Stochastic Modeling

Analysis & Simulation

Barry L. Nelson

Northwestern University

DOVER PUBLICATIONS, INC.
Mineola, New York

Bibliographical Note

This Dover edition, first published in 2002, is an unabridged republication of the work originally published by McGraw-Hill, Inc., New York, in 1995.

Library of Congress Cataloging-in-Publication Data

Nelson, Barry L.
   Stochastic modeling : analysis & simulation / Barry L. Nelson.
      p. cm.
   Originally published: New York : McGraw-Hill, c1995, in series: McGraw-Hill series in industrial engineering and management science.
   Includes bibliographical references and index.
   ISBN 0-486-42569-X (pbk.)
   1. Stochastic processes. I. Title.

QA274 .N46 2002
519.2'3—dc21

2002029916

Manufactured in the United States of America
Dover Publications, Inc., 31 East 2nd Street, Mineola, N.Y. 11501

ABOUT THE AUTHOR

Barry L. Nelson is the James N. and Margie M. Krebs Professor in the Department of Industrial Engineering and Management Sciences at Northwestern University, and is Director of the Master of Engineering Management Program there. His teaching and research centers on the design and analysis of computer simulation experiments with applications in manufacturing, services and transportation. His work has been consistently funded by the National Science Foundation, and he has published numerous papers and two books. Dr. Nelson received his B.S. in Mathematics from DePauw University, and his M.S. and Ph.D. in Industrial Engineering from Purdue University. He has served the profession as the Simulation Area Editor of *Operations Research* and President of the INFORMS (then TIMS) College on Simulation. He has held many positions for the annual Winter Simulation Conference, including Program Chair in 1997, and is currently a member of its Board of Directors representing INFORMS.

To Jeanne for now
To Kyle for the future

CONTENTS

PREFACE

A student who had taken my simulation course called me from work. The student's engineering group thought it had a problem for which simulation might be the solution. It was trying to estimate the expected lead time to produce a particular product, and the group had some data on this lead time, but not much. The student wondered if the group could fit a distribution to the data, simulate a larger sample of lead times, and obtain a better estimate of the expected lead time from this larger sample.

To the beginner it may not be obvious what is wrong here. A short answer is that the proposed simulation merely represents what was already observed, so it cannot add anything beyond what was observed. Was the confusion due to poor training of the student? I conjecture (in self-defense), no. In fact, the student understood several important concepts: Real phenomena that are subject to uncertainty (the lead times) can be modeled using the language of probability (a probability distribution). The probability model can be parameterized or "fit" using real data. Given a probability model, simulation is one way to analyze it (generate samples from the probability model). And the more data you generate, the better your estimates of performance measures (such as the expected lead time) will be. The confusion arose in how these concepts fit together.

I do not think that this confusion is surprising. We tend to expect either too much or too little from first courses in stochastic modeling and simulation. We expect too much when we start with the difficult language and notation of probability and expect students to have any intuition about the kinds of physical processes it describes. We expect too little when we teach a collection of formulas without the mathematical structure that supports them. Sometimes students are misled when we fail to make the important distinction between probability models and the mathematical, simulation, and numerical tools that are available to analyze them. And the dual role of statistics, to parameterize stochastic models and to analyze them when we simulate, is rarely clear.

In this book I attempt to provide a unified presentation of stochastic modeling, analysis and simulation that encompasses all of these aspects. The unifying principle is the discrete-event-sample-path view of stochastic processes, which is the basis for discrete-event simulation, but can also be exploited to make sense of many mathematically tractable processes. A primary goal is to use simulation to help the student develop intuition that will serve as a faithful guide when the mathematics gets tough. Another goal is to ensure that the student has nothing to relearn when progressing to the next level of mathematical rigor in a second course. In other words, I have tried to be correct without being formal. I take as my starting point the cumulative distribution function of a random variable and do not describe the underlying probability space, although one is certainly implied by the way we generate sample paths.

A course from this book should precede, or be the first part of, a course in simulation modeling and analysis that emphasizes the particulars of a simulation language and how to model complex systems. This allows the student to develop some intuition about simulation from simulating the simple processes in this book and also to gain an understanding of the supporting probability structure that is hidden by most simulation languages. The simple Markovian processes emphasized here also provide approximating models for more complex systems, allowing students to obtain rough-cut estimates prior to simulating.

I presume linear algebra, calculus, computer programming and an introductory course in probability and statistics (that typically emphasizes statistics) as prerequisites. The concepts of random variables, probability distributions and sample statistics need to be familiar, but not necessarily comfortable. The only stochastic process about which I assume any knowledge is a sequence of independent and identically distributed random variables. Modeling and analysis tools are introduced as needed, with no extras "just for completeness."

The book is designed to be read in chapter sequence, and there are many backward references from later chapters to earlier chapters. However, it is possible to omit certain sections within chapters, especially the more difficult derivations (a euphemism for "proofs") that are collected in their own skipable sections, and the "Fine Points" at the end of most chapters. The chapters are structured around miniature cases that are introduced at the beginning and carried throughout to illustrate key points.

All of the book can be covered in a semester course (45 class hours), with the possible exception of the last chapter, Chapter 9, "Topics in Simulation of Stochastic Processes." For those who have the misfortune, as I have, of cramming the material into a one-quarter course (30 class hours), I offer three compromises:

1. If fundamentals are more important than the particular application of queueing, then cover portions of Chapters 1–7. These chapters contain simulation, arrival-counting processes, discrete-time Markov chains and continuous-time Markov processes.

2. If queueing is important, as it is in most industrial engineering curriculums, then cover portions of Chapters 1–6, skip Chapter 7, "Continuous-Time Processes," and instead do Chapter 8, "Queueing Processes." I have included a "quick start" introduction to continuous-time Markov processes in Chapter 8 to make this possible.

3. An alternative that puts more of a burden on the instructor, but works well, is to cover Chapters 1–3, skip Chapter 4, "Simulation," and go directly to Chapters 5–8. This approach is feasible because the simulations in Chapters 5–8 can be understood without the generic introduction to simulation in Chapter 4 if a little in-class support is provided.

In any case it is important not to spend too much time on the probability and statistics review material in Chapter 3. This is prerequisite material given a slightly different slant to match the perspective of the book, and I wrote it assuming students would be required to read it without much in-class discussion. If time is critical, then the random variate generation section, Section 3.3, can also be treated lightly, but it should not be skipped entirely. And no matter what path you follow through the book, I recommend finishing the course with Section 9.2, "Rough-Cut Modeling," because "quick-and-dirty" analysis is one of the most important uses of Markovian models.

I assume that every instructor will have favorite models or discipline-specific topics to add or will want more depth on some topics than I provide. Rather than try to anticipate all of these needs, I have endeavored to provide a foundation that will not get in the instructor's way when following their own instinct. In a few instances some obvious omissions are covered in the exercises.

I strongly recommend that instructors employ the same simulation language that the students will use in the simulation course, when applicable. This guarantees that the connection between stochastic processes and simulation analysis will not be lost when simulating complex systems. I do not have the students code any of the simulations themselves (remember I only have 30 class hours), but instead they exercise and modify my simulation programs and then study the sample paths (ideally graphically). The examples in the book have been coded in several languages and are available from the publisher. The book can be used without ever actually simulating anything, but if that approach is adopted, then at least one demonstration and one of the manual simulations from the exercises in Chapter 2 is advisable.

To support a course from this book, a good statistical-analysis package or electronic spreadsheet should be available so that students can "muck around" with the data they generate. A symbolic calculation program is also useful, especially for manipulating the summations and performing the matrix operations that are often required to compute performance measures. Many of the exercises simply require too much calculation to do by hand, which is also a feature of real problems.

My perspective on stochastic processes and how to teach the subject has been significantly influenced by two books: Ravindran, Philips and Solberg (1987), who emphasize intuition and modeling, and Çinlar (1975), who emphasizes sample paths and underlying structure. Although I did not formally adopt the generalized semi-Markov process view of simulation, I also found inspiration from Glasserman (1991).

Several reviewers provided critical suggestions about content and balance, including M. Jeya Chandra, David Goldsman, Carl Harris, Sheldon Jacobson, Maria Rieders, Gennady Samorodnitsky, Lee Schruben and Frank Wolf. Many colleagues offered valuable feedback, discussion and encouragement, including Ben Fox, David Kelton, Bruce Schmeiser, Mike Taaffe and Laurel Travis. David worked through one draft of the book at the subatomic level, helping me more than I can explain. Mike offered his standard unreserved (but insightful) critique of the queueing chapter. I am especially grateful to David and Mike for arranging my sabbatical at the University of Minnesota, which is when the first draft was written. The sabbatical was supported by the Operations and Management Science Department of the Carlson School of Management, the Minnesota Supercomputer Institute, and The Ohio State University. Discussions and debates with my Ohio State colleagues Gordon Clark, Jane Fraser, Chuck Reilly and Marc Posner have indirectly influenced this book and directly influenced my thinking about modeling and analysis. Most importantly, the (former) Department of Industrial and Systems Engineering at Ohio State, under the direction of George Smith, Jr., has consistently allowed me to follow my own instincts, to experiment and to learn. This is a debt I can never fully repay.

Special thanks go to my editor, Eric Munson, who was willing to take a chance on doing something really different. And finally to my wife Jeanne, who continues to support me despite the *big lie* ("I won't be as busy after tenure").

Barry L. Nelson
July 1994

CHAPTER
1

WHY ARE
WE HERE?

This book describes basic tools for modeling, analysis and simulation of systems that evolve dynamically over time and whose behavior is uncertain. One class of examples is queueing systems, which are systems of customers waiting to use service resources. Our goal is to describe these tools in a way that exploits your common sense and intuition but also enables you to use the mathematics, probability and statistics at your disposal to perform a detailed analysis. At the end of the day you should know a bit more about probability and statistics too.

A brief list of the settings in which these tools have been useful includes:

- Manufacturing applications, such as capacity planning, inventory control and evaluation of process quality
- Health-care applications, such as hospital staffing and medical decision making
- Computer applications, such as designing hardware configurations and operating-system protocols
- Communication applications, such as sizing message buffers and evaluating network reliability
- Economic applications, such as portfolio management and forecasting industrial relocation
- Business applications, such as consumer behavior and product-distribution logistics

1

- Military applications, such as combat strategy and predicting ordnance effectiveness
- Biological applications, such as population genetics and the spread of epidemics

There are many more.

Our approach is to model systems in terms of how we would *simulate* them, and then recognize situations in which a mathematical analysis can take the place of a simulation. There are three reasons for this approach: Simulation is intuitive, because it describes how a system physically responds to uncertainty; the fundamentals of simulation do not change from application to application, while mathematical analysis depends profoundly on characteristics of the model; and simulation provides a common modeling paradigm from which either a mathematical analysis or a simulation study can be conducted. The primary drawback of simulation is that it produces only *estimates* of system performance, whereas mathematical analysis provides *the right answer*, up to the fidelity of the model. Other advantages that favor mathematical analysis are described later in the book.

The next chapter illustrates how we propose to think about dynamic systems that are subject to uncertainty. Chapters 3–4 then provide the basic constructs for modeling such systems and simulating them. Standard methods of mathematical analysis, and when they apply, appear in Chapters 5–8. Finally, Chapter 9 describes some critical issues that arise when simulating models that are not amenable to mathematical analysis.

CHAPTER

2

SAMPLE PATHS

This chapter introduces *sample-path decomposition*, which is one way to think about dynamic systems that evolve over time. A service-system example is used to illustrate the approach, but sample-path decomposition can characterize the behavior of many types of systems, including those described in Chapter 1. Sample-path decomposition is also a convenient way to formulate mathematical models of systems, models that can be used to evaluate how changes will affect existing systems or predict the performance of systems that do not yet exist. Formulation and analysis of these models is the topic of this book, and sample-path decomposition is the perspective that we employ throughout.

2.1 THE CASE OF THE COPY ENLARGEMENT

Case 2.1. A national chain of small photocopying shops, called The Darker Image, currently configures each store with one photocopying machine and one clerk. Arriving customers stand in a single line to wait for the clerk. The clerk completes the customers' photocopying jobs one at a time, first-come-first-served, including collecting payment for the job.

Business is good enough that the parent company of The Darker Image plans to enlarge some stores by adding a second photocopying machine and hiring a second clerk. The second copier could be operated by the new clerk, but since some customers with small copying jobs have complained about having to wait a long time, the company is considering installing the second copier for self-service copying only. The company wants to know which option will deliver better service.

Case 2.1 describes a system that is quite simple compared to the complex systems that engineers and management scientists study every day. Nevertheless, the answer to the company's question is not obvious. Certainly some additional information is needed to perform a thorough analysis. For instance, how many customers will use a self-service copier? Can a self-service copier provide all the services that a customer wants (such as collating, stapling, or attaching covers)? Should a page limit be placed on self-service copying? Are there differences from store to store, or is a single solution good for all stores?

One important characteristic of the system described in Case 2.1 is that neither option will be the best option all of the time. The system's performance will be subject to some uncertainty: The number of customers that arrive and when they arrive each day are variable. The nature of the customers' copying jobs—large or small, simple or requiring special handling—is also unknown. These factors, and others, will affect the service delivered on a particular day, so that a system design that seems superior on one day may be inferior on another day. Therefore, the definition of "better service" is not straightforward.

Modeling and analysis of systems that are subject to uncertainty is the topic of this book. The mathematical models that we develop to analyze such systems are called **stochastic processes**. *"Stochastic" is a synonym for random.*

The words used in the previous paragraph were carefully chosen. *System* refers to a physical entity, such as The Darker Image, which may be real or conceptual (yet to be constructed). Systems are subject to *uncertainty*, which we cannot completely control. A *model* is a mathematical representation of a system; it is always distinct from the system it is intended to model, and it is never a perfect representation. A stochastic process is a particular type of model that represents the uncertainty in a dynamic system using the language of probability.

We return to Case 2.1 after reviewing some of the supporting tools we will use to analyze it.

2.2 NOTATION AND REVIEW

The second section of most chapters in this book introduces notation and reviews the analysis tools that support the remainder of that chapter. The tools should be familiar, and the presentation is concise so that you can easily refer back to these sections to find a result. The analysis tools used in this chapter are two descriptive statistics: the sample average and the histogram.

Suppose that we have some numerical data, such as the amount of time that was required to complete each photocopying job during a day. Let m denote the number of values, and let $s_1, s_2, \ldots, s_m$ denote the values themselves.

One way to summarize the data is the *sample average*,

$$\bar{s} = \frac{1}{m} \sum_{i=1}^{m} s_i$$

The "bar" (" ¯ ") notation is a standard way to indicate an average. The sample average is a measure of the center of the data. A more complete representation of the distribution of the data is a *histogram*. A histogram is formed by fixing a set of numbers, $c_1, c_2, \ldots, c_{k-1}$, such that

$$-\infty < c_1 < c_2 < \cdots < c_{k-1} < \infty.$$

Let c_0 and c_k denote $-\infty$ and ∞, respectively, and form the k intervals, $(c_0, c_1]$, $(c_1, c_2], \ldots, (c_{k-1}, c_k]$, where "(" indicates that the endpoint is not included in the interval, while "]" indicates that the endpoint is included.

A histogram is a plot of the height h_j above the interval $(c_{j-1}, c_j]$, where h_j is the fraction of the data values $s_1, s_2, \ldots, s_m$ that fall in the interval. More precisely,

$$h_j = \frac{1}{m} \sum_{i=1}^{m} \mathcal{I}\left(s_i \in (c_{j-1}, c_j]\right) \tag{2.1}$$

for $j = 1, 2, \ldots, k$. The function $\mathcal{I}$ is called an *indicator function*; it takes the value 1 if its argument is true, and 0 otherwise. In this case the argument is $s_i \in (c_{j-1}, c_j]$, which is read, "The data value s_i is contained in the interval $(c_{j-1}, c_j]$." Thus, $\mathcal{I}\left(s_i \in (c_{j-1}, c_j]\right)$ adds a 1 to the summation for each data value in the jth interval. What we call a "histogram" some books refer to as a *relative frequency distribution*, due to the division by m in (2.1).

For example, if the data values are $\{7, 16, 3, 11, 3, 8, 6, 4\}$ and $c_1 = 0$, $c_2 = 4$, then

$$h_2 = \frac{1}{8} \sum_{i=1}^{8} \mathcal{I}\left(s_i \in (0, 4]\right) = \frac{0+0+1+0+1+0+0+1}{8} = \frac{3}{8}$$

since three of the eight values are greater than 0 but less than or equal to 4. In some contexts the height h_2 may be interpreted as the probability of a value falling in this interval. Using an indicator function to represent a histogram may seem unnecessarily complicated, but the simple idea of an indicator function turns out to be a powerful representation that we exploit later in the book.

Figure 2.1 in Section 2.3 shows several histograms where the intervals are plotted along the horizontal axis and the heights h_j are plotted along the vertical axis. The intervals at each end, $(c_0, c_1]$ and $(c_{k-1}, c_k]$, are often omitted from a histogram when they are empty, and the widths of the other intervals, $c_j - c_{j-1}$, are typically identical.

2.3 SAMPLE-PATH DECOMPOSITION

An important first step in any modeling and analysis problem is to understand how the system of interest works, or will work. Table 2.1 displays some time-study data collected one morning at a location of The Darker Image. Whenever a customer arrived (denoted by "customer arrived") or received the finished copying job (denoted by "customer finished"), the time was recorded. The customers were also assigned an identification number by their order of arrival. And if there was anything special about the customer's job then a comment was appended to the entry. A total of 20 customers were observed.

The parent company of The Darker Image thinks that any customer whose job does not require special handling (collate, staple, covers, two-sided copies or special paper) could potentially have used a self-service copier. Of the 20 customers observed, 12 (60%) had jobs that required no special handling. A reasonable first question to ask is, Are the *service times*—the times required to complete the copying jobs—different for the potential self-service customers' jobs compared to the other customers' jobs? The company's conjecture is that the potential self-service customers tend to have shorter service times, and therefore might consider a long delay to be unfair.

We can deduce from Table 2.1 that customer 1 spent $9:12 - 9:19 = 7$ minutes being served. Customer 2 did not begin service until customer 1 finished at 9:19, so customer 2 spent $9:21 - 9:19 = 2$ minutes being served.

To generalize this calculation, let a_i denote the arrival time, and let d_i denote the departure (finish) time, of customer i, for $i = 1, 2, \ldots, 20$. The service time, denoted s_i, is the difference between the time customer i departed and the time customer i reached the front of the queue. A *queue* is the collection of customers waiting for service in a service system. If customer i arrived to find the shop empty, then the customer reached the front of the queue at the time of arrival, a_i. Otherwise, the customer reached the front of the queue at the time the customer ahead, customer $i - 1$, departed, which is d_{i-1}. Thus, the service time for customer i was

$$s_i = d_i - \max\{a_i, d_{i-1}\} \tag{2.2}$$

for $i = 1, 2, \ldots, 20$, provided we define $d_0 = 0$. For example, the service time of customer 5 was

$$s_5 = d_5 - \max\{a_5, d_4\} = 9:57 - \max\{9:39, 9:41\}$$
$$= 9:57 - 9:41 = 16 \text{ minutes.}$$

TABLE 2.1
Time-study data from The Darker Image

Time-study record		
Time	**What happened**	**Comment**
9:00		shop opens
9:12	customer 1 arrived	
9:14	customer 2 arrived	
9:17	customer 3 arrived	
9:19	customer 1 finished	collate, staple
9:21	customer 2 finished	
9:22	customer 3 finished	
9:38	customer 4 arrived	
9:39	customer 5 arrived	
9:41	customer 4 finished	
9:43	customer 6 arrived	
9:45	customer 7 arrived	
9:52	customer 8 arrived	
9:57	customer 5 finished	special paper
9:58	customer 9 arrived	
10:00	customer 6 finished	
10:01	customer 7 finished	
10:08	customer 8 finished	
10:11	customer 9 finished	collate, covers
10:13	customer 10 arrived	
10:14	customer 11 arrived	
10:16	customer 12 arrived	
10:19	customer 13 arrived	
10:24	customer 10 finished	two-sided, staple
10:26	customer 11 finished	
10:28	customer 14 arrived	
10:36	customer 12 finished	
10:36	customer 15 arrived	
10:38	customer 13 finished	
10:39	customer 14 finished	
10:42	customer 15 finished	collate, staple
10:45	customer 16 arrived	
10:47	customer 16 finished	
10:48	customer 17 arrived	
10:50	customer 17 finished	
10:53	customer 18 arrived	
10:58	customer 19 arrived	
11:00	customer 20 arrived	
11:01	customer 18 finished	special paper
11:07	customer 19 finished	two-sided, staple
11:09	customer 20 finished	collate, staple

TABLE 2.2
Service times for customers in the time study (An * indicates special handling.)

Customer number, i	Service time, s_i
1*	7
2	2
3	1
4	3
5*	16
6	3
7	1
8	7
9*	3
10*	11
11	2
12	10
13	2
14	1
15*	3
16	2
17	2
18*	8
19*	6
20*	2

Applying Equation (2.2) to the data in Table 2.1 gives the values in Table 2.2, where an asterisk indicates a job that required special handling. Figure 2.1 shows histograms of these service-time data in total, and separated into those that involved special handling (* in the table) and those that did not (the potential self-service customers). We can also compute the sample average of the service times: 3 minutes for self-service customers and 7 minutes for full-service customers.

Based on this statistical analysis, it appears that the potential self-service customers may have been delayed unfairly long. Their sample-average service time of 3 minutes is less than the corresponding 7 minutes for jobs requiring special handling. And the histograms show that the distribution of the service times for jobs requiring special handling included some extremely long times. May we therefore conclude that the company should make the new copier a self-service copier?

Unfortunately, our analysis so far has some shortcomings. One is that it concentrates on the wrong performance measure, the service time. The addition of a second copier, either self-service or full-service, will likely have no effect on the service times, unless self-service increases the service time because inexperienced customers make mistakes. More relevant is a performance measure such as the time a customer spends waiting in queue, called the *delay*. Adding a self-service

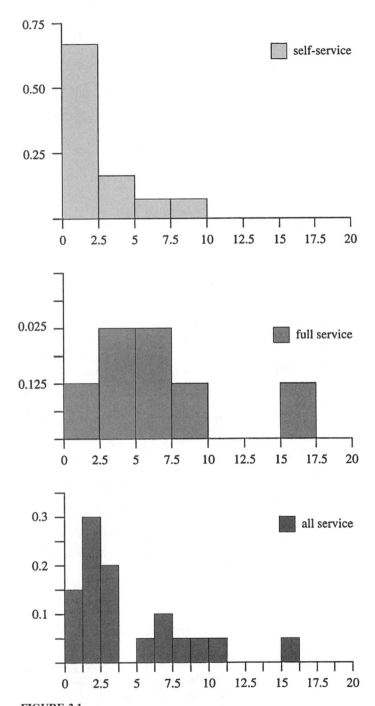

FIGURE 2.1
Histograms of service-time data from The Darker Image.

copier will certainly decrease the delay for self-service customers, but will it decrease the delay more than adding a second full-service copier? And if so, what will be the effect on full-service customers?

Answering these questions is not easy since The Darker Image is a dynamic system in which what happens to one customer can depend on the actions of other customers. The dynamics of the system will change considerably with the addition of a second copier, and we cannot predict the effect on a performance measure such as delay without considering these dynamics.

The following observation is fundamental to the modeling and analysis approach in this book. The time-study data in Table 2.1 describes a *sample path* for The Darker Image that is composed of two parts: The characteristics of the customers (arrival times and service times), which *are not* under the control of The Darker Image, and the procedure by which The Darker Image serves customers (one at a time, first-come-first-served), which *is* under their control. These two parts are the system *inputs* and system *logic*, respectively. Given a set of inputs (arrival times and service times), the system logic implies a sample path (such as Table 2.1).

The Darker Image is proposing to change the system logic by either adding a new self-service or full-service copier. However, it cannot change the system inputs, except via changes outside the scope of our case, such as reducing its prices or increasing its advertising. This suggests that we subject each proposed change in system logic to the system inputs we obtained in Table 2.1 and compare the results. Stated differently, we can create the sample paths that *would have occurred* if these customers had encountered each of the proposed systems. This provides an assessment of which proposed system would have performed better for these customers, an assessment that includes the dynamics of the system.

A **sample path** is a record of the time-dependent behavior of a system. **Sample-path decomposition** represents a sample path as inputs and logic. **Simulation** generates new sample paths without building the new system. And **sample-path analysis** extracts system performance measures from sample paths.

We simulate the self-service and full-service systems and analyze the resulting sample paths in the next two sections.

There are two conventions that make it easier to simulate the passage of time in a dynamic system: We always start time at 0, and we keep track of time in a standard unit, such as 1 minute. Thus, 9:00 becomes time 0, 9:17 becomes time 17, and 10:37 becomes time 97. In addition, we keep track of the time gap *between* customer arrivals rather than the actual times that customers arrive. Let

TABLE 2.3
Interarrival-time gaps for customers in the time study (An * indicates special handling.)

Customer number, i	Interarrival-time gap, g_i
1*	12
2	2
3	3
4	21
5*	1
6	4
7	2
8	7
9*	6
10*	15
11	1
12	2
13	3
14	9
15*	8
16	9
17	3
18*	5
19*	5
20*	2

g_i denote the time gap between the arrival of customers $i - 1$ and i. Then

$$g_i = a_i - a_{i-1}$$

for $i = 1, 2, \ldots, 20$, provided we define $a_0 = 0$. These values are displayed in Table 2.3. Tables 2.2 and 2.3 are the inputs for our simulations.

2.3.1 Simulating the Self-Service System

To generate a sample path for the self-service system, we need to define its system logic. Since the self-service system does not exist, we have to draw on our own experience and the experience of people who will design and operate the system. This is part of modeling.

The parent company of The Darker Image agrees to the following logical characterization:

- There will be two queues for customers, one for full service and one for self service. Customers will not change queues after they join one.
- Self-service customers will always join the self-service queue. Full-service customers (those whose jobs require special handling) will always join the full-service queue.

- Service times for self-service customers will be the same as they would have been in the original one-copier system.

The performance measure we use as a basis for comparison is *delay*, which is the time that a customer spends waiting to begin copying. Delay can be 0 if a customer begins copying immediately upon arrival.

The time study in Table 2.1 recorded two types of changes in the system, the arrival of a customer and finishing a copying job. We refer to these system changes as *system events*. The corresponding system events for the proposed self-service system are the arrival of a customer, finishing a job at the full-service copier and finishing a job at the self-service copier.

Figure 2.2 is a graphical representation of The Darker Image beginning at time 0, then updated just after the system changes. For instance, the first update occurs just after time 12 when the first customer (full service) arrives; customers are represented by a circle that encloses their identification number. We conduct the simulation by updating this graphical representation every time there is a system event. During the course of the simulation it is convenient to keep track of the time the next system event of each type is due to occur. This information is also displayed in Figure 2.2.

The system inputs needed to generate the sample path are contained in Tables 2.2–2.3. For example, the first system event is the arrival of the first

CURRENT TIME: 0

Full Service		Self-Service	

NEXT SYSTEM EVENT	TIME
customer arrival	0 + 12 = 12
full-service finish	
self-service finish	

CURRENT TIME: 12

Full Service		Self-Service	
①			

NEXT SYSTEM EVENT	TIME
customer arrival	12 + 2 = 14
full-service finish	12 + 7 = 19
self-service finish	

CURRENT TIME: 14

Full Service		Self-Service	
①		②	

NEXT SYSTEM EVENT	TIME
customer arrival	14 + 3 = 17
full-service finish	19
self-service finish	14 + 2 = 16

CURRENT TIME: 16

Full Service		Self-Service	
①			

NEXT SYSTEM EVENT	TIME
customer arrival	17
full-service finish	19
self-service finish	

FIGURE 2.2
Simulation of the proposed self-service system, times 0–16 minutes.

customer at time $0 + g_1 = 0 + 12 = 12$; this full-service customer is finished at time $12 + s_1 = 12 + 7 = 19$. Notice that customer 1 had a delay of 0 minutes, since the system was empty when the customer arrived. When the first customer arrives at time 12, we can determine that the next arrival occurs at time $12 + g_2 = 12 + 2 = 14$.

The next system event after time 12 occurs at time 14 minutes, when customer 2 arrives, not at time 19 when customer 1 finishes, because the simulation progresses just like the real system: first things first. To continue the simulation, we select the system event that is due to occur next; advance the system time to the time of the next system event; update the status of the system depending upon the type of system event and current system status; and repeat the procedure.

The first nonzero delay occurs at time 45, when customer 7 arrives; see Figure 2.3. Customer 7 is a self-service customer, and therefore enters the self-service queue behind customer 6, who is already in service. At time 46, when customer 6 finishes service, customer 7 begins service having experienced a delay of 1 minute. You should fill in the details between times 16–43, and 46–79, paying particular attention to time 76 when two system events occur simultaneously.

Figure 2.4 shows the system from times 79 through 86. Customer 13 arrives at time 79. At time 84 full-service customer 10 finishes, but self-service

CURRENT TIME: 43

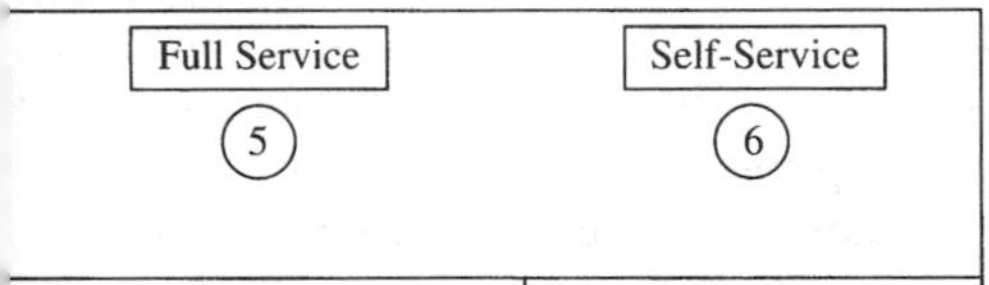

Full Service	Self-Service
⑤	⑥

NEXT SYSTEM EVENT	TIME
customer arrival	$43 + 2 = 45$
full-service finish	55
self-service finish	$43 + 3 = 46$

CURRENT TIME: 45

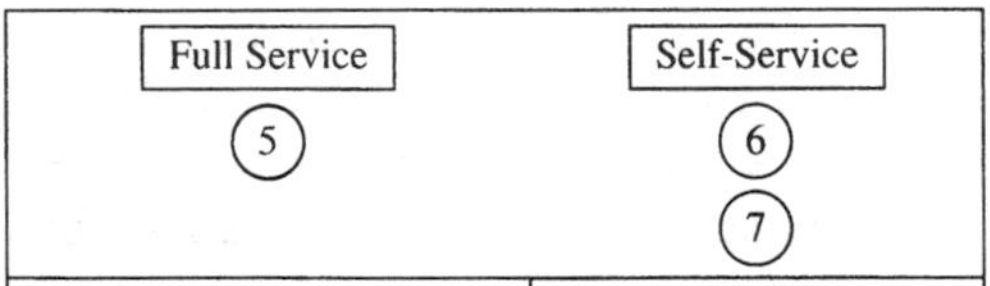

Full Service	Self-Service
⑤	⑥
	⑦

NEXT SYSTEM EVENT	TIME
customer arrival	$45 + 7 = 52$
full-service finish	55
self-service finish	46

CURRENT TIME: 46

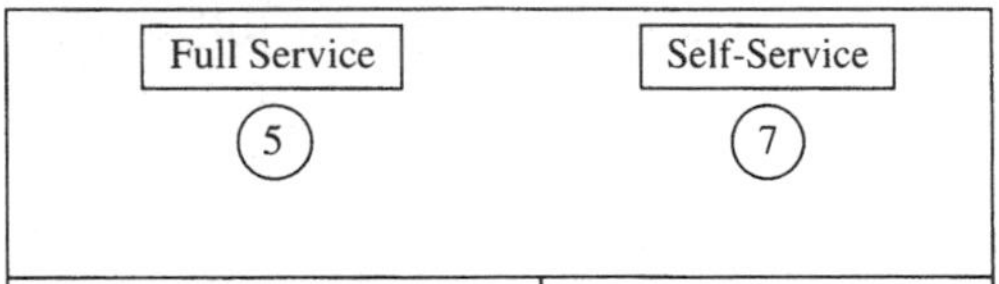

Full Service	Self-Service
⑤	⑦

NEXT SYSTEM EVENT	TIME
customer arrival	52
full-service finish	55
self-service finish	$46 + 1 = 47$

FIGURE 2.3
Simulation of the proposed self-service system, times 43–46 minutes.

CURRENT TIME: 79

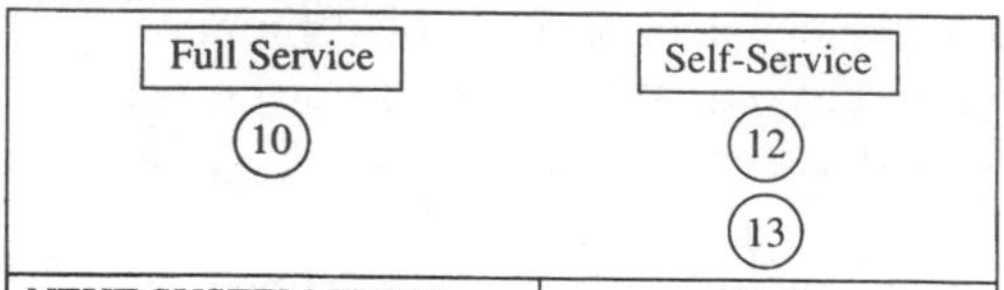

Full Service	Self-Service
(10)	(12)
	(13)

NEXT SYSTEM EVENT	TIME
customer arrival	79 + 9 = 88
full-service finish	84
self-service finish	86

CURRENT TIME: 84

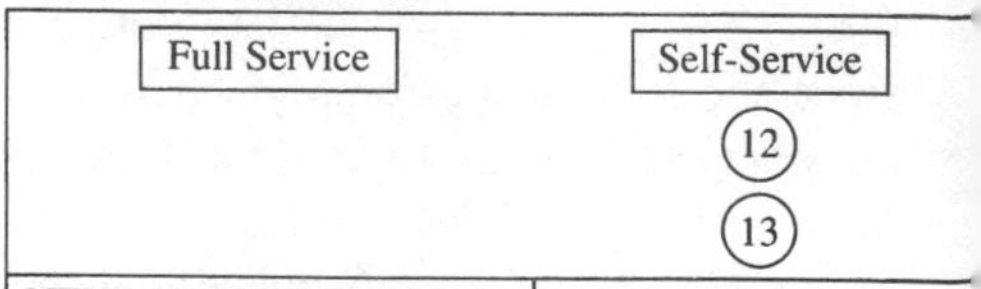

Full Service	Self-Service
	(12)
	(13)

NEXT SYSTEM EVENT	TIME
customer arrival	88
full-service finish	
self-service finish	86

CURRENT TIME: 86

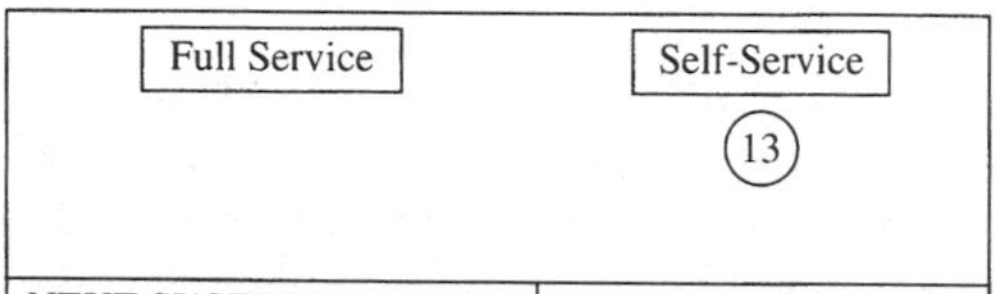

Full Service	Self-Service
	(13)

NEXT SYSTEM EVENT	TIME
customer arrival	88
full-service finish	
self-service finish	86 + 2 = 88

FIGURE 2.4
Simulation of the proposed self-service system, times 79–86 minutes.

customer 13 does not begin service because customers do not change queues. Customer 13 begins service at time 86, after an $86 - 79 = 7$-minute delay. Notice that we do not need to know what happened prior to time 79 to continue the simulation, as long as we know which customers are present and when the next system events are due to occur after time 79.

You should now complete the simulation until all 20 customers are served. The simulation ends at time 129 minutes (11:09 a.m.). You should find that self-service customers 7 and 13 experience delays of 1 and 7 minutes, respectively, while full-service customers 19 and 20 experience delays of 3 and 7 minutes, respectively; all other customers have delays of 0 minutes. We will see how this performance compares to the proposed full-service system in the next section.

2.3.2 Simulating the Full-Service System

The parent company of The Darker Image agrees to the following logical characterization of the proposed full-service system:

- There will be one queue of customers with service delivered first-come-first-served by the next available clerk (copier).

CURRENT TIME: 0

Full Service	Full Service

NEXT SYSTEM EVENT	TIME
customer arrival	
full-service finish (left)	
full-service finish (right)	

FIGURE 2.5
Simulation of the proposed full-service system.

The corresponding system events for the proposed full-service system are the arrival of a customer, finishing a job at the left full-service copier and finishing a job at the right full-service copier. This information is displayed in Figure 2.5.

You should complete this simulation, which ends at time 124 and looks similar to the simulation of the self-service system except at those times when a customer is delayed: Delayed customers wait for the first available copier, rather than staying in a queue dedicated to one copier. You should find that self-service customers 7 and 13 experience delays of 1 and 5 minutes, respectively, while full-service customer 20 experiences a delay of 1 minute; all other customers have delays of 0 minutes.

2.3.3   Discussion

Based on the simulations in Sections 2.3.1 and 2.3.2, the full-service system is superior for both types of customers since it leads to delays that are the same or smaller than the self-service system (see Table 2.4). This is unexpected. How much confidence should we have in these conclusions?

Perhaps these 20 customers are representative of the customers that The Darker Image serves every day, but 20 customers do not even constitute a full day's work. And the particular pattern of inputs observed (times and types of service) will likely never be repeated, and may even differ from store to store. The fact that full-service customers have longer service times must *sometimes* imply an advantage for self-service customers if they have a dedicated copier.

The problem is that we generated just *one* possible sample path for this system. To gain confidence in our conclusions, we need to know more about the *distribution* of performance over *many* possible sample paths. One solution is to obtain additional time studies and perform more simulations. The next two chapters develop a different approach that allows us to generate sample paths that are longer than the time-study data we have available, or to generate sample paths when we have no data at all. The remainder of the book shows that in some cases we can analyze sample paths without actually generating them.

TABLE 2.4
Customer delays from the simulations (An * indicates special handling.)

Customer number, i	Delays	
	Self-service system	Full-service system
1*	0	0
2	0	0
3	0	0
4	0	0
5*	0	0
6	0	0
7	1	1
8	0	0
9*	0	0
10*	0	0
11	0	0
12	0	0
13	7	5
14	0	0
15*	0	0
16	0	0
17	0	0
18*	0	0
19*	3	0
20*	7	1

2.4   EXERCISES

Although the exercises in this section can be completed by hand, some are more easily done on a computer using spreadsheet software. The specifics differ depending on the software, but the central idea is illustrated in Exercise 6. While spreadsheets are useful for simple simulations, they are not well suited for simulations with complex system-event structure. A more general approach is presented in Chapter 4.

2.1. Complete the simulations of the proposed self-service and full-service systems.

2.2. Compute the sample average and plot a histogram of the interarrival-time gaps in Table 2.3. Does the histogram resemble any familiar probability distribution?

2.3. In the proposed self-service system for The Darker Image, it may not be appropriate to assume that self-service customers will never join the full-service queue or will never switch queues. Simulate the self-service system allowing self-service customers to select the shortest queue when they arrive and to switch queues when the full-service copier becomes idle. Observe how this changes the results. What other assumptions of both proposed systems seem troublesome, and why?

2.4. Propose an alternative system logic for using a second copier that is different from the two that the parent company of The Darker Image proposed. Simulate it.

2.5. The time-study data below were collected from observing four parts being processed on a single machine in their natural order of arrival. What would the *makespan* (time to complete all parts) be if the a new machine could process parts 10% faster? Identify the inputs and logic; then conduct the simulation necessary to determine the answer.

Time	What happened
0	start observing
5	part 1 arrived
6	part 2 arrived
8	part 1 finished
10	part 3 arrived
11	part 2 finished
15	part 4 arrived
17	part 3 finished
20	part 4 finished

2.6. A company needs to replace its high-use photocopying machine. Should the company purchase a model similar to the one it currently has or purchase a slightly more expensive one that promises to be 20% faster on jobs that involve collating but 10% slower on jobs that do not involve collating?

Time-study data for 10 jobs are given below (times are in hours). Perform sample-path decomposition; then generate the sample path that would

Time	System event
0.00	begin observing
0.09	arrival
0.35	arrival
0.54	arrival
0.92	arrival
2.27	arrival
2.33	finish collate job
3.09	finish collate job
3.53	arrival
3.66	arrival
4.39	finish no-collate job
4.52	finish collate job
4.62	finish no-collate job
4.63	arrival
4.64	finish no-collate job
4.80	finish collate job
4.86	finish collate job
7.33	arrival
7.61	arrival
8.24	finish collate job
8.50	finish no-collate job

result from installing the more expensive copier. Compare the two sample paths in terms of *makespan*, which is the time required to finish a fixed number of jobs, 10 in this case.

A spreadsheet program can be used to perform the sample-path decomposition and the simulation for this problem. Set up columns A–G as shown below:

A	B	C	D	E	F	G
cust no.	arrival	finish	collate?	copy time	new copy time	new finish

Number the customers from 0 to 10 in column A, where customer 0 is a phantom customer included to make the calculations work. Enter the customer arrival times in column B; the finish times in column C; and a 0.8 or 1.1 in column D for collate and no-collate jobs, respectively. Column D is the factor by which the customer's copying time would increase or decrease if the new machine was installed.

Column E, the time the customer spends copying, is determined by the formula

$$E[i] = C[i] - MAX(B[i], C[i-1])$$

as described in Section 2.3, where `[i]` indicates the row number in the spreadsheet. The new copying times are `F[i] = D[i]*E[i]`. Finally, the new finish time for customer `i` is the maximum of the time customer `i-1` finished and the time customer `i` arrived, plus the new copy time of customer `i`; that is,

$$G[i] = MAX(G[i-1], B[i]) + F[i]$$

Notice that this approach hides the two system events, the arrival of a customer and finishing a job, which is not possible in more complex simulations.

2.7. A fast-food restaurant receives supplies from its distributor every other day just prior to opening for business. The manager places an order for new supplies just prior to opening for business on the days between receiving orders. To clarify the situation, we number the days sequentially, 1, 2, 3, ..., and specify that orders can only be placed on odd-numbered days and orders are received at the beginning of the following (even-numbered) day.

The manager decides how much of each item to order by checking the inventory on hand just prior to ordering. One item that he orders is frozen hamburger patties. If he finds 600 or fewer patties in inventory on the day of an order, then he orders 1000 patties minus the number of patties in inventory. If there are more than 600 patties on hand, then he does not order any more. Notice that if he chooses not to order, then it will be at least 3 days before receiving another shipment, since his next order cannot be placed for 2 more days.

Listed below is the number of hamburgers on hand at the beginning of each day for a 14-day study period. The inventory on hand on a day that an order is received includes the order.

Day	Inventory	Comment
1	415	place order
2	704	receive order
3	214	place order
4	856	receive order
5	620	no order placed
6	353	
7	172	place order
8	976	receive order
9	735	no order placed
10	433	
11	217	place order
12	860	receive order
13	598	place order
14	833	receive order

Because it is expensive to keep supplies on hand, the manager would like to reduce the reorder point from 600 patties to 500 patties. Can he do this without running out of hamburger patties? Perform sample-path decomposition; then generate the sample path that would result from the new reorder point. If he runs out, record how many hamburger sales that he loses. (Hint: The system input is the number of hamburgers demanded each day.)

2.8. There are some limitations to the approach proposed in this chapter, as the following example illustrates:

The Charles H. Reilly Dome (an indoor stadium) has 230 high-intensity flood-lights in its ceiling. All of these lights are replaced annually, on June 1, whether or not they are burned out. The cost of replacing the lights, including labor, special equipment and the lights themselves, is $50,000.

Good lighting is necessary for television broadcasts from the stadium. If at any time 10 or more lights have burned out, then they must be replaced immediately. The cost for these interim replacements is $10,000 plus $500 per light. Replacement data since 1988 are given in the table below:

Date	Description
June 1, 1988	annual replacement
June 1, 1989	annual replacement
February 7, 1990	replace 10 lights
April 23, 1990	replace 10 lights
June 1, 1990	annual replacement
June 1, 1991	annual replacement
June 1, 1992	annual replacement
December 21, 1992	replace 11 lights
June 1, 1993	annual replacement
June 1, 1994	annual replacement

Since annual replacement is so expensive, the management of the Reilly Dome is considering replacing all lights every 18 months, rather than annually. For the given sample path, how would the new policy change the cost of lighting? Can you create the new sample path? If not, why not?

2.9. The following design is proposed for an automated manufacturing cell: A new job will arrive at the cell at precisely 30-minute intervals, and jobs will be processed first-come-first-served. The jobs can be classified into six types, depending upon the amount of work required to complete the job, as shown in the following table:

Job type	Time to complete (minutes)
1	20
2	50
3	23
4	17
5	38
6	26

Since not all jobs can be completed in 30 minutes, the design decision is how much "buffer space" (queue) to allocate to hold waiting jobs.

The system does not yet exist, so it is not possible to obtain a sample path. However, it is believed that a reasonable model of arriving jobs is that they are equally likely to be of each type. Therefore, the job-type inputs to a simulation of this system can be generated by rolling a fair, six-sided die.

Simulate 25 jobs arriving at this manufacturing cell, and note the maximum number of jobs ever waiting for the cell. Have everyone in the class report his or her observed maximum, and discuss how this information could be used to allocate buffer space.

2.10. Jobs submitted to a proposed computer system will be distributed to two identical central-processing units (CPUs). Because of the overhead involved in checking the load on each CPU, a front-end processor will distribute incoming jobs randomly; that is, an incoming job will have a 50% chance of being assigned to CPU A, and a 50% chance of being assigned to CPU B. The design question is whether the random allocation strategy can lead to a severe imbalance in processor load.

For simplicity, suppose that a new computer job arrives every 40 microseconds, a CPU takes exactly 70 microseconds to process a job, and jobs are processed first-come-first-served (this obviously ignores realities such as different size jobs, job priorities, the need for input/output resources, time slicing, etc.). The system does not yet exist, so it is not possible to obtain a sample path. However, since the job routing is equally likely, routing inputs can be generated by flipping a fair coin.

Simulate 30 jobs arriving to this computer system, and note the maximum absolute difference in the number of jobs waiting for each CPU. Have everyone in the class report his or her observed maximum absolute difference, and discuss whether it appears that the random routing strategy is sound.

2.11. Select a simple, dynamic system that is easily observable (such as a vending machine, fast-food restaurant, ticket window, or check-out counter) and record appropriate time-study data. Decompose the system into inputs and logic. Then propose a change in the system logic that should improve performance. Generate the new sample path, and analyze it in terms of some useful performance measure.

2.12. Read the article by D. L. Martell, R. J. Drysdale, G. E. Doan and D. Boychuk. 1984. An evaluation of forest fire initial attack resources. *Interfaces* **14**, 5, 20–32. Identify the inputs and logic in this simulation.

2.13. Read the article by M. E. Cebry, A. H. deSilva and F. J. DiLisio. 1992. Management science in automating postal operations: facility and equipment planning in the United States Postal Service. *Interfaces* **22**, 1, 110–130. Identify the inputs and logic in this simulation.

CHAPTER
3

BASICS

The purpose of modeling is to deduce statements about the performance of a real or conceptual system.

In this book we are interested in the performance of dynamic systems that are subject to uncertainty. For any system about which we would like to make statements, we start by formulating a model of that system and then we deduce statements from the model. Sample-path decomposition is our primary tool for model formulation. If the model is a faithful representation of the system, then the statements deduced from the model will also apply to the system.

The distinction between model and system is important. A model is always an approximate representation of a system, and it is never perfect. We can deduce provably correct statements about models, but not about systems.

For example, we should *not* say that the time required for a pizza to be delivered has an exponential distribution, but we *might* say that the time until delivery is *well approximated* as a random variable with an exponential distribution having mean $\frac{1}{2}$ hour. The phrase "random variable with an exponential distribution having mean $\frac{1}{2}$ hour" is a model of the system from which we can derive probability statements. For instance, we may derive the probability that delivery will take more than 40 minutes, which may be an important performance measure if a free pizza is promised when delivery takes longer than 40 minutes.

Statements about stochastic models are deduced using *probability* or *statistics*. We make a distinction between probability and statistics that is useful in our context.[1]

Probability = look forward in time

> *Statistical inference often looks backward in time, while probability statements typically look forward in time.*

Inference = backward in time

Given data, we use statistics to infer statements about the model or system that generated the data. When the data are simulation output, it is clear what we mean by "generated." When the data are obtained by observing a real system, then we can treat the data *as if* they were simulation output. There is no need to differentiate "real" data from the output of a simulation.

Statistics are useful when our model is not completely specified because the values of some parameters are unknown. Probability, on the other hand, is useful for deriving statements about the data that a completely specified model *might generate if it were simulated*. Thus, probability addresses future data.

The next two sections are brief reviews of probability and statistics. They are not a substitute for previous exposure to these topics, but rather they summarize important ideas from the perspective of this book. Section 3.3 connects probability and statistics through the idea of random-variate generation, which is assumed to be a new topic. We then apply these ideas to Case 2.1 in Chapter 2.

3.1 PROBABILITY

This section provides a brief review of probability via examples. Additional concepts will be introduced in later chapters as they are required.

3.1.1 Random Variables

In professional baseball players are sometimes traded from team to team. Occasionally, trades are made without naming all of the players that will be traded, in which case the trade involves "a player to be named later." In other words, the final result is uncertain. Since we do not know which player will eventually be traded, but we may want to refer to that player, we could substitute a symbol, X,

---

[1]This principle is not always appropriate, since probability and statistics are also used in contexts where there is no explicit accounting of time. But in the limited context of this book—modeling and analysis of dynamic systems—the statement provides a useful contrast.

for the name of the unknown player. In this book we use capital letters, such as X, Y or S, to denote unknown future results ("to be named later"), and we refer to them as *random variables*. We require that random variables take only numerical values. Thus, in the baseball example, we refer to players by an identification number rather than by name.

In any trade of players there is some set of eligible "players to be named later." Suppose that there are four of them and we assign them identification numbers $1, 2, 3, 4$. To discuss the result that player 2 is the one who is ultimately traded, we write the *event* $\{X = 2\}$. The curly braces $\{\}$ are important. We will never (again) write $X = 2$ when X is a random variable; when we refer to a random variable X, we are referring to the future, and we do not know what the future will be. The curly braces and the term "event" emphasize this point. Notice that the term *event* in probability refers to an unknown future result, while the term *system event* in Chapter 2 refers to a change in the status of a system.

Probability is a way to attach a likelihood to an event. We write $\Pr\{X = 2\}$ for the probability that the event $\{X = 2\}$ occurs. Recall that:

- Probabilities take values in the interval $[0, 1]$.
- The probability that an event occurs is 1 minus the probability that is does not occur.

A random variable X is characterized by its probability distribution. One form of the probability distribution is the *cumulative distribution function* (cdf), F_X, defined as

$$\boxed{F_X(a) \equiv \Pr\{X \le a\}} \tag{3.1}$$

for any real number $-\infty < a < \infty$. The notation "$\equiv$" means "is defined to be," and the notation "$\Pr\{X \le a\}$" is read, "The probability of the event that X is less than or equal to a." The notation F is generic for a cdf, and we use subscripts on F to distinguish the cdfs of different random variables.

The cdf may seem awkward for the baseball example in which we are interested in events like "player 2 is traded" rather than events like "the index of the player traded is less than or equal to 2." But for any random variable X and constants $a < b$,

$$\boxed{\Pr\{a < X \le b\} = \Pr\{X \le b\} - \Pr\{X \le a\} = F_X(b) - F_X(a)} \tag{3.2}$$

In the baseball example the only result greater than 1 but less than or equal to 2 is $\{X = 2\}$. Therefore,

$$\Pr\{X = 2\} = \Pr\{1 < X \le 2\} = F_X(2) - F_X(1)$$

and we can recover the probability for individual players from the cdf.

The reason we emphasize the cdf is that Definition (3.1) does not depend on whether X refers to the (discrete) identification numbers of players ($a = 1, 2, 3, 4$) or the (continuous) time it takes to have a pizza delivered ($a \geq 0$), or some mixture of discrete and continuous values. This is not true of other forms of the probability distribution, such as density functions and mass functions; however, these other forms are useful so we review them later in this section.

One possible cdf for the random variable X that models the baseball trade is

$$F_X(a) = \begin{cases} 0, & a < 1 \\ 0.1, & 1 \leq a < 2 \\ 0.4, & 2 \leq a < 3 \\ 0.9, & 3 \leq a < 4 \\ 1, & 4 \leq a \end{cases} \tag{3.3}$$

which is plotted in Figure 3.1. Notice that $\Pr\{X = 2\} = F_X(2) - F_X(1) = 0.4 - 0.1 = 0.3$.

If we let Y be the random variable that models the time to deliver a pizza, then one possible cdf for Y is

$$F_Y(a) = \begin{cases} 0, & a < 0 \\ 1 - e^{-2a}, & 0 \leq a \end{cases} \tag{3.4}$$

where the delivery time a is measured in hours. This cdf is plotted in Figure 3.2.

All cdfs F have the following properties:

- Their domain is $-\infty < a < \infty$ and range is $0 \leq F(a) \leq 1$, since $F(a)$ is a probability.
- They are nondecreasing, which means that as the value of a increases, $F(a)$ never decreases. In other words, the probability of being less than a cannot get smaller if a gets larger.

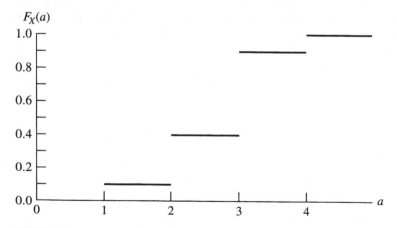

FIGURE 3.1
A cdf for the baseball-trade example.

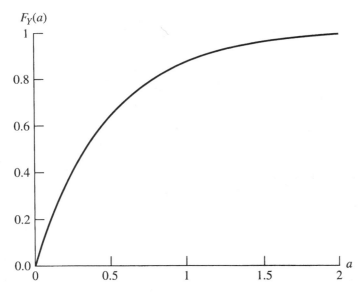

FIGURE 3.2
A cdf for the pizza-delivery example.

- They are right-continuous, which means that if they have a discontinuity at a point a, then $F(a)$ is determined by the piece of the function on the right-hand side of the discontinuity. For example, there is a discontinuity at $a = 3$ in Figure 3.1, so $F_X(3) = 0.9$, not 0.4.

Some cdfs are so useful that they are named. The cdf (3.4) is one instance of the *exponential distribution*, which has the general form

$$F_Y(a) = \begin{cases} 0, & a < 0 \\ 1 - e^{-\lambda a}, & 0 \le a \end{cases} \tag{3.5}$$

for some fixed value of the parameter $0 < \lambda < \infty$. Another important probability distribution is the *uniform distribution* on the interval $[0, 1]$. If U is a random variable having the uniform distribution on $[0, 1]$, then

$$F_U(a) = \begin{cases} 0, & a < 0 \\ a, & 0 \le a < 1 \\ 1, & 1 \le a \end{cases} \tag{3.6}$$

The uniform distribution has the property that $F_U(a) = a$ for $0 \le a \le 1$.

Suppose X is some discrete-valued random variable with range $b_1 < b_2 < \cdots$ (possibly infinite) and cdf F_X. Let a be some fixed real number. Then the *mass function* of X is the function p_X with the property that

$$F_X(a) = \sum_{\{i:b_i \le a\}} p_X(b_i) \tag{3.7}$$

The summation is over all of the indices i such that b_i is less than or equal to a. If none of the b_i are less than or equal to a, then we define the summation to be 0. This definition implies that

$$\Pr\{X = b_j\} = F_X(b_j) - F_X(b_{j-1}) = \sum_{\{i:b_i \le b_j\}} p_X(b_i) - \sum_{\{i:b_i \le b_{j-1}\}} p_X(b_i) = p_X(b_j)$$

provided we define $b_0 = -\infty$. In other words, the mass function provides the probability of individual values. All discrete-valued random variables have an associated mass function. For instance, in the baseball trade example $p_X(2) = 0.3$.

Suppose that Y is a continuous-valued random variable with cdf F_Y. Some continuous-valued random variables have an associated *density function*, f_Y, a function with the property that

$$F_Y(a) = \int_{-\infty}^{a} f_Y(b)\, db \tag{3.8}$$

Although Definitions (3.7) and (3.8) are similar, a density function is different from a mass function in that $f_Y(a) \ne \Pr\{Y = a\}$. In fact, $\Pr\{Y = a\} = 0$ for random variables that have a density function. This is a mathematical necessity, since $\int_a^a f_Y(b)\, db = 0$. But events such as $\{Y \le \frac{1}{2}\}$ can be assigned probabilities, and that is sufficient.

When a random variable Y has a density, the density function is the derivative of the cdf F_Y; that is, $f_Y(a) = dF_Y(a)/da$. Thus, in the pizza-delivery example

$$f_Y(a) = \frac{d}{da}(1 - e^{-2a}) = 2e^{-2a}$$

for $a \ge 0$, and $f_Y(a) = 0$ for $a < 0$.

3.1.2   Joint Distributions

Returning to the baseball example, suppose that three teams are involved in the trade, the Cubs, the Reds and the Giants, and each one will name a player later to complete the trade. Let X, Y and Z be the random variables that model the player ultimately traded by the Cubs, the Reds and the Giants, respectively, and let F_X, F_Y and F_Z denote their respective cdfs.

Since each team wants to obtain a player that is as valuable as the one it gives up, it is likely that the player ultimately traded by the Cubs will affect the player traded by the Reds and the Giants. Stated in the language of probability, the random variables X, Y and Z are *dependent* on each other. This dependence is not captured in the *marginal* cdfs F_X, F_Y and F_Z, which only apply to a specific team. Therefore, we need the concept of a *joint probability distri-*

bution for (X, Y, Z) together. One representation is the joint cdf $F_{XYZ}(a, b, c)$ defined as

$$F_{XYZ}(a, b, c) \equiv \Pr\{X \le a, Y \le b, Z \le c\} \qquad (3.9)$$

for any real numbers $-\infty < a, b, c < \infty$. The joint event $\{X \le a, Y \le b, Z \le c\}$ is interpreted to mean that the component events $\{X \le a\}$ *and* $\{Y \le b\}$ *and* $\{Z \le c\}$ all occur. This idea generalizes to any finite number of random variables.

Joint probability distributions are often difficult to formulate and to manipulate. One easy case is when the random variables are *independent*, meaning that knowledge of the values of, say, X and Y, does not change the probability distribution for the value of Z. In the baseball example independence might be plausible if none of the trades interact (for example, the Cubs are trading with the Pirates, the Reds are trading with the Dodgers, and the Giants are trading with the Red Sox, and none of these teams are trading with each other). If X, Y and Z are independent, then

$$F_{XYZ}(a, b, c) = F_X(a)F_Y(b)F_Z(c) = \Pr\{X \le a\}\Pr\{Y \le b\}\Pr\{Z \le c\}$$

Notice that $\Pr\{Z \le c\}$ on the right-hand side is not a function of a or b; this is what we mean by independence of Z from X and Y. Stated succinctly:

A joint probability statement about any finite number of independent random variables decomposes into the product of the marginal probability statements. This fundamental decomposition applies to joint cdfs, joint density functions, joint mass functions and joint probability statements in general.

Although we exploit the simplification of independence when it is appropriate, characterizing the dependence among random variables is central to the study of stochastic processes.

3.1.3 Expected Value

The cdf of a random variable X captures everything that there is to know about X. However, it is sometimes useful to work with summary measures of F_X rather than the entire distribution, just as it is sometimes convenient to work with summary measures of data rather than all of the data. The *expected value* of X, denoted E[X], is a summary measure indicating the center of the distribution F_X, and it is sometimes called the *mean* of X. The "E" is called the *expectation operator*.

If X is a discrete-valued random variable taking values $a_1, a_2, \ldots$ (possibly infinite) with mass function p_X, then

$$E[X] \equiv \sum_{\text{all } i} a_i \, p_X(a_i)$$

a weighted average of the possible values of X with each value weighted by its probability of being generated. For example, if X has a *Bernoulli distribution* with parameter $0 < \gamma < 1$, then its mass function is

$$p_X(b) = \begin{cases} 1 - \gamma, & \text{if } b = 0 \\ \gamma, & \text{if } b = 1 \\ 0, & \text{otherwise} \end{cases}$$

and the

$$E[X] = 0 \cdot p_X(0) + 1 \cdot p_X(1) = \gamma$$

If Y is a continuous-valued random variable with density f_Y, then

$$E[Y] \equiv \int_{-\infty}^{\infty} a f_Y(a) \, da$$

the integral of the possible values of Y with each value weighted by the density f_Y. For example, if Y has an exponential distribution with parameter $0 < \lambda < \infty$, then its density function is

$$f_Y(a) = \begin{cases} 0, & a < 0 \\ \lambda e^{-\lambda a}, & 0 \le a \end{cases}$$

and the

$$E[Y] = \int_0^{\infty} a \lambda e^{-\lambda a} \, da = \frac{1}{\lambda}$$

(integration by parts is required to derive this result).

The expected value of a random variable is the most used (and abused) summary measure of a distribution. It provides a single number to represent all of the possible values a random variable might take, and it is often treated as a "best guess" of what will happen in the future. But it is not always a meaningful summary measure. Recall the random variable X that represents the baseball-player trade described in Section 3.1.1. The range of X is $\{1, 2, 3, 4\}$, a set of index numbers that correspond to particular players that might be traded. Direct calculation gives $E[X] = 1(0.1) + 2(0.3) + 3(0.5) + 4(0.1) = 2.6$. Although the expected value of this random variable is a well-defined summary measure, it has no physical meaning in this example. There is no player 2.6, and a weighted average of the players' indices has no useful interpretation. The term "expected value" does *not* mean the most likely value, and it may not even be a possible value.

The concept of the expected value of a random variable can be extended to the expected value of a function of a random variable. For a function g that is defined over the range of X, the $E[g(X)]$ can be computed as follows: If X is a discrete-valued random variable taking values $a_1, a_2, \ldots$ (possibly infinite) with mass function p_X, then

$$E[g(X)] = \sum_{\text{all } i} g(a_i) p_X(a_i)$$

If Y is a continuous-valued random variable with density f_Y, then

$$E[g(Y)] = \int_{-\infty}^{\infty} g(a) f_Y(a) \, da$$

An important function of a random variable X is $g(X) = (X - E[X])^2$, the squared deviation of X from its expected value. For simplicity, let $\mu \equiv E[X]$. The $E\left[(X - \mu)^2\right]$ is called the *variance* of X, and is denoted $\text{Var}[X]$. The $\sqrt{\text{Var}[X]}$ is called the *standard deviation* of X. Both are summary measures of the spread of the distribution F_X around the expected value μ. The larger the standard deviation (or variance) of X is, the farther values of X tend to stray from μ.

For example, if X has a Bernoulli distribution with parameter $0 < \gamma < 1$, then

$$\text{Var}[X] = (0 - \gamma)^2 \cdot p_X(0) + (1 - \gamma)^2 \cdot p_X(1) = \gamma^2(1 - \gamma) + (1 - \gamma)^2 \gamma = \gamma(1 - \gamma)$$

Notice that the variance is largest when $\gamma = \frac{1}{2}$, which is the distribution often used to model the outcome of flipping a fair coin. As a second example, if Y has an exponential distribution with parameter λ, then

$$\text{Var}[Y] = \int_0^{\infty} \left(a - \frac{1}{\lambda}\right)^2 \lambda e^{-\lambda a} \, da = \frac{1}{\lambda^2}$$

Notice that the variance increases as λ decreases.

The variance and standard deviation are measures of the spread of the distribution of a random variable around its expected value. Whether or not this spread is large, relative to the expected value, is captured by the *coefficient of variation* of X,

$$CV[X] \equiv \frac{\sqrt{\text{Var}[X]}}{|E[X]|} \tag{3.10}$$

The coefficient of variation expresses the standard deviation in units of (the absolute value of) the expected value. Thus, it is a measure of relative variability.

For example, if X has a Bernoulli distribution then

$$\text{CV}[X] = \frac{\sqrt{\gamma(1-\gamma)}}{\gamma} = \sqrt{\frac{1-\gamma}{\gamma}}$$

which decreases as γ goes to 1. If Y has an exponential distribution then

$$\text{CV}[Y] = \frac{\sqrt{\frac{1}{\lambda^2}}}{\frac{1}{\lambda}} = 1$$

for all values of λ. A coefficient of variation of 1 is an important standard: Random variables with coefficient of variation 1 or greater are considered highly variable, relative to their expectation.

Another important function of a random variable X is $g(X) = \mathcal{I}(X \leq b)$. Recall that $\mathcal{I}$ is the indicator function that takes the value 1 if its argument is true, and 0 otherwise. For discrete-valued random variable X,

$$\text{E}[\mathcal{I}(X \leq b)] = \sum_{\text{all } i} \mathcal{I}(a_i \leq b) p_X(a_i)$$

$$= \sum_{\{i:a_i \leq b\}} 1 \, p_X(a_i) - \sum_{\{i:a_i > b\}} 0 \, p_X(a_i)$$

$$= \text{Pr}\{X \leq b\} = F_X(b)$$

An analogous result holds for continuous-valued random variable Y: $\text{E}[\mathcal{I}(Y \leq b)] = \text{Pr}\{Y \leq b\}$. Therefore, *probabilities can be expressed as the expected value of an indicator function.*

A direct application of the definition of expected value establishes that $\text{E}[aX + b] = a\text{E}[X] + b$, for any constants a and b (Exercise 24 asks you to show this). But in general $\text{E}[g(X)] \neq g(\text{E}[X])$, although this substitution is often tempting. As a reminder, notice that if we define $g(X) = (X - \mu)^2$, then assuming (incorrectly) that $\text{E}[g(X)] = g(\text{E}[X])$ implies that $\text{E}[g(X)] = \text{E}[(X - \mu)^2] = g(\mu) = (\text{E}[X] - \mu)^2 = 0$, showing (incorrectly) that the variance of every random variable is 0!

3.1.4 Conditional Probability

The final concept that we need to review is *conditional probability*. Let V and W be two random variables with joint cdf F_{VW} and marginal cdfs F_V and F_W. Conditional probability addresses the question, how should we revise our probability statements about W given that we have some knowledge of the value of V? This type of question arises frequently in the study of stochastic processes.

Let a and b be constants, and suppose that the marginal probability $\text{Pr}\{V \leq a\} > 0$. We define the conditional probability of the event $\{W \leq b\}$, given that the event $\{V \le a\}$ has occurred, to be

$$Pr\{W \le b|V \le a\} \equiv \frac{Pr\{W \le b, V \le a\}}{Pr\{V \le a\}} = \frac{F_{VW}(a, b)}{F_V(a)} \tag{3.11}$$

In an event like $\{W \le b|V \le a\}$, the event appearing to the right of the "|" is called a *condition* or *conditional event*, meaning that this event is assumed to have occurred.

Equation (3.11) revises the probability of the event $\{W \le b\}$ given the conditional event $\{V \le a\}$. The revised probability is the probability of the joint event $\{W \le b, V \le a\}$ relative to the probability of the conditional event $\{V \le a\}$. This revision is in fact quite reasonable.

If V and W are independent, then

$$\frac{Pr\{W \le b, V \le a\}}{Pr\{V \le a\}} = \frac{Pr\{W \le b\}\,Pr\{V \le a\}}{Pr\{V \le a\}} = Pr\{W \le b\}$$

This makes sense, because independence implies that the distribution of W does not depend on V in any way. If V and W are perfectly dependent (that is, $V \equiv W$) and $b \ge a$ then

$$\frac{Pr\{W \le b, V \le a\}}{Pr\{V \le a\}} = \frac{Pr\{W \le b, W \le a\}}{Pr\{W \le a\}} = \frac{Pr\{W \le a\}}{Pr\{W \le a\}} = 1$$

This also makes sense, because the fact that the event $\{W \le a\}$ has occurred should make the event $\{W \le b\}$ certain for any $b \ge a$.

How does this definition behave as a function of b? It is certainly nondecreasing in b, since $Pr\{W \le b, V \le a\}$ cannot get smaller if b gets larger. If we let $b \to \infty$, then

$$\frac{Pr\{W \le b, V \le a\}}{Pr\{V \le a\}} \to \frac{Pr\{W \le \infty, V \le a\}}{Pr\{V \le a\}} = \frac{Pr\{V \le a\}}{Pr\{V \le a\}} = 1$$

since the event $\{W \le \infty\}$ is certain. Similarly, if we let $b \to -\infty$, then

$$\frac{Pr\{W \le b, V \le a\}}{Pr\{V \le a\}} \to \frac{Pr\{W \le -\infty, V \le a\}}{Pr\{V \le a\}} = \frac{0}{Pr\{V \le a\}} = 0$$

Thus, as a function of b, $Pr\{W \le b|V \le a\}$ behaves like a cdf, and therefore defines a *conditional cdf* for W. Let

$$F_{W|V \le a}(b) \equiv Pr\{W \le b|V \le a\}$$

denote the conditional cdf of W given $\{V \le a\}$. Notice that different values of a imply different conditional cdfs. We can similarly define a conditional density or mass function, conditional expected values, and so on.

The conditional probability of more general events than $\{W \le b\}$ is defined in the following way: Let $\mathcal{A}$ and $\mathcal{B}$ be collections of possible values for V and W, respectively (for instance, $\mathcal{A}$ could be the interval $(5, 27]$, and $\mathcal{B}$ could be the set of values $\{44, 73\}$). Then the conditional probability that W takes a value in $\mathcal{B}$, given that V takes a value in $\mathcal{A}$, is

$$\Pr\{W \in \mathcal{B} | V \in \mathcal{A}\} \equiv \frac{\Pr\{W \in \mathcal{B}, V \in \mathcal{A}\}}{\Pr\{V \in \mathcal{A}\}} \tag{3.12}$$

provided $\Pr\{V \in \mathcal{A}\} > 0$.

Definition (3.12) implies that we can decompose joint probabilities into products of conditional probabilities and marginal probabilities; specifically

$$\Pr\{W \in \mathcal{B}, V \in \mathcal{A}\} = \Pr\{W \in \mathcal{B} | V \in \mathcal{A}\} \Pr\{V \in \mathcal{A}\} \tag{3.13}$$

We will frequently encounter situations in which Equation (3.13) is the best way to obtain a joint probability.

Pressing this idea further, we can also decompose marginal probabilities. Suppose that V is a discrete-valued random variable taking values $a_1, a_2, \ldots$ (possibly infinite). The marginal probability of the event $\{W \in \mathcal{B}\}$ can be written as

$$\Pr\{W \in \mathcal{B}\} = \sum_{\text{all } i} \Pr\{W \in \mathcal{B}, V = a_i\}. \tag{3.14}$$

In words, the marginal probability of the event $\{W \in \mathcal{B}\}$ is the sum of the probabilities of the joint events $\{W \in \mathcal{B}, V = a_i\}$ over all possible values of V, since some value of V must occur in conjunction with W. Combining (3.13) and (3.14) gives

$$\begin{aligned}
\Pr\{W \in \mathcal{B}\} &= \sum_{\text{all } i} \Pr\{W \in \mathcal{B}, V = a_i\} \\
&= \sum_{\text{all } i} \Pr\{W \in \mathcal{B} | V = a_i\} \Pr\{V = a_i\}
\end{aligned} \tag{3.15}$$

Equation (3.15) is one form of the *law of total probability*, which is an extremely powerful decomposition. We will often encounter situations in which it is easier to build up the marginal probabilities we want from conditional probabilities that are readily available, rather than derive the marginal probabilities directly.

When V is a continuous-valued random variable that has a density function f_V, a second form of the law of total probability is

$$\Pr\{W \in \mathcal{B}\} = \int_{-\infty}^{\infty} \Pr\{W \in \mathcal{B} | V = a\} f_V(a)\, da \tag{3.16}$$

provided $\Pr\{W \in B|V = a\}$ is well defined. Notice that since $\Pr\{V = a\} = 0$ when V has a density, the definition of $\Pr\{W \in B|V = a\}$ is problem-dependent and is not (3.12).

To illustrate these concepts, suppose that a company that makes "in-line skates" (roller skates with the wheels configured in a straight line) sells three different types of replacement wheels ("cool," "hot" and "ouch!") and two different types of replacement bearings ("warp 2" and "warp 9"). Wheels and bearings must be ordered as a set, but customer's can decide which combination of wheel type and bearing type that they want. Last year 4000 replacement sets were sold in the following quantities:

		wheel			
		cool	hot	ouch!	total
bearing	warp 2	800	400	400	1600
	warp 9	200	1600	600	2400
	total	1000	2000	1000	4000

To model the sale of replacement sets, define two random variables V and W to represent the type of bearing and wheel, respectively, in a replacement set. The joint mass function p_{VW} of (V, W) can be estimated by dividing the entries in the table above by 4000, giving

p_{VW}		W			
		1	2	3	p_V
V	1	2/10	1/10	1/10	4/10
	2	1/20	8/20	3/20	12/20
	p_W	5/20	10/20	5/20	

where numerical values have been assigned to the names of the wheel and bearing types. The marginal mass functions p_V and p_W are given by the row sums and column sums, respectively.

The conditional distribution of wheel type given the bearing type is obtained from the definition of conditional probability,

$$\Pr\{W = b|V = a\} = \frac{\Pr\{V = a, W = b\}}{\Pr\{V = a\}} = \frac{p_{VW}(a, b)}{p_V(a)}$$

For example, the probability that a customer will order "hot" wheels after ordering "warp 2" bearings is

$$\Pr\{W = 2|V = 1\} = \frac{\Pr\{V = 1, W = 2\}}{\Pr\{V = 1\}} = \frac{1/10}{4/10} = \frac{1}{4}$$

The conditional distribution of W for each possible value of V is

	1	2	3
	b		
$\Pr\{W = b \mid V = 1\}$	2/4	1/4	1/4
$\Pr\{W = b \mid V = 2\}$	1/12	8/12	3/12

Each row of the table corresponds to a conditional mass function $p_{W \mid V = a}(b)$.

Suppose that the revenue from selling "cool," "hot" and "ouch!" wheels is \$40, \$60 and \$100, respectively. We can represent the revenue from wheel sales as a function $g(W)$ where $g(1) = 40$, $g(2) = 60$ and $g(3) = 100$. Then the expected wheel revenue, given that the customer purchased "warp 2" bearings, is

$$E[g(W) \mid V = 1] = \sum_{b=1}^{3} g(b) p_{W \mid V = 1}(b) = (40)\frac{2}{4} + (60)\frac{1}{4} + (100)\frac{1}{4} = \$60$$

Notice also that marginal probabilities for wheel type can be recovered from the conditional probabilities by using the law of total probability. For example

$$\Pr\{W = 2\} = \sum_{a=1}^{2} \Pr\{W = 2 \mid V = a\} \Pr\{V = a\}$$

$$= \Pr\{W = 2 \mid V = 1\} \Pr\{V = 1\} + \Pr\{W = 2 \mid V = 2\} \Pr\{V = 2\}$$

$$= \frac{1}{4}\frac{4}{10} + \frac{8}{12}\frac{12}{20} = \frac{10}{20}$$

3.1.5 Limit Distributions

Up to this point in our review of probability we have focused on making statements about random variables with known probability distributions. This section reviews two results that are useful for making statements about random variables whose distributions are not known. These results provide a probabilistic justification for many statistical procedures and therefore are a bridge to the next section.

Suppose that $X_1, X_2, \ldots$ is an infinite sequence of random variables that are mutually independent and have a common cdf F_X. Therefore, they also have a common expected value $\mu = E[X_i]$ and common variance $\sigma^2 = \text{Var}[X_i]$. Let $\bar{X}_n \equiv \frac{1}{n}\sum_{i=1}^{n} X_i$, the average of the first n of the X_i's. This defines a new sequence of random variables $\bar{X}_1, \bar{X}_2, \ldots$. Further, let

$$\bar{Z}_n = \frac{\bar{X}_n - \mu}{\sigma/\sqrt{n}}$$

the standardized average of the first n of the X_i's, which also defines a new sequence of random variables $\bar{Z}_1, \bar{Z}_2, \ldots$. Even without knowing F_X, we can say quite a bit about $\bar{X}_n$ and $\bar{Z}_n$ as n gets large:

With probability 1,
$$\lim_{n \to \infty} \bar{X}_n = \mu$$
This result is called the **Strong Law of Large Numbers**.

The
$$\lim_{n \to \infty} \Pr\{\bar{Z}_n \le a\} = \int_{-\infty}^{a} \frac{1}{\sqrt{2\pi}} e^{-b^2/2} \, db$$
This result is called the **Central Limit Theorem**. *The expression on the right-hand side is the cdf of the standard normal distribution.*

When the sequence $X_1, X_2, \ldots$ is used to model simulation output data or data collected from a real system, the Strong Law of Large Numbers provides a justification for using the sample average to estimate the expected value (mean) of the process. It states that as the sample size gets larger, the sample average must inevitably converge to the true mean. The Central Limit Theorem goes even further. It provides a probability distribution for $\bar{Z}_n$ as n gets larger, permitting us to make statements about just how far $\bar{X}_n$ might be from μ.

3.2 STATISTICS

This section provides a brief review of the use of statistics via an example. Additional concepts will be introduced in later chapters as they are required. Since basics statistics is assumed as background, the emphasis in this section is on certain pitfalls that arise in the routine application of statistics.

The mean grade-point averages (GPAs) for students in the College of Liberal Arts at the University of Minnesota for the academic years 1987–1992 were 2.66, 2.69, 2.75, 2.79, 2.83 and 2.78, out of a possible 4.00.[2] Is it likely that the GPA in the year 2010 will be above 2.90?

To answer a question such as the one posed above, we need a probability model of the GPA system (recall that we only make statements about models). Here is one possible model: *The GPA for College of Liberal Arts students at the University of Minnesota in any year can be represented as a random variable, X, that has cdf F_X. Further, the GPA in any year is independent of the GPA in any other year.* If this model is a good representation of the GPA system and we can

---

[2]Source: *The Minnesota Daily*, University of Minnesota, September 30, 1992.

find an appropriate cdf F_X, then we can make a probability statement about GPA in the year 2010, or any year in the future.

What are the implications of adopting this model? First, stating that the probability distribution F_X does not depend on the year implies that there is no trend in student GPAs. *The Minnesota Daily* argued that there is an upward trend. When the distribution of a process that evolves over time does not depend on time, the process is *time stationary*. This concept will be defined more precisely later in the book. In the GPA example time stationarity implies that it is reasonable to use the 1987–1992 data to determine a distribution F_X that also applies to the year 2010. If there is an upward trend, however, then a more sophisticated model is required to predict future GPAs.

Second, the conjecture that the GPAs in each year are independent implies that knowledge of the GPA in another year, say year 2009, would not cause us to change our probability distribution for the year 2010 GPA. In other words, we would still use F_X to make probability statements about the GPA in the year 2010, even if it was currently the year 2009 and we knew the GPA. Under independence only the marginal distribution for the year 2010 matters (see the discussion in Section 3.1.2). But independence is not necessarily a good approximation here because some of the same students and instructors that contribute to the year-2009 GPA will contribute to the year-2010 GPA. On the other hand, the year-2010 GPA may be (nearly) independent of the 1987–1992 GPAs, since so much time will have passed.

For the purpose of illustration, suppose that we adopt the independent, time-stationary model despite its obvious shortcomings. Then we must develop an appropriate F_X from the 1987–1992 data. This is a problem in statistics, while using F_X to make a statement about the year 2010 is a problem in probability.

In statistics problems we first propose a *model for the data*—which we may later reject—and then use statistical reasoning to infer statements about that model. *Parameterizing*, or "fitting" probability distributions from data, is a common statistical problem, and it is essential for practical application of stochastic modeling, analysis and simulation. There are many specific ways to do it, each with advantages and disadvantages, but for our purposes the following general method will suffice:

1. Based on known properties of the system or through exploratory analysis of the data, select a candidate family of distributions with some undetermined parameters.
2. Determine values of the parameters so that the parameterized distribution closely matches the distribution of the data.
3. Evaluate the adequacy of the resulting distribution relative to the distribution of the data.

When we parameterize a distribution, we assume that there are in fact correct values for the parameters. The values we ultimately assign to these parameters, based on a sample of data, are called *estimates,* and the process of determining the values is called *estimation*. We use the convention that a "^" (called a "hat") over a quantity indicates that its value has been estimated.

For example, suppose that we decide that F_X to represent the GPA data is from the family of normal distributions. Notice that this choice of distribution family *might* be supported by the Central Limit Theorem (Section 3.1.5) because the GPA for the College of Liberal Arts is the average of a large number of individual students' GPAs. The normal family has two parameters, its mean (denoted μ) and variance (denoted σ^2). The *sample average* and *sample variance* of the GPA data are *estimates* of these parameters. The sample average was defined in Section 2.2. The sample variance of m numbers, $d_1, d_2, \ldots, d_m$, is

$$\frac{1}{m-1} \sum_{i=1}^{m} (d_i - \bar{d})^2 = \frac{1}{m-1} \left(\sum_{i=1}^{m} d_i^2 - \frac{(\sum_{j=1}^{m} d_j)^2}{m} \right) \tag{3.17}$$

where $\bar{d}$ is the sample average of the data. The sample variance is the average of the *squared deviations* of each data value from the sample mean. The square root of the sample variance is called the *sample standard deviation*, and it is (roughly) the average deviation of the data values from the sample mean.

For the GPA data, the sample average, variance and standard deviation are 2.750, 0.004 and 0.064, respectively. Therefore, one way to parameterize the normal distribution based on this data is to set $\hat{\mu} = 2.75$ and $\hat{\sigma}^2 = 0.004$ so that our normal-distribution model has mean and variance that match the sample average and sample variance of the data. For X having such a normal distribution,

$$\Pr\{X > 2.90\} = 1 - \Pr\{X \leq 2.90\} = 1 - F_X(2.90) \approx 1 - 0.990 = 0.010$$

meaning it is unlikely that we will observe a GPA greater than 2.90, provided our model is a good representation of the GPA system. *Notice that we used statistics to parameterize a model with unspecified parameters (the normal distribution with unspecified μ and σ^2), and then used probability to make statements about future results based on the (now completely specified) model.*

The step in our procedure that often causes uneasiness is selecting a family of distributions; we selected the normal distribution in the case of the GPA data. One way to avoid this choice is to use the cdf implied by the data itself, which is called the *empirical cdf*. The empirical cdf has the pleasing property that if the data really are a time-stationary sample with marginal cdf F_X, then as more and more data are collected, the empirical cdf gets closer and closer to F_X. We do not obtain such convergence if we select the wrong family of distributions to parameterize.

The empirical cdf, denoted $\widehat{F}_X$, is constructed as follows: Let $d_1, d_2, \ldots, d_m$ be the observed data values. Then for any $-\infty < a < \infty$, the empirical cdf is

$$\widehat{F}_X(a) \equiv \frac{1}{m} \sum_{i=1}^{m} \mathcal{I}(d_i \leq a) \tag{3.18}$$

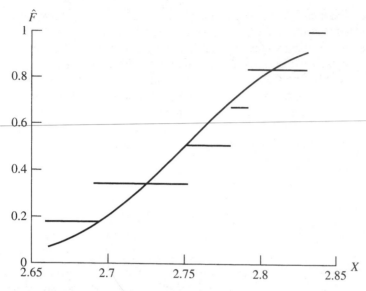

FIGURE 3.3
Empirical cdf and parameterized cdf for the GPA data.

That is, $\widehat{F}_X(a)$ is the fraction of the m observed values that are less than or equal to a.

The step function in Figure 3.3 is the empirical cdf of the GPA data. Notice that there is a probability jump of $1/6$ (in general $1/m$) at each observed data value. The empirical cdf is the distribution of a discrete-valued random variable, assigning probability $1/6$ (in general $1/m$) to each observed value. A simple algorithm for plotting the empirical cdf is the following:

algorithm plot_empirical_cdf

1. Sort the data so that $d_1 \le d_2 \le \cdots \le d_m$.
2. Plot i/m versus d_i for $i = 1, 2, \ldots, m$ as a step function.

A disadvantage of the empirical cdf is that it only assigns probabilities to the values that were observed, whereas a parameterized distribution can fill in the gaps between and beyond the observed values. For example, the empirical cdf of the GPA data gives $\widehat{Pr}\{X > 2.90\} = 1 - \widehat{F}_X(2.90) = 1 - 1 = 0$, since all the observed values were less than 2.90.

One important use of the empirical cdf is for evaluating the fit of a parameterized distribution. The smooth curve in Figure 3.3 is the cdf of a normal distribution with mean 2.75 and variance 0.004. A histogram is also useful for evaluating the fit of a parameterized distribution, because a histogram is the empirical version of the density or mass function. The height h_j over the interval

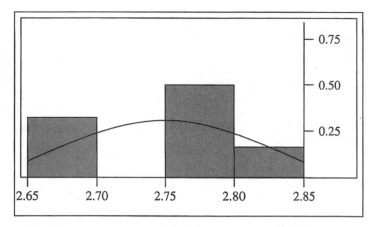

FIGURE 3.4
Histogram and parameterized density function for the GPA data.

$(c_{j-1}, c_j]$ is the fraction of the m observed values that fell in the interval; therefore, h_j estimates $\Pr\{c_{j-1} < X \le c_j\}$. Figure 3.4 is a histogram of the GPA data
with the parameterized density function of the normal distribution superimposed
over it.

A more subtle reason to be uneasy about using either a parameterized distribution or the empirical cdf is the effect of sample size. For instance, the
six available observations of GPA are a very small sample from which to parameterize a distribution and an absurdly small sample from which to assess
whether or not we have a good fit. From Figures 3.3–3.4 it is impossible to
confirm that the normal distribution represents the data. But we should also be
uncomfortable with the empirical cdf because it assigns probability to only the
six observed GPAs when we know that any GPA between 0 and 4 is achiev-
able.

At the other extreme, if we have a very large sample of data—as is often the
case when a computer automatically collects data—then we are likely to encounter
a situation in which no parameterized distribution seems to fit. This problem arises
because a parameterized distribution is an *imperfect representation* of the data, and
when we have too much data, the imperfections become obvious. Unfortunately,
an empirical cdf may be impractical in this case if there are, say, thousands of
data values in its specification.

The sensitivity of our distribution models to the available sample of data
indicates that it is important to evaluate the sensitivity of our conclusions to
the distribution we select. When we use a parameterized distribution, we should,
at a minimum, assess the effect of the error in the parameter estimates. If $\widehat{\theta}$
is an estimator of a parameter θ, then it is almost certainly in error. Fortunately, in many contexts the data used to form $\widehat{\theta}$ can also be used to estimate its
error.

Prior to obtaining a sample of data, we may think of $\widehat{\theta}$ as a random variable. The *standard error (se)* of $\widehat{\theta}$ as an estimator of θ is defined to be

$$\text{se} \equiv \sqrt{\text{Var}[\widehat{\theta}]}$$

which is simply the standard deviation of $\widehat{\theta}$. The se is (roughly) the average deviation of an estimator from the quantity it is to estimate. For appropriate constant c, $\widehat{\theta} \pm c \, \text{se}$ is often used as a range within which the true value of θ may lie. Typically the se is also unknown, but an estimate of it, denoted $\widehat{\text{se}}$, can often be obtained from the following result:

> The se of a sample average of m independent, time-stationary random variables, each with marginal variance σ^2, is $\sigma/\sqrt{m}$.

For further discussion of the se of a sample average see Chapter 9 or any basic statistics text.

Specializing this result to the case of the mean GPA μ, an estimate of the se of $\widehat{\mu}$ is

$$\widehat{\text{se}} = \sqrt{\frac{\widehat{\sigma}^2}{6}} = \sqrt{\frac{0.004}{6}} \approx 0.026$$

where we have substituted the estimate $\widehat{\sigma}^2$ for the unknown variance σ^2. We might now want to investigate how much our conclusion about the year-2010 GPA changes if we use $\widehat{\mu}$ as large as $2.75 + c\,0.026$ or as small as $2.75 - c\,0.026$. Whenever we present a parameter estimator in this book, we always provide a corresponding estimator of its standard error that can be used for sensitivity analysis.

The choice $c = 2$ is common because of the intuitive appeal of plus-or-minus two standard errors. In some contexts we can select c so that we form a *confidence interval* for θ. One frequently used confidence interval is

$$\widehat{\theta} \pm t_{1-\alpha/2,\,\nu}\,\widehat{\text{se}}$$

where $t_{1-\alpha/2,\,\nu}$ is the $1 - \alpha/2$ quantile of the t-distribution with ν degrees of freedom. The "confidence" in a confidence interval is the probability that the procedure used to construct the interval will yield an interval that contains θ. For the interval above, the probability is $1 - \alpha$, so we refer to it as a $(1 - \alpha)100\%$ confidence interval. The width of a confidence interval is often used as a measure of sampling error.

A situation in which this confidence interval is valid is when $\widehat{\theta}$ is a sample average of m independent, time-stationary, normally distributed random variables.

In that case $\nu = m - 1$ degrees of freedom. For example, in the case of the mean GPA μ, the estimator $\widehat{\mu}$ is a sample average of $m = 6$ observations. Therefore, a 95% confidence interval for μ is

$$2.75 \pm 0.067$$

since $t_{0.975,5} \, \widehat{se} = 2.57(0.026) \approx 0.067$.

3.3 RANDOM-VARIATE GENERATION

At the beginning of the chapter we stated that probability is used to derive statements about the data that a model might generate, while statistics are used to infer statements about the model or system that generated a sample of data. We have been intentionally vague about what we mean by a model generating data. This section and Chapter 4 make the concept explicit, thereby providing a connection between probability and statistics.

Our goal in this section is as follows: Given a cdf, F, generate m numbers in such a way that, prior to generating them, probability statements about their values can be derived by treating them as independent and time-stationary random variables with marginal cdf F, and after generating them, statistical analysis supports the hypothesis that they are an independent, time-stationary sample from the model F (for instance, their empirical cdf is close to F).

To accomplish this goal, we need to have a way to generate values from one specific probability distribution, the uniform distribution on the interval $[0, 1]$, denoted by F_U (see (3.6)). Suppose that we have been blessed with a function, `random()`, that will yield such numbers. We reveal more about the nature of `random()` later.

We first show that we can define any random variable as a function of another random variable U that has the uniform distribution on $[0, 1]$. Since `random()` generates values for U, we can therefore generate values for any random variable. Before offering a proof, we develop intuition from the baseball-trade example.

Define the random variable X, which represents the baseball player that will be traded, in terms of the random variable U as follows:

$$X \equiv \begin{cases} 1, & \text{if } F_X(-\infty) \le U \le F_X(1) \\ 2, & \text{if } F_X(1) < U \le F_X(2) \\ 3, & \text{if } F_X(2) < U \le F_X(3) \\ 4, & \text{if } F_X(3) < U \le F_X(\infty) \end{cases} \tag{3.19}$$

$$= \begin{cases} 1, & \text{if } 0 \le U \le 0.1 \\ 2, & \text{if } 0.1 < U \le 0.4 \\ 3, & \text{if } 0.4 < U \le 0.9 \\ 4, & \text{if } 0.9 < U \le 1 \end{cases}$$

What is the probability distribution of this random variable? For example, the probability that player 2 is traded is

$$\Pr\{X = 2\} = \Pr\{F_X(1) < U \le F_X(2)\} \qquad \text{(Equation (3.19))}$$

$$= \Pr\{U \le F_X(2)\} - \Pr\{U \le F_X(1)\} \quad \text{(Equation (3.2))}$$

$$= F_U(F_X(2)) - F_U(F_X(1)) \qquad \text{(definition of cdf)}$$

$$= F_X(2) - F_X(1) \qquad \text{(properties of uniform cdf)}$$

$$= 0.4 - 0.1 = 0.3$$

exactly as desired. The probabilities of the other results are also consistent with (3.3).

Figure 3.5 illustrates what we have done. The cdf F_X has been used to partition the range $[0, 1]$ of U into intervals with the correct probabilities. A large interval, like $(F_X(2), F_X(3)] = (0.4, 0.9]$, corresponds to a value of X with large probability, $\Pr\{X = 3\} = 0.9 - 0.4 = 0.5$. A small interval, like $(0, F_X(1)] = (0, 0.1]$, corresponds to a value of X with small probability, $\Pr\{X = 1\} = 0.1 - 0.0 = 0.1$.

This idea generalizes to any random variable X whose range is a discrete set of values, $a_1 < a_2 < a_3 < \cdots$ (possibly infinite): Set X equal to the smallest value of a_i such that $F_X(a_{i-1}) < U \le F_X(a_i)$, provided we define $a_0 \equiv -\infty$. Then it follows that

$$\Pr\{X = a_i\} = \Pr\{F_X(a_{i-1}) < U \le F_X(a_i)\} = F_X(a_i) - F_X(a_{i-1}) = p_X(a_i)$$

from Equation (3.2).

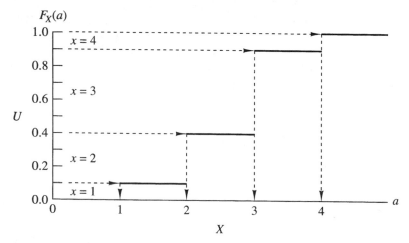

FIGURE 3.5
Defining X in terms of U for the baseball-trade example.

The algorithm below generates values from any discrete cdf given a function `random()` to return values of U from the uniform distribution on $[0, 1]$. In algorithms, an $\leftarrow$ indicates that a value is assigned to a variable.

algorithm discrete_inverse_cdf

1. $U \leftarrow$ `random()`
   $i \leftarrow 1$
2. until $U \leq F_X(a_i)$
   do
        $i \leftarrow i + 1$
   enddo
3. $X \leftarrow a_i$
4. return X

For example, suppose F_X is the cdf for the baseball trade, so that $a_1 = 1$, $a_2 = 2$, $a_3 = 3$ and $a_4 = 4$ representing the players eligible to be traded. If `random()` returns the value 0.61719902, then Step 2 will be executed three times, until $0.61719902 \leq F_X(3) = 0.9$, and Step 3 will assign $X \leftarrow 3$. Thus, player 3 will be traded.

When we have a random variable Y that has a continuous range, such as the exponential distribution (3.4) that models the pizza-delivery time, there are no jumps in F_Y to use to partition $[0, 1]$. But, by analogy, we want to set Y equal to that value a such that $F_Y(a) \leq U \leq F_Y(a)$. That is, we want $F_Y(Y) = U$. For the pizza-delivery example

$$F_Y(Y) = 1 - e^{-2Y} = U$$
$$e^{-2Y} = 1 - U$$
$$-2Y = \ln(1 - U)$$
$$Y = -\ln(1 - U)/2$$

where ln is the natural logarithm function. Figure 3.6 illustrates how Y is defined in terms of U for this distribution.

For any continuous, increasing cdf, let the solution of $F_Y(Y) = U$ in terms of Y be denoted by $Y = F_Y^{-1}(U)$, the *inverse cdf*. The following derivation shows that the inverse cdf defines a random variable Y with cdf F_Y:

$$\Pr\{Y \leq a\} = \Pr\left\{F_Y^{-1}(U) \leq a\right\} \qquad \text{(definition of } Y\text{)}$$
$$= \Pr\left\{F_Y\left[F_Y^{-1}(U)\right] \leq F_Y(a)\right\} \quad \text{(since } F_Y \text{ increasing)}$$
$$= \Pr\{U \leq F_Y(a)\} \qquad \text{(inverse of inverse is the identity)}$$
$$= F_Y(a) \qquad \text{(property of uniform cdf)}$$

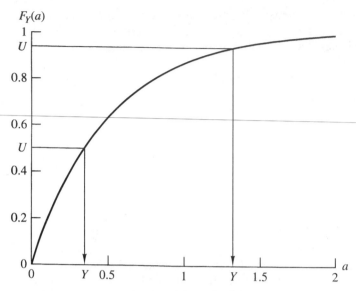

FIGURE 3.6
Defining Y in terms of U for the pizza-delivery example.

The following algorithm generates values from any continuous, increasing cdf:

algorithm continuous_inverse_cdf

1. $U \leftarrow$ random()
2. $Y \leftarrow F_Y^{-1}(U)$
3. return Y

For example, suppose F_Y is the cdf for the pizza-delivery time. If random() returns the value 0.41171980, then Step 2 will assign $Y \leftarrow -\ln(1 - 0.41171980)/2 \approx 0.27$. Thus, the pizza is delivered in 0.27 hours, or approximately 16 minutes.

Provided that repeated calls to the function random() return independent results, we can achieve our goal of generating m independent, time-stationary values from F by repeated use of algorithm discrete_inverse_cdf or continuous_inverse_cdf, whichever is appropriate. This idea, called *random-variate generation*, is one way that a probability model (specifically a cdf) can generate values. In Chapter 4 we extend this idea to generating sample paths of a stochastic process.

Figure 3.7 shows a histogram of $m = 100$ pizza-delivery times, and another of $m = 500$ pizza-delivery times, generated by algorithm continuous_inverse_cdf with $Y = -\ln(1 - U)/2$. As more and more values of Y are generated, the histogram conforms more and more closely to the shape of the density function, which is also displayed in the figure.

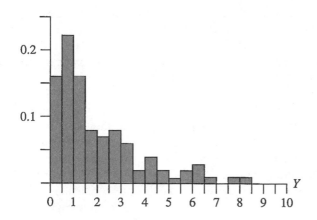

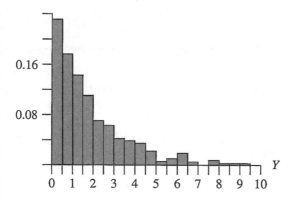

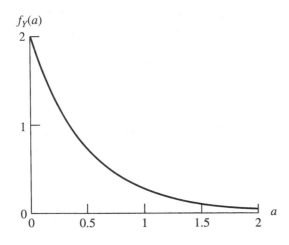

FIGURE 3.7
Histogram of 100 and 500 pizza-delivery times, and the density of the exponential distribution.

If you are already familiar with stochastic simulation, you know that the inverse-cdf method of random-variate generation is not necessarily the best method for all cdfs: It can execute very slowly for discrete-valued random variables with a large range, and F_X^{-1} may not be easily computed for some continuous distributions, such as the normal distribution. Nevertheless, since we can always define X in terms of the inverse cdf, and we can always (perhaps with great numerical effort) generate observations via the inverse cdf, the inverse-cdf method is sufficient to develop intuition.[3]

To our knowledge there is no such function random(). But there are functions that appear to do what random() is supposed to do, and they do it so well that it makes no sense to quibble. Such functions are called *pseudorandom-number generators*, and we will use them as if they do indeed generate values from F_U.

A useful way to think about pseudorandom numbers is as a large, ordered table, where the number of entries in the table is typically greater than 2×10^9. Smaller tables appear in some books, and a very small table is shown in Figure 3.8. Given a starting position in the table, random() returns the pseudorandom numbers until the simulation is completed. If the end of the table is encountered, then numbers starting from the beginning of the table are used. Most functions like random() do not actually reference a table of numbers, but rather the numbers are generated by a recursion of the form

next random number = function(previous random numbers).

In any case, the table of pseudorandom numbers is a good physical representation of how random() works.

Although the (conceptual) table of pseudorandom numbers is ordered (for example, we might use the pseudorandom numbers in Figure 3.8 one row at

```
0.14971781 0.2140427 0.85765977 0.4272876 0.8080653
0.57792245 0.4877184 0.41339374 0.5251962 0.6381426
0.06875780 0.1676558 0.56041437 0.6709518 0.8346895
0.03736677 0.4933928 0.33699042 0.3089899 0.8364343
0.01385732 0.8602028 0.07936855 0.4776284 0.4002407
0.68897327 0.7535193 0.28217973 0.9859376 0.4575937
0.71841285 0.2029749 0.10450996 0.7519042 0.2870101
0.66305717 0.8961229 0.65070454 0.5802081 0.1773571
0.72322937 0.4513504 0.02232056 0.3868480 0.2090379
0.20906964 0.4027947 0.35705923 0.5123724 0.5768044
```

FIGURE 3.8
A small table of pseudorandom numbers.

---

[3] Any algorithm for random-variate generation can be used to implement the simulations in this book, and another one is developed in Exercises 31–33.

a time, from left to right), the order does not matter. As long as the numbers are used without replacement, they can be taken in any manner or starting from any position in the table and still appear to be a sample of independent, time-stationary values from F_U. An important feature of functions like random() is that they permit control of the pseudorandom numbers through *seeds* or *streams*. The seeds or streams are nothing more than different starting positions in the table, typically spaced far apart. For example, stream 1 might correspond to entering the table at the 121, 567th random number, or in row 2, column 3 of Figure 3.8. One obvious advantage of pseudorandom-number generators over truly random numbers is that the values are repeatable by starting over from the same position in the table. Seeds and streams permit this control. We will represent different seeds or streams by giving random() an integer argument, such as random(3). Any random-variate-generation algorithm that uses random(3) will start at the same position in the (conceptual) pseudorandom-number table.

3.4 THE CASE OF THE COPY ENLARGEMENT, REVISITED

Recall the simulations of the proposed changes to the photocopying shop, The Darker Image, in Chapter 2, Case 2.1. They were limited by the rather short sample path available to us, specifically the limited number of system inputs: interarrival-time gaps, customer types (self-service or full service), and service times. Since these inputs are responsible for the uncertainty in the system, it is natural to represent them using probability models. And given a probability model, random-variate generation provides a way to produce a (conceptually) unlimited sample of inputs.

We begin by developing a probability model for customers arriving at The Darker Image, which means defining random variables. Let G_i be the interarrival-time gap between customers $i - 1$ and i in minutes, where the arrival time of customer 0 is defined to be 0. Let X_i be the service time of customer i in minutes, and let B_i be the "type" of customer i, where the event $\{B_i = 0\}$ means customer i is a (potential) self-service customer, and $\{B_i = 1\}$ means that customer i is a full-service customer. Notice that G and X are naturally represented as continuous-valued random variables, while B is a discrete-valued random variable with two possible values.

The inputs to our simulation are characterized by the joint distribution of the collection of random variables

$$(G_1, X_1, B_1, G_2, X_2, B_2, G_3, X_3, B_3, \ldots, G_m, X_m, B_m) \qquad (3.20)$$

where m is the total number of customers we want to model. Directly parameterizing this joint distribution is clearly difficult. We know that independence allows us to decompose the joint distribution, so we should use our knowledge of the system to determine which random variables in (3.20) can be modeled as independent.

Are the inputs associated with different customers independent? More precisely, are (G_i, X_i, B_i) and (G_j, X_j, B_j) independent when $i \neq j$? In most cases customers choose their time of arrival and the type of service they need without consulting other customers. This argues for independence. In some cases customers might arrive as a group—which argues against independence—but if these groups are a small portion of The Darker Image's business, then treating (G_i, X_i, B_i) and (G_j, X_j, B_j) as independent is a reasonable approximation. This is a significant decomposition because it means that we can concentrate on the joint distribution of (G_i, X_i, B_i), for $i = 1, 2, \ldots, m$, and multiply their joint distributions to obtain the overall joint distribution of (3.20).

Which of the random variables (G_i, X_i, B_i) are independent? The analysis in Figure 2.1 showed that the potential self-service customers tend to have shorter service times, which argues that X_i and B_i are dependent. But it seems unlikely that the customer characteristics are related to the interarrival-time gap. Therefore, we have argued that G_i is independent of (X_i, B_i).

Thus, we have simplified our problem to characterizing two independent collections of random variables:

$$G_1, G_2, \ldots, G_m$$

and

$$(X_1, B_1), (X_2, B_2), \ldots, (X_m, B_m)$$

It is tempting to model the interarrival-time gaps as independent random variables with common cdf F_G, but this assumes that $G_1, G_2, \ldots, G_m$ are also time stationary (have the same cdf). Does G_1, the time until the first customer arrives after the shop opens, have a different distribution from the gaps between the other customers? Does the shop experience any "rush hour" in which the gaps tend to be smaller? A yes answer to either of these questions argues against time stationarity.

Time stationarity is often violated because most systems change over time. Later in the book we develop some tools for modeling nonstationarity. For now, suppose that treating $G_1, G_2, \ldots, G_m$ as independent and time-stationary random variables with cdf F_G is a reasonable approximation. Exercise 23 asks you to develop an appropriate cdf F_G from the available data.

We also suppose that it is reasonable to approximate $(X_1, B_1), (X_2, B_2)$, $\ldots, (X_m, B_m)$ as time stationary. In other words, all pairs have the same joint cdf F_{XB}. However, since X_i and B_i are clearly dependent, we must model the dependence in some way. Here is one approach: First model B, the random variable that determines customer type, as a random variable with marginal mass function p_B. Then model the conditional distribution of X given that the value of B is known. In other words, develop two marginal cdfs for X: one for potential self-service customers, $F_{X|B=0}$, and one for full-service customers, $F_{X|B=1}$. Recall that we have some data on the service times of each type of customer. This approach, known as *conditioning*, is equivalent to determining the joint distribution of X and B, since by the decomposition in Equation (3.13)

$$\Pr\{X \leq a, B = i\} = \Pr\{X \leq a | B = i\} \Pr\{B = i\} = F_{X|B=i}(a)\, p_B(i)$$

Exercise 23 asks you to develop appropriate cdfs for X and B.

3.5 FINE POINTS

Our characterization of independent and time-stationary random variables in Section 3.2 is the same as the characterization of independent and identically distributed (i.i.d.) random variables in many statistics texts. We chose the term "time stationary" rather than "identically distributed" because the dynamic processes we study in this book are always time-ordered. Also, it introduces the concept of stationarity that underlies much of stochastic modeling.

To define any random variable X in terms of a uniformly distributed random variable U, we must address the cases of mixed discrete and continuous-valued random variables and random variables whose cdfs are not strictly increasing (which occur when there are gaps in the range of the random variable). A general definition of the inverse cdf that works for *any* cdf is

$$X \equiv \min\{a : F_X(a) \geq U\} \tag{3.21}$$

which is read, "Set X equal to the minimum number a such that F_X evaluated at a is greater than or equal to U." This is not a practical definition to actually implement, but if we let $X = F_X^{-1}(U)$ mean (3.21), then we can refer to the "inverse cdf" of any random variable, discrete valued, continuous valued, or mixed.

In many books the definition of a random variable X begins with a probability space for an experiment (sample space and probability measure) and then defines a random variable as a function from the sample space to the real numbers. We started with random variables and ignored the underlying probability space. But notice that random-variate generation completes the description: The sample space of the experiment is the interval $[0, 1]$, the probability measure is the uniform distribution F_U, and the function from the sample space to the real numbers is $X = F_X^{-1}(U)$.

3.6 EXERCISES

3.1. Calculate the following probabilities from the cdf (3.3) for the baseball-trade example. In each case begin by writing the appropriate probability statement, for example $\Pr\{X = 4\}$; then calculate the probability.
(a) The probability that player 4 will be traded.
(b) The probability that player 2 will not be traded.
(c) The probability that the index of the player traded will be less than 3.
(d) The probability that the index of the player traded will be larger than 1.

3.2. Calculate the following probabilities from the cdf (3.4) for the pizza-delivery example. In each case begin by writing the appropriate probability statement, for example $\Pr\left\{X \leq \frac{1}{2}\right\}$; then calculate the probability.
(a) The pizza-delivery company promises delivery within 40 minutes or the pizza is free. What is the probability that it will have to give a pizza away for free?
(b) The probability that delivery takes longer than 1 hour.

(c) The probability that delivery takes between 10 and 40 minutes.

(d) The probability that delivery takes less than 5 minutes.

3.3. A random variable X has the following cdf:

$$F_X(a) = \begin{cases} 0, & a < 0 \\ \frac{a^2}{\delta^2}, & 0 \le a \le \delta \\ 1, & \delta < a \end{cases}$$

Answer the following questions about this random variable.

(a) What is the density function of X?

(b) What is the maximum possible value that X can take?

(c) What is the expected value of X?

3.4. A random variable X has the following density function:

$$f_X(a) = \begin{cases} 0, & a < 0 \\ \frac{3a^2}{\beta^3}, & 0 \le a \le \beta \\ 0, & \beta < a \end{cases}$$

Answer the following questions about this random variable.

(a) What is the cdf of X?

(b) What is the minimum possible value that X can take?

(c) What is the expected value of X?

3.5. A random variable Y has the following density function:

$$f_Y(a) = \begin{cases} 0, & a < 0 \\ \frac{3}{16}a^2 + \frac{1}{4} & 0 \le a \le 2 \\ 0 & 2 < a \end{cases}$$

Answer the following questions about this random variable.

(a) What is the expected value of Y?

(b) What is the cdf of Y?

(c) For this random variable, which is more likely: a value near $\frac{1}{2}$ or a value near $1\frac{1}{2}$?

3.6. Calculate the following conditional probabilities from the cdf (3.3) for the baseball-trade example.

(a) The probability that player 4 will be traded, given that player 1 is not traded.

(b) The probability that player 4 will be traded, given that players 1 and 2 are not traded.

(c) The probability that player 2 will be traded, given that player 1 or 2 is traded.

3.7. Calculate the following conditional probabilities from the cdf (3.4) for the pizza-delivery example.

(a) The pizza-delivery company promises delivery within 40 minutes or the pizza is free. What is the probability that we will get a free pizza, given that we have already waited 30 minutes?

(b) The probability that delivery takes longer than 1 hour, given it takes longer than 40 mintutes.

(c) The probability that delivery takes less than 10 minutes, given it takes less than 40 minutes.

3.8. Recall the joint distribution of replacement-wheel type and replacement-bearing type presented in Section 3.1.4. Answer the following questions:

(a) What is the probability that a customer will order "ouch!" wheels if the customer ordered "warp 9" bearings?

(b) What is the probability that a customer will order "cool" or "hot" wheels if the customer ordered "warp 2" bearings?

(c) What is the probability that a customer will not order "hot" wheels?

3.9. A machine produces components that are either acceptable or defective. From the observation of 100 pairs of components produced in sequence, the following data were collected:

		second component	
first component		acceptable	defective
	acceptable	75	5
	defective	10	10

Let X_1 represent the first component and X_2 represent the second component where a value of 0 corresponds to an acceptable component and a value of 1 corresponds to a defective component. Estimate a joint distribution for these two random variables; then answer the following questions:

(a) What is the probability that the second component will be defective if the first component is not defective? What is the probability that the second component will be defective if the first component is defective? Are X_1 and X_2 dependent?

(b) If we make a profit of $100 for each acceptable component but lose $20 for each defective component produced, what is the expected profit for the second component given that the first component was defective? What does this suggest that we should do whenever we observe a defective component?

3.10. In Section 3.2 we parameterized a normal-distribution model for the random variable X representing GPA with $\widehat{\mu} = 2.75$ and $\widehat{\sigma}^2 = 0.004$. Under this model $\Pr\{X > 2.90\} \approx 0.010$. Determine how sensitive this result is to the error in our estimate of $\widehat{\mu}$ by calculating the probability when μ is $\widehat{\mu} + \widehat{se}$ and when μ is $\widehat{\mu} - \widehat{se}$.

3.11. Industrial Engineers have observed the time to inspect five components. The observed times, in seconds, were 12.2, 11.1, 7.7, 3.5, 5.9.

(a) Compute the sample average and sample standard deviation of these data.

(b) Compute the standard error of the sample average as an estimator of the expected time to inspect.

(c) Construct the empirical cdf of these data.

3.12. The times a doctor spent with 20 patients were 10.2, 11.3, 22.2, 10.7, 8.7, 6.8, 9.8, 8.2, 19.5, 4.1, 13.2, 13.0, 3.8, 13.2, 18.0, 8.3, 10.0, 10.2, 13.1, 5.2 (times are in minutes).

(*a*) Compute the sample average and sample standard deviation of these data.

(*b*) Compute the standard error of the sample average as an estimator of the expected time spent with patients.

(*c*) Construct the empirical cdf of these data.

(*d*) Find a probability distribution that provides a good fit to these data.

3.13. The following data were collected on the number of compact discs (CDs) purchased by customers entering a record store: 1, 0, 1, 0, 0, 0, 2, 1, 2, 0, 2, 2, 1, 0, 0, 0, 2, 0, 0, 1, 3, 0, 0, 0, 1, 1, 1, 0, 1, 2.

(*a*) Compute the sample average and sample standard deviation of these data.

(*b*) Compute the standard error of the sample average as an estimator of the expected number of CDs purchased.

(*c*) Construct the empirical cdf of these data.

(*d*) Write the mass function corresponding to your empirical cdf.

3.14. The diameters of 25 grapefruit obtained from a particular grower were 5.98, 5.85, 6.17, 6.55, 5.53, 5.90, 5.78, 5.98, 6.44, 5.95, 6.18, 6.42, 5.88, 6.13, 6.07, 5.98, 6.05, 5.90, 6.11, 5.87, 6.27, 5.99, 6.33, 5.90, 6.19 (in inches).

(*a*) Compute the sample average and sample standard deviation of these data.

(*b*) Compute the standard error of the sample average as an estimator of the expected diameter of a grapefruit from this grower.

(*c*) Construct the empirical cdf of this data.

(*d*) Find a probability distribution that provides a good fit to these data.

3.15. Consider the following situation: A penny (or other coin) is placed on its edge on a table. The table is then slapped, causing the coin to fall either head-side or tail-side up. We are interested in making a statement about which side comes up.

(*a*) Propose a probability model of this situation, and parameterize it by collecting data.

(*b*) What factors do you think affect the model? The particular table or coin? How hard the table is slapped? Do you think that different trials are independent?

3.16. Show that $Y = -\ln(U)/\lambda$ defines an exponentially distributed random variable with parameter λ, when U is a random variable having the uniform distribution on $[0, 1]$.

3.17. Plot the cdfs of the following distributions. Write an algorithm involving the function `random()` to generate values from each distribution. In each case, what value is generated if `random()` returns the values $0.1, 0.5$ and 0.9?

(*a*) The uniform distribution on the interval $[\alpha, \beta]$

$$F(a) = \begin{cases} 0, & a < \alpha \\ \frac{a-\alpha}{\beta-\alpha}, & \alpha \le a \le \beta \\ 1, & \beta < a \end{cases}$$

Use $\alpha = 0$ and $\beta = 4$.

(b) The Weibull distribution with scale parameter $\alpha > 0$ and shape parameter $\beta > 0$ (when $\alpha = 1$ this is the exponential distribution)

$$F(a) = \begin{cases} 0, & a \le 0 \\ 1 - e^{-(a/\beta)^{\alpha}} & 0 < a \end{cases}$$

Use $\alpha = 1/2$ and 2, and $\beta = 1$.

(c) A discrete distribution

$$F(a) = \begin{cases} 0, & a < 1 \\ 0.3, & 1 \le a < 2 \\ 0.4, & 2 \le a < 3 \\ 0.7, & 3 \le a < 4 \\ 0.95, & 4 \le a < 6 \\ 1, & 6 \le a \end{cases}$$

(d) The Bernoulli distribution with success probability $0 < \gamma < 1$

$$F(a) = \begin{cases} 0, & a < 0 \\ 1 - \gamma, & 0 \le a < 1 \\ 1, & 1 \le a \end{cases}$$

Use $\gamma = 1/4$.

(e) The geometric distribution with success probability $0 < \gamma < 1$

$$F(a) = \begin{cases} 0, & a < 1 \\ 1 - (1 - \gamma)^{\lfloor a \rfloor}, & 1 \le a \end{cases}$$

The floor function $\lfloor a \rfloor$ means "the largest integer in a" (in other words round down). Use $\gamma = 1/4$. (Hint: Begin by writing an expression for $\Pr\{X = a\}$ for the geometric distribution.)

3.18. Derive the density or mass functions for the cdfs in Exercise 17 and plot them.

3.19. Derive the expected values of random variables having the cdfs in Exercise 17. (Hint for part (b): Consider the substitution $u = (a/\beta)^{\alpha}$. The solution will include a gamma function. The gamma function Γ is defined to be

$$\Gamma(c) = \int_0^{\infty} a^{c-1} e^{-a}\, da$$

for $c > 0$. One property of the gamma function that will help simplify the answer is that $\Gamma(c + 1) = c\Gamma(c)$.)

3.20. Derive the variances of the random variables having the cdfs in Exercise 17.

3.21. Generate 100 values from the distributions in Exercise 17. Plot the empirical cdf and histogram, and compare them to the actual cdf and density or mass function. Increase the number of values until they conform closely. How many does it take? If you have algorithms available for other distributions, generate values and do the same plots for those distributions.

3.22. Random-variate generation allows us to sample values from any probability distribution we specify. However, the empirical distribution of any finite (small) sample of random variates may have characteristics that are quite different from the specified distribution. A spreadsheet program can be used to gain some understanding of this phenomenon.

In a spreadsheet, let column A be a column of 100 pseudorandom numbers. Most spreadsheet programs have a function like RAND() or @RAND that generates pseudorandom numbers; copy this function into 100 rows of column A.

Let column B be the random variates corresponding to the pseudorandom numbers in column A. For example, to generate exponentially distributed random variates with parameter 5, the function in column B, row 1 would be something like B[1] = -LN(1 - A[1])/5, based on the inverse-cdf method of random-variate generation. Copy this formula (or whatever one you are using) into the other 99 rows of column B.

In column C create cells that contain the sample average, sample standard deviation, maximum and minimum of the values in column B. Most spreadsheet programs have built-in functions for sample average, standard deviation (or variance, in which case you can take the square root), maximum and minimum. If your spreadsheet program has the capability to create a histogram of the values in column B, then do this also.

Each time you issue the "recalculate" command in the spreadsheet, new pseudorandom numbers are generated in column A. Therefore, new random variates are also generated in column B, and new sample statistics are generated in column C. Each recalculation corresponds to an independent replication of the experiment.

For each distribution specified below, recalculate the spreadsheet 10 times. After each recalculation enter the sample average from the replication into column D (do not "copy" the elements in column C or else they will also change each time you recalculate the spreadsheet). When you have finished all 10 replications, compute the sample average and sample standard deviation of the 10 sample averages in column D, placing it in column E. How does this overall average compare to the *true expected value* of the distribution (which is 2 for each of the distributions listed below)?

(a) The exponential distribution with parameter $\lambda = 1/2$.

(b) The uniform distribution with parameters $\alpha = 0$ and $\beta = 4$.

(c) The Weibull distribution with parameters $\alpha = 2$ and $\beta = 2.257$.

3.23. Develop probability models (parameterized cdfs) for the interarrival-time gaps; the service time conditional on the customer being a potential self-service customer; the service time conditional on the customer being a full-service customer; and the customer type. Also develop a model for all the customer service times, without conditioning on the customer type. Compare the empirical cdfs to the parameterized cdfs. (Hint: Try the exponential distribution for the time data and the Bernoulli distribution for the customer type.)

3.24. Show that $E[aX + b] = aE[X] + b$ for a discrete-valued or a continuous-valued random variable X.

3.25. Show that $Var[X] = E[X^2] - (E[X])^2$.

3.26. Suppose that X is a discrete-valued random variable taking values $0, 1, 2, \ldots$. Show that

$$E[X] = \sum_{i=0}^{\infty} \Pr\{X > i\}$$

(Hint: Start with the standard definition of expected value.) This result sometimes provides a more convenient way to compute an expected value.

3.27. Suppose that X is a Bernoulli random variable with parameter γ. Derive $E\left[X^m\right]$ for $m = 2, 3, \ldots$.

3.28. Recall the empirical cdf defined by (3.18). Let $X_1, X_2, \ldots, X_m$ be independent, time-stationary random variables with common cdf F_X that represent the data prior to sampling. Then the empirical cdf, prior to sampling, is

$$\widehat{F}_X(a) \equiv \frac{1}{m} \sum_{i=1}^{m} \mathcal{I}(X_i \leq a)$$

Show that $E\left[\widehat{F}_X(a)\right] = F_X(a)$.

3.29. Suppose that Y is a continuous-valued random variable with density function f_Y, and let $\mathcal{A}$ be some set of values. Let $g(Y) = \mathcal{I}(Y \in \mathcal{A})$. Show that

$$E\left[g(Y)\right] = \Pr\{Y \in \mathcal{A}\}$$

This result demonstrates that a probability can always be expressed as the expected value of an indicator function of a random variable.

3.30. The *q-quantile* of a random variable X is the smallest value η such that $\Pr\{X \leq \eta\} \geq q$. If the cdf of X is continuous, then $\eta = F_X^{-1}(q)$. Expressed as a percentage, η is called the $100q$th *percentile*. Some special quantiles are the 0.5 quantile (or fiftieth percentile), called the *median*, and the 0.25 and 0.75 quantiles, called the *quartiles*. Like the expected value and variance, quantiles (or percentiles) are summary measures of random variables.

(a) Suppose X has an exponential distribution with parameter λ. Give an expression for the median of X, and compare it to the $E[X]$. What quantile is the $E[X]$ (in other words, what is $F_X(E[X])$?).

(b) Suppose that X has the uniform distribution with parameters $\alpha < \beta$. Give an expression for the median of X, and compare it to the $E[X]$.

3.31. You have at your disposal a fair, six-sided die, which implies that by throwing the die, you can generate values for the discrete-valued random variable Z with mass function:

$$p_Z(a) = \frac{1}{6} \text{ for } a = 1, 2, 3, 4, 5, 6$$

Suppose that what you really need is to generate values for the random variable Y with mass function

$$p_Y(b) = \frac{1}{5} \text{ for } b = 1, 2, 3, 4, 5$$

Here is an easy approach: Roll the die. If a "6" results, then ignore it and roll again. Show that this approach works. (Hint: $\Pr\{Y = b\} = \Pr\{Z = b | Z \neq 6\}$.)

3.32. (Continuation of Exercise 31.) Suppose that we want to generate observations of Y, a discrete-valued random variable taking values in a set $\mathcal{A}$ and with mass function p_Y. Let Z be another discrete-valued random variable taking values in a set $\mathcal{B}$, such that $\mathcal{A}$ is contained in $\mathcal{B}$ (that is, every value in $\mathcal{A}$ is also in $\mathcal{B}$), and with mass function p_Z. If it helps to make the problem concrete, take $\mathcal{A} = \mathcal{B} = \{1, 2, 3, \ldots, m\}$.

Suppose that there is a constant $1 < c < \infty$ such that $c p_Z(a) \geq p_Y(a)$ for all a in $\mathcal{A}$. Then values of Y can be generated in the following way:

algorithm discrete_acceptance_rejection
1.   generate a value of Z by any method, including inverse cdf
2. $V \leftarrow$ random()
3.   if $V \leq p_Y(Z)/(c p_Z(Z))$ then
         return $Y \leftarrow Z$
     else
         goto step 1
     endif

This algorithm is known as the *acceptance-rejection method* of random-variate generation.

(a) Show that this algorithm works. Hint: Use the law of total probability.

(b) Show that the expected number of trials until acceptance is c.

(c) Develop an acceptance-rejection algorithm for the random variable Y having the cdf in Exercise 17(c). Let Z have the mass function

$$p_Z(a) = \frac{1}{6} \text{ for } a = 1, 2, 3, 4, 5, 6$$

Compute the expected number of trials for your algorithm.

3.33. (Continuation of Exercise 31.) Suppose we want to generate observations of a continuous-valued random variable Y, but F_Y is difficult to invert. Let F_Y have a density function f_Y. Suppose that there is another random variable Z with cdf F_Z that is easy to invert, and F_Z has density function f_Z. If there is a constant $1 < c < \infty$ such that $c f_Z(a) \geq f_Y(a)$ for all a, then the following algorithm will generate values of Y:

algorithm continuous_acceptance_rejection
1. $U \leftarrow$ random()
2. $Z \leftarrow F_Y^{-1}(U)$
3. $V \leftarrow$ random()
4.   if $V \leq f_Y(Z)/(c f_Z(Z))$ then
         return $Y \leftarrow Z$
     else
         goto step 1
     endif

This algorithm is known as the *acceptance-rejection method* of random-variate generation.

(a) Show that this algorithm works. Hint:

$$\Pr\{Y \le a\} = \Pr\left\{Z \le a | V \le \frac{f_Y(Z)}{cf_Z(Z)}\right\}$$

then use the law of total probability.
(b) Show that the expected number of trials until acceptance is c.
(c) Develop an acceptance-rejection algorithm for the random variable Y having density function $f_Y(a) = 6a(1 - a)$, for $0 \le a \le 1$. Compute the expected number of trials for your algorithm.

3.34. The right-triangular distribution with parameters $\alpha < \beta$ has density function

$$f(a) = \begin{cases} 0, & a < \alpha \\ \frac{2(a-\alpha)}{(\beta-\alpha)^2}, & \alpha \le a \le \beta \\ 0, & \beta < a \end{cases}$$

(a) Derive an inverse-cdf algorithm for generating values from this distribution.
(b) Derive an acceptance-rejection algorithm for generating values from this distribution. Use the uniform distribution on $[\alpha, \beta]$ for f_Z.

CHAPTER
4

SIMULATION

Chapters 2 and 3 provided the tools needed to formulate models of dynamic systems that are subject to uncertainty. Sample-path decomposition (Chapter 2) separates system behavior into inputs (the uncertain part) and logic (how the system reacts to the inputs). Chapter 3 proposed random variables as models of the inputs. When data are available, statistics parameterize the distributions of these random variables. Numerical values of the inputs can be obtained via random-variate generation.

We are now ready to combine these tools to formulate models of dynamic systems. This chapter presents a generic stochastic-process model, including an algorithm for generating sample paths. This algorithm is our logical representation of stochastic processes throughout the book. We analyze the sample paths of such models to obtain system performance measures.

4.1 THE CASE OF THE LEAKY BIT BUCKET

Case 4.1. Bit Bucket Computers[1] specializes in installing and maintaining highly reliable computer systems. One of its standard configurations is to install a primary computer, an identical backup computer that is idle until needed, and provide a service contract that guarantees complete repair of a failed computer within 48 hours (if it has not fixed a computer within 48 hours, then it replaces the computer).

---

[1] Many thanks to Jane Fraser of The Ohio State University.

Computer systems are rated in terms of their "time to failure" (TTF). The engineers at Bit Bucket Computers have developed a probability model for the TTF of the individual computers and a probability model for the time required to complete repairs, but they would like to have a TTF rating for the entire system. A failure of the system is when both computers are not working ("down") simultaneously.

Developing a TTF rating for a highly reliable system by actually testing the system is often impractical because excessive time is required to observe even a small number of failures. By generating sample paths from a model of the system via simulation, however, a large number of *simulated* failures can be observed in a short period of time. This is what the engineers would like to do.

4.2   NOTATION AND REVIEW

When we refer to a collection of numbers ordered by subscript (but not necessarily by value), like $a_1, a_2, \ldots, a_m$, it is sometimes convenient to represent them as a vector or matrix. We use boldface to denote vectors or matrices. For example

$$\mathbf{a} \equiv \begin{pmatrix} a_1 \\ a_2 \\ \vdots \\ a_m \end{pmatrix}$$

is an $m \times 1$ column vector. Vectors are column vectors unless otherwise specified. A prime ("$\prime$") attached to a vector or matrix indicates the transpose operation; for instance, $\mathbf{a}' = (a_1, a_2, \ldots, a_m)$ is a $1 \times m$ row vector. A 1×1 matrix is called a *scalar*.

Section 2.2 defined the sample average of m numbers, $b_1, b_2, \ldots, b_m$, denoted $\bar{b}$. We can also define the sample average of a function $g(t)$ of a continuous variable t. The mean-value theorem from calculus establishes that if $\int_a^b g(t)\,dt$ is defined, then there is a number $\bar{g}$ such that $(b - a)\bar{g} = \int_a^b g(t)\,dt$. When t represents time (as it does in this book), we call $\bar{g}$ the *time-average* value of $g(t)$ between a and b; it is computed as

$$\bar{g} = \frac{\int_a^b g(t)\,dt}{b - a} \tag{4.1}$$

4.3   STOCHASTIC PROCESSES

Our models of dynamic systems that are subject to uncertainty are called *stochastic processes*. A stochastic process is nothing more than a sequence of random variables ordered by an index set.

For example, the random variables $S_0, S_1, S_2, \ldots$ form a stochastic process ordered by the *discrete index set* $\{0, 1, 2, \ldots\}$. Another way to denote this stochastic process is $\{S_n; n = 0, 1, 2, \ldots\}$. We also consider stochastic processes in which the index set is continuous, for instance, the stochastic process $\{Y_t; t \geq 0\}$, which is ordered by the *continuous index set* $\{t \geq 0\}$.

The indices n and t are often referred to as "time," so that S_n is a *discrete-time process* and Y_t is a *continuous-time process*. By convention the index set of a stochastic process is always infinite.

The range (possible values) of the random variables in a stochastic process is called the *state space* of the process. We consider both discrete and continuous-state processes. The state space may also be multidimensional, in which case the stochastic process is an ordered sequence of random vectors. For example, an m-dimensional stochastic process is $\{\mathbf{S}_n; n = 0, 1, 2, \ldots\}$ where

$$\mathbf{S}_n \equiv \begin{pmatrix} S_{1,n} \\ S_{2,n} \\ \vdots \\ S_{m,n} \end{pmatrix}$$

and each element $S_{i,n}$ is a scalar random variable.

Examples of stochastic-process models we will formulate later in the book include:

- $\{S_n; n = 0, 1, 2, \ldots\}$, where the state space of S_n is $\{1, 2, 3, 4\}$ representing which of four types of transactions a person submits to an on-line data-base service, and time n corresponds to the number of transactions submitted.
- $\{S_n; n = 0, 1, 2, \ldots\}$, where the state space of S_n is $\{1, 2\}$ representing whether an electronic component is acceptable or defective, and time n corresponds to the number of components produced.
- $\{Y_t; t \geq 0\}$, where the state space of Y_t is $\{0, 1, 2, \ldots\}$ representing the number of accidents that have occurred at an intersection, and time t corresponds to weeks.
- $\{Y_t; t \geq 0\}$, where the state space is $\{0, 1, 2, \ldots, s\}$ representing the number of copies of a software product in inventory, and time t corresponds to days.
- $\{Y_t; t \geq 0\}$, where the state space of Y_t is $\{0, 1, 2, \ldots\}$ representing the number cars parked in a parking garage at a shopping mall, and time t corresponds to hours.

A stochastic process is a sequence of random variables, so it is characterized by the joint distribution of (any finite collection of) its component random variables. Unfortunately, characterizing a stochastic process in this way presents some technical difficulties, including the fact that there are an infinite number of random variables to characterize. One way to finesse this problem is to define a stochastic process via an algorithm for generating its sample paths, rather than specifying the joint distribution of its component random variables directly. The next section illustrates this approach for the Bit Bucket example, and we formalize it in Section 4.5.

4.4 SIMULATING THE LEAKY BIT BUCKET

The engineers at Bit Bucket Computers, Case 4.1, have defined two random variables to model the uncertainty in their system: The TTF for a computer, X, with cdf F_X, and the time to repair a computer, R, with cdf F_R. These random variables are the system inputs in their model.

The distribution F_X is from the Weibull family (Section 3.6, Exercise 17) with parameters $\alpha = 2$ and $\beta = 812$. These parameters imply that the expected TTF of a computer is approximately 720 hours (30 days) with standard deviation 376 hours (16 days). The Weibull family is often used to model the time until failure of a component because its flexible shape can describe components whose likelihood of failure increases or decreases with age.

Let X_i denote the TTF of the ith computer in service. The engineers believe that $X_1, X_2, \ldots$ form an independent, time-stationary sequence of random variables with marginal cdf F_X. This characterization implies that the distribution of TTF is the same for a new computer and one that has just been repaired, which may be plausible if repair typically means replacing a failed component with a new one.

Based on service records, F_R is chosen as the uniform distribution on [4, 48] hours (Section 3.6, Exercise 17). Therefore, the expected repair time is 26 hours, with standard deviation approximately 13 hours. If R_i is the time required to repair the ith computer failure, then the engineers believe that $R_1, R_2, \ldots$ form an independent, time-stationary sequence of random variables with marginal cdf F_R. Further, they believe that the repair times are independent of the time to failure, X. This characterization implies that the repair time is not affected by the number of times a computer has been repaired and also that the time until a computer fails is unrelated to the time it took to repair it.

The random variable of interest to Bit Bucket is the time until the entire system fails, denoted by D. Clearly D is a function of the X_i and R_i, but not a simple one. Since we can generate values of X_i and R_i via random-variate generation (Section 3.3), we can simulate values of D provided we have a model of the system logic. The observed values of D can then be used to compute summary performance measures, to construct the empirical cdf of D, or to parameterize a distribution for D (Section 3.2).

The engineers at Bit Bucket agree to the following model of the system logic:

- After a system is installed, the primary computer is started. When it fails, the backup computer is immediately started and a service call is made to Bit Bucket.

- If the primary computer is repaired before the backup computer fails, then the primary computer becomes the backup computer, the former backup computer remains the primary computer, and the system continues to operate in this fashion. In other words, the distinction between primary and backup computer is irrelevant, since whichever one is running is the primary computer and the other one is the backup.

- If at any time neither computer is available, then the entire system fails.

- Only one computer can be repaired at a time. If at any time both computers have failed, repairs continue on the one that failed first until it can be put back in service.

Thus, sample paths of this system are a time-ordered sequence of failure and repair times.

Based on this description of the system logic, there are two types of system events of interest: the failure of a computer and the completion of a computer repair. Recall that system events are the changes in the system that we incorporate into our model. Let e_1 denote the system event that a computer fails, and let e_2 denote the system event that a repair is completed on a computer.

The critical variable that characterizes system status is the number of computers that are down (not available). Let S_n denote the number of computers that are down just after the nth system event (failure or repair). The stochastic process $\{S_n; n = 0, 1, \ldots\}$ characterizes the status of the system. Notice that S_n is a discrete-valued random variable with three possible values: 0 (both primary and backup computers are available), 1 (one computer is working but the other is being repaired) and 2 (both computers are down). In other words, the state space of S_n is $\{0, 1, 2\}$.

Associated with system event e_i is a *clock* C_i that represents the time the next system event of type e_i is due to occur. When no event e_i is pending, then the value of C_i is set to ∞. Let $\{T_n; n = 0, 1, \ldots\}$ be the time at which the nth system event occurs; T_n is called the nth *event epoch*, and it is a continuous-valued random variable that depends on the clock settings.

Two things can happen at the time of the $(n+1)$st system event, time T_{n+1}: The system state can change and the clocks can be reset. The system events are functions that map the current system state into the next system state and reset clock values. Here is the system event e_1 representing a computer failure:

$e_1()$ **(computer failure)**

1. $S_{n+1} \leftarrow S_n + 1$ (one more computer down)

2.   if $\{S_{n+1} = 1\}$ then   (if one up and one down)

                 $C_2 \leftarrow T_{n+1} + F_R^{-1}(\text{random}())$ (set clock for end of repair)

                 $C_1 \leftarrow T_{n+1} + F_X^{-1}(\text{random}())$ (set clock for computer TTF)

      endif

When a failure occurs, Step 1 increases the number of computers that are down. Step 2 resets the clocks when necessary: If the computer that just failed is the only computer down ($\{S_{n+1} = 1\}$), then a repair clock for that computer is set to the current time, T_{n+1}, plus the time to complete the repair, and a failure clock is set to the current time plus the time until the backup computer fails. If, on the other hand, both computers are down ($\{S_{n+1} = 2\}$), then no clocks are set because there must already be one computer under repair and there is no other computer available to fail. The event $\{S_{n+1} = 0\}$ is logically impossible just after a computer failure. Notice that clocks are set using the inverse-cdf method of random-variate generation with the function `random()` supplying values from the uniform distribution on $[0, 1]$ (Section 3.3).

Here is the function e_2 representing the end of a computer repair. Notice that clocks only need to be set if $\{S_{n+1} = 1\}$ occurs, implying that just prior to this system event both computers were down.

$e_2()$ **(end of computer repair)**

1. $S_{n+1} \leftarrow S_n - 1$ (one fewer computer down)

2.   if $\{S_{n+1} = 1\}$ then   (if one up and one down)

       $C_1 \leftarrow T_{n+1} + F_X^{-1}(\text{random}())$ (set clock for computer TTF)

       $C_2 \leftarrow T_{n+1} + F_R^{-1}(\text{random}())$ (set clock for end of repair)

   endif

We also include a system event e_0 to set the initial values for the system state and clocks. Think of e_0 as representing the installation of a new computer system.

$e_0()$ **(installation)**

1. $S_0 \leftarrow 0$ (initially no computers down)

2. $C_1 \leftarrow F_X^{-1}(\text{random}())$ (set clock for first computer TTF)

       $C_2 \leftarrow \infty$ (indicate no pending repair)

As we observed in Chapter 2, a simulation proceeds from system event to system event, with the next system event determined by the pending event with the smallest clock value. The following algorithm formalizes this for the Bit Bucket model.

algorithm Bit_Bucket_simulation

1. $n \leftarrow 0$ (initialize system-event counter)
       $T_0 \leftarrow 0$ (initialize event epoch)
       $e_0()$ (execute initial system event)

2. $T_{n+1} \leftarrow \min\{C_1, C_2\}$ (advance time to next pending system event)
       $I \leftarrow \text{argmin}\{C_1, C_2\}$ (find index of next system event)

3. $C_I \leftarrow \infty$ (event I no longer pending)

4. $e_I()$ (execute system event I)
   $n \leftarrow n + 1$ (update event counter)

5. repeat 2

After initializing the system in Step 1, Step 2 advances time to the smallest clock value. The statement $I \leftarrow \operatorname{argmin}\{C_1, C_2\}$ is a formal way of stating that we set I equal to the subscript of the clock with the smallest value. In Step 3, event I is marked as no longer pending by setting $C_I \leftarrow \infty$. Step 4 calls system event function e_I to update the state and reset clocks. Finally, the event counter is incremented and the simulation continues. This algorithm is a precise statement of what we did informally in the simulation of The Darker Image in Chapter 2.

The system events, along with algorithm Bit_Bucket_simulation, define the random variables in the stochastic process in a *constructive* manner. To generate sample paths, we select streams for random() and execute the algorithm. For example, suppose that the first five TTF values generated via $F_X^{-1}(\text{random(1)})$ are 877, 1041, 612, 36 and 975 hours, and the first four repair-time values generated via $F_R^{-1}(\text{random(2)})$ are 17, 8, 39 and 9 hours (these values would not, in general, be integer, but we use integer values to make the example easier to follow). Table 4.1 shows the sample path generated by algorithm Bit_Bucket_simulation given these values.

The $n = 0$ line of Table 4.1 is determined by Step 1 of the algorithm Bit_Bucket_simulation: It sets C_1, the clock for the first computer failure, to 877 and C_2, the clock for the next computer repair, to ∞ since both computers are initially available. Step 2 finds T_n, the time of the first event epoch, to be the smallest clock value of 877, which is the value on clock $I = 1$. Step 4 executes system event e_1, incrementing the value of S_1 to 1 and setting the

TABLE 4.1
Simulation of the Bit Bucket Computers model

Event counter n	System event I	Time T_n	State S_n	Failure clock C_1	Repair clock C_2
0		0	0	877	∞
1	1	877	1	877+1041=1918	877+17=894
2	2	894	0	1918	∞
3	1	1918	1	1918+612=2530	1918+8=1926
4	2	1926	0	2530	∞
5	1	2530	1	2530+36=2566	2530+39=2569
6	1	2566	2	∞	2569
7	2	2569	1	2569+975=3544	2569+9=2578
8	2	2578	0	3544	∞

clocks for the time of the next failure and repair events. Then the cycle repeats. It is important that you work through the remainder of this example in detail.

The stochastic process $\{S_n; n = 0, 1, \ldots\}$ represents the number of failed computers, and it is indexed by the *number* of system events, n, rather than the *time* the events occur. The stochastic process $\{T_n; n = 0, 1, 2, \ldots\}$ represents the time the system events occur, but not what happens when they do occur. These are the stochastic processes that are most naturally generated by a simulation. But we are usually interested in how a system evolves over time, so it is worthwhile to derive another stochastic process that combines both types of information.

For all $t \geq 0$, let

$$Y_t \leftarrow S_n \text{ for all } T_n \leq t < T_{n+1}$$

The stochastic process $\{Y_t; t \geq 0\}$ is a record of the state of the system at *continuous time* t. This is a formal definition of a simple idea, which is illustrated by the plot in Figure 4.1 that is based on the sample path in Table 4.1. Notice that Y_t is piecewise constant (flat), taking the values $S_0, S_1, \ldots$, but only changing values at the times $T_0, T_1, T_2, \ldots$. The process Y_t has a value at all $t \geq 0$. For example, Y_{2000} is 0 in this sample path, indicating that at time 2000 hours there were no computers down.

The process $\{Y_t; t \geq 0\}$ contains information about the state of the system and how long the system visited a particular state. Since a sample path of Y_t is a function of t, we can compute its sample time average using Equation (4.1). Specifically, the time average of Y_t up to the nth event epoch is

$$\bar{Y} \equiv \frac{\int_{T_0}^{T_n} Y_t \, dt}{T_n - T_0}$$

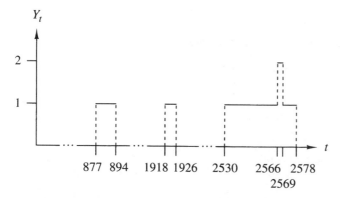

FIGURE 4.1
Plot of the system state Y_t of Bit Bucket Computers (t axis not to scale).

Using the data in Table 4.1, the integral (area under the function) from time 0 to time 2578 is computed by adding the areas of all the rectangles (some with height 0) as follows:

$$\text{area} = 0(877 - 0) + 1(894 - 877) + 0(1918 - 894)$$
$$+1(1926 - 1918) + 0(2530 - 1926) + 1(2566 - 2530)$$
$$+2(2569 - 2566) + 1(2578 - 2569)$$
$$= 76$$

Therefore, the time-average number of failed computers is $76/(2578 - 0) \approx 0.03$, a very small value as we might expect for a highly reliable system.

The random variable D, representing the time of total system failure, can be defined in terms of the random variable Y_t as

$$D \equiv \min\{t : Y_t = 2\}$$

Using this definition, the value of D generated in Table 4.1 is 2566 hours (approximately 107 days). Notice that the simulation algorithm is written to continue forever, because the index set of a stochastic processes is always infinite. But in practice we stop the simulation the first time the event $\{S_n = 2\}$ occurs, indicating that both computers are down. A value of D can be generated by adding Step 4.5 to algorithm Bit_Bucket_simulation:

4.5 if $\{S_n = 2\}$ then   (both computers down)
         $D \leftarrow T_n$ (record time of system failure)
         stop   (terminate simulation)
    endif

Of course, one observation of D does not tell us much about the distribution of D. An advantage of random-variate generation, as opposed to using time-study data we did as in Chapter 2, is that we can repeat or *replicate* the simulation as many times as we desire, using different pseudorandom numbers to obtain independent results. This can be accomplished by embedding algorithm Bit_Bucket_simulation in the following loop:

for $r = 1$ to m
do
    algorithm Bit_Bucket_simulation
enddo

Using a computer, $m = 500$ replications were generated to obtain values for $D_1, D_2, \ldots, D_{500}$, where D_i is the time of system failure on the ith replication. The sample average of the data was $551,606$ hours, or approximately 63 years, which makes the system seem virtually failure free! The sample average of the observed times to failure is often reported as the "mean time to failure" of a system.

More information is obtained by looking at the distribution of the data. The empirical cdf and a histogram of the 500 values are shown in Figure 4.2, with the scale changed from hours to years. A close-up of a portion of the empirical cdf from 0 to $2\frac{1}{2}$ years is also shown. Although the TTF was typically quite long, there were some replications on which it was short. For instance, the estimated probability of system failure within 2 years is $\widehat{F}_D(2) = 10/500 = 0.02$, because on 10 replications the TTF was less than or equal to 2 years. Depending on the mission of the computer system this probability could be considered acceptable or unacceptable, but 2 chances in 100 is certainly *not* failure free.

Case 4.1 demonstrates the power of modeling and simulation: There is simply no way to estimate the distribution of system TTF by experimenting with the real system. By the time even a small sample of failures were observed, the computer system would be obsolete. This case also illustrates why we examine the empirical distribution of a random variable and not just a summary measure like the sample average: The sample average characterizes the center of a distribution, but gives little information about extreme ("tail") behavior, such as the probability of failure within 2 years.

4.5   A GENERIC STOCHASTIC-PROCESS MODEL

The Bit Bucket example illustrates our generic stochastic-process model. In this section we define it more generally. You should think of this as a reference section that gathers all of the important definitions without providing the intuition contained in the examples. A second example follows this section to clarify the definitions further.

Our generic stochastic-process model consists of the following components:

- $\{S_n; n = 0, 1, 2, \ldots\}$, the *state-change* process, is a discrete-time, possibly mixed discrete and continuous state, process. The state-change process represents all the relevant information about the status of the system that is included in the model.
- $\{T_n; n = 0, 1, \ldots\}$, the *event-epoch* process, is defined by $T_{n+1} \equiv \min\{C_i\}$; it is a discrete-time, continuous-state process where T_n is the time of the nth system event.
- $\{Y_t; t \geq 0\}$, the *output* process, is defined by $Y_t \leftarrow S_n$ when $T_n \leq t < T_{n+1}$; it is a continuous-time process with the same state space as S_n, and it connects the state changes to the times that they occur.
- The *simulation algorithm*. The state-change, event-epoch and output processes are defined by k system events $e_1, e_2, \ldots, e_k$, an initial system event e_0, the *clocks* $\mathbf{C} = (C_1, C_2, \ldots, C_k)'$, and the generic simulation algorithm below.

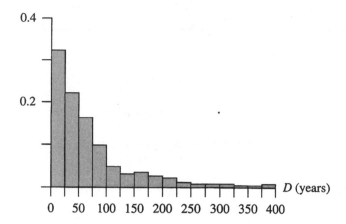

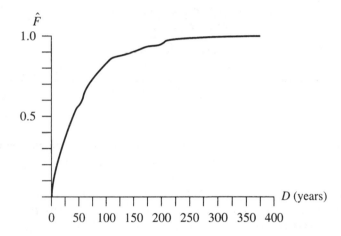

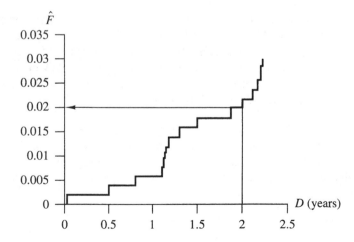

FIGURE 4.2
Empirical cdf and histogram of D values from Bit Bucket Computers (time in years).

---

algorithm simulation

1. $n \leftarrow 0$ *(initialize system-event counter)*
    $T_0 \leftarrow 0$ *(initialize event epoch)*
    $e_0()$ *(execute initial system event)*

2. $T_{n+1} \leftarrow \min\{C_i\}$ *(advance time to next pending system event)*
    $I \leftarrow \operatorname{argmin}\{C_i\}$ *(find index of next system event)*

3. $\mathbf{S}_{n+1} \leftarrow \mathbf{S}_n$ *(temporarily maintain previous system state)*
    $C_I \leftarrow \infty$ *(event I no longer pending)*

4. $e_I()$ *(execute system event I)*
    $n \leftarrow n + 1$ *(update event counter)*

5. *repeat 2*

- Comments:

  - The system-event functions, e_i, perform two tasks: They update the new system state $\mathbf{S}_{n+1}$ from the previous system state $\mathbf{S}_n$, and they reset the clocks, $\mathbf{C}$, if necessary.
  - The statement $\mathbf{S}_{n+1} \leftarrow \mathbf{S}_n$ in Step 3 is included for convenience. By initially defining the state of the system at event epoch $n + 1$ to be *the same as* at event epoch n, the system-event routines need only specify the *changes* in the system state from one event epoch to the next.
  - Our simulation algorithm has been defined so that the only information needed to advance from $(\mathbf{S}_n, \mathbf{C})$ to $(\mathbf{S}_{n+1}, \mathbf{C})$ are the values of $\mathbf{S}_n$ and $\mathbf{C}$. Nothing about the process at event epochs $0, 1, \ldots, n-1$ is required. This is efficient, because we do not have to save the entire sample path up to event epoch n to generate the sample path beyond epoch n, and it is also a fundamental property that we will exploit to analyze sample paths.

Figure 4.3 shows a possible sample path for our generic stochastic-process model when S_n is a scalar. The jumps in Y_t correspond to state changes of S_n, which occur only at the event epochs, $T_0, T_1, \ldots$. Since Y_t is always piecewise constant, the area under Y_t is composed of rectangles of height S_{n-1} and width $T_n - T_{n-1}$. Therefore,

$$\int_{T_0}^{T_n} Y_t \, dt = \sum_{j=1}^{n} \left(T_j - T_{j-1}\right) S_{j-1} \tag{4.2}$$

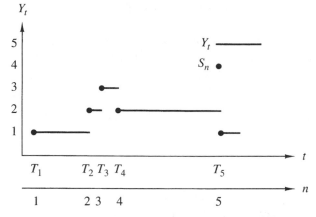

FIGURE 4.3

Sample path of the stochastic processes in our generic model.

This relationship is useful when we want to compute the time average of Y_t.

The distinction between a stochastic process and a sample path of that process is important. We can derive statements about how a process *will* behave from a stochastic-process model. A sample path is a record of how a process actually *did* behave in one instance. Sample paths are generated by executing algorithm simulation with specific seeds or streams for the pseudorandom-number generator.

---

*A sample path is a collection of **time-ordered data** describing what happened to a dynamic process in one instance. A stochastic process is a probability model describing a collection of **time-ordered random variables** that represent the possible sample paths.*

Our simulation algorithm has a natural computer implementation, and it forms the basis for many computer-simulation languages (although they are implemented more efficiently and use more powerful data structures, as explained in Section 4.7). In the next section we show how this approach can be applied to another modeling-and-analysis problem, Case 2.1 described in Chapter 2.

4.6 SIMULATING THE COPY ENLARGEMENT

In this section we formulate a stochastic-process model for the photocopying shop, The Darker Image, as it currently operates. Refer to Chapter 2 for descriptions of the input random variables: G, the interarrival-time gap, and X, the customer service time. Models of the proposed full-service and self-service systems are presented in Chapter 8.

The system events are the arrival of a customer (e_1, with clock C_1) and finishing a customer (e_2, with clock C_2). Let the state of the system be $\mathbf{S}_n = (Q_n, A_n)'$, where just after the nth event epoch

Q_n is the number of customers waiting in the Queue, not including the customer being served if any and

A_n is 0 if the copier is Available and is 1 if it is in use.

The system starts each day with no customers present and the copier is available. These initial conditions are set in the system event e_0:

$e_0()$ (**start of day**)

$\quad Q_0 \leftarrow 0$
$\quad A_0 \leftarrow 0$ (initialize state)

$\quad C_1 \leftarrow F_G^{-1}(\texttt{random()})$ (set clock for first arrival)
$\quad C_2 \leftarrow \infty$ (indicate no finish pending)

When a customer arrives, the state of the system changes. However, the components of $\mathbf{S}_{n+1}$ that are different from $\mathbf{S}_n$ depend on whether or not the copier is in use.

$e_1()$ (**customer arrives**)

$\quad$ if $\{A_n = 0\}$ then   (if copier is available)
$\quad\quad A_{n+1} \leftarrow 1$ (put copier into use)
$\quad\quad\quad C_2 \leftarrow T_{n+1} + F_X^{-1}(\texttt{random()})$ (set clock for next finish)
$\quad$ else   (else copier is not available)
$\quad\quad\quad Q_{n+1} \leftarrow Q_n + 1$ (one more customer waiting)
$\quad$ endif

$\quad C_1 \leftarrow T_{n+1} + F_G^{-1}(\texttt{random()})$ (set clock for next arrival)

System event e_2 updates the state and the clocks when a customer finishes. What happens depends on whether or not there is another customer waiting.

$e_2()$ (**customer finishes**)

$\quad$ if $\{Q_n = 0\}$ then   (if no customers waiting)
$\quad\quad A_{n+1} \leftarrow 0$ (make copier available)

else   (else a customer is waiting)

$$Q_{n+1} \leftarrow Q_n - 1$$ (one fewer customer waiting)

$$C_2 \leftarrow T_{n+1} + F_X^{-1}(\texttt{random()})$$ (set clock for next finish)

endif

Let the output process be

$$\mathbf{Y}_t = \begin{pmatrix} Y_{1,t} \\ Y_{2,t} \end{pmatrix} \leftarrow \begin{pmatrix} Q_n \\ A_n \end{pmatrix} \quad \text{for all } T_n \leq t < T_{n+1}$$

Several system performance measures can be extracted from the sample paths of this output process. For example, suppose a working day for The Darker Image is 6 hours in length, which is 360 minutes. Let

$$\bar{Y}_i \equiv \frac{\int_0^{360} Y_{i,t}\, dt}{360}$$

for $i = 1, 2$ be the time average of each component of the output process over one day. Clearly, $\bar{Y}_1$ is the time-average number of customers waiting for service. Since $Y_{2,t}$ is 1 when the copier is in use and 0 when it is idle, $\bar{Y}_2$ is the time-average number of copiers in use. Because there is only one copier, this is also the fraction of time that the copier is in use, which is called the *utilization* of the copier.

These two performance measures are valuable, but we are most interested in customer delay, a quantity that is not explicitly included in the model. Delay can be directly recorded for each customer (see Section 4.7), or we can estimate the delay indirectly from $Y_{1,t}$, the number of customers waiting for service at time t, as follows.

At time t there are $Y_{1,t}$ customers waiting for service. If there are no arrivals or departures in the next small time increment Δt, then each of these customers is delayed Δt time units, and altogether they are delayed $Y_{1,t}\Delta t$ time units. Now divide the entire time interval from 0 to t into small increments Δt. Summing $Y_{1,t}\Delta t$ over each Δt interval is an approximation of the total accumulated delay during the interval. The smaller the Δt, the better the approximation. Let Δt go to 0, and the summation becomes the integral $\int_0^t Y_{1,t}\, dt$, the total accumulated delay up to time t. To compute the average delay per customer, we divide this quantity by the total number of customer arrivals by time t, say N_t, to obtain

$$\frac{\int_0^{360} Y_{1,t}\, dt}{N_{360}}$$

Thus, we can obtain delay information indirectly from information about the number of customers waiting, $Y_{1,t}$.

Relationship (4.2) can then be used to replace the integral by a summation that can be computed by updating the area after each system event; specifically,

$$\text{area} \leftarrow \text{area} + (T_n - T_{n-1})Q_{n-1}$$

4.7 SIMULATION PROGRAMMING

There are many computer programming languages available for simulating stochastic processes, and you might use one of them to perform the simulations in this book. For reasons of efficiency in large-scale simulations, simulation languages generate sample paths in a manner that is equivalent to, but different from, our algorithm simulation. While our approach is useful for teaching modeling, analysis and simulation, it is not efficient to implement. The purpose of this section is to explain the connection between the two approaches so that you can easily translate between them. We also present a widely used alternative representation of stochastic processes. Complete simulation programs for The Darker Image—in several languages—are given in Appendix A.

A stochastic process is an ordered sequence of random variables, for instance $\{S_n; n = 0, 1, 2, \ldots\}$. To make statements about the possible sample paths of a process, it is necessary to uniquely identify each random variable in the sequence, which is the role of the index n. However, when we use a programming language and a computer to generate a sample path, the computer memory required to maintain a unique name for each random variable could be excessive, particularly if we generate a long sample path or have a multidimensional process. Therefore, instead of a process like $\{S_n; n = 0, 1, 2, \ldots\}$, we maintain a single variable name, like S, to represent the process. A statement such as

$$S_{n+1} \leftarrow S_n + 1$$

becomes

```
S = S + 1
```

in a programming language. If the entire sample path is required, then the value of S is saved or printed after each update.

We have chosen a very simple data structure to keep track of the pending events and clock values: a unique random variable for each clock. This data structure illustrates the essential idea of keeping track of future system events, but it is inefficient in a large-scale simulation with perhaps hundreds of pending system events. An efficient data structure is needed to set clocks and find the next system event quickly. In simulation languages the pending system events and their clock values are typically maintained in some sort of linked list that is ranked in increasing order of clock value. And since "scheduling" pending events is done frequently, simulation languages have a built-in function to do it. For example, a function such as

```
SCHEDULE(2, FX)
```

performs operations equivalent to our statement

$$C_2 \leftarrow T_{n+1} + F_X^{-1}(\text{random}())$$

Most simulation languages maintain their own algorithm simulation so that it does not have to be provided by the user. The user provides system event

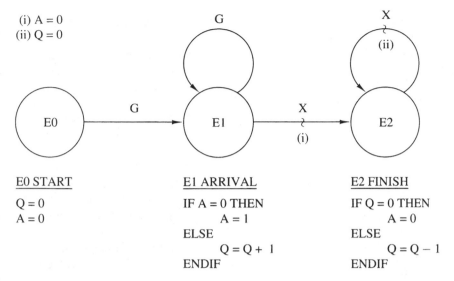

(i) A = 0
(ii) Q = 0

G

X

E0

E1

E2

E0 START	E1 ARRIVAL	E2 FINISH
Q = 0	IF A = 0 THEN	IF Q = 0 THEN
A = 0	A = 1	A = 0
	ELSE	ELSE
	Q = Q + 1	Q = Q − 1
	ENDIF	ENDIF

FIGURE 4.4
Event graph representation of The Darker Image.

functions and an experiment design and then releases control of the execution of the simulation to the language's internal algorithm simulation.

An intuitive graphical representation of a discrete-event simulation is the *event graph* (Schruben 1992). The event graph corresponding to the simulation of The Darker Image in Section 4.6 is shown in Figure 4.4. In the graph, each system event is represented by a node ◯ that has appropriate system logic associated with it. When one system event schedules another system event, a directed edge ⟶ connects them. For example, the edge from E1 to itself indicates that whenever an arrival occurs, another arrival is also scheduled. The time until a system event occurs is indicated by a random variable above the edge, and any conditions that must be satisfied prior to scheduling are indicated by a ~ and a numbered condition below the edge. For example, the edge from E1 to E2 indicates that a customer finishes X time units after an arrival, provided the condition (i) A = 0 is in force when the customer arrives (in other words, the copier is available). There are no edges entering E0, which indicates that E0 is the initial system event executed at time 0 to get the simulation started.

We close this section by describing a popular representation of stochastic processes that is different from the discrete-event view adopted in this book. To illustrate the representation, we use the simulation of The Darker Image in Section 4.6.

The discrete-event view of stochastic processes represents a system as a collection of state variables and system events, where the state variables describe the status of the system and the system events define when and how the state variables change. The simulation proceeds by advancing time chronologically from system event to system event, updating the state variables and system-event clocks whenever a system event occurs.

An equivalent description of The Darker Image can be constructed from the perspective of an individual customer, rather than from the perspective of the system. Whenever a customer arrives, the customer joins a queue of others waiting for service. If the copier is available and the customer is at the front of the queue, then the customer leaves the queue and begins service; otherwise the customer is delayed in the queue. When service is completed, the customer exits the system.

The change in perspective is that we describe the system in terms of the activities of a customer that passes through it, rather than in terms of the how the system changes as it serves customers. We refer to this new perspective as the *process view* and to the customers that pass through the system as *entities* or *transactions*.

Figure 4.5 displays a process-view model of The Darker Image. The model describes the activities of an entity (representing a customer) from its arrival until it exits from the shop. The syntax of the language is

```
STATEMENT, OPTIONS; comments
```

Although this is not an example of any existing simulation language, it is similar to many commercial languages (see Appendix A).

The ARRIVE statement specifies that new entities will be introduced into the system with an interarrival-time gap generated from an exponential distribution with mean 6 using random number stream 1. Arriving entities will join a queue to wait for the copier. The QUEUE statement automatically maintains a first-come-first-served list of the entities waiting for the resource COPIER. When the COPIER becomes available, the first entity in the QUEUE will leave the QUEUE and ACQUIRE the copier; when the copying job is finished, the entity

```
ARRIVE,GAP=EXPONENTIAL(MEAN=6,STREAM=1);        generate custome:
SET,ATTRIBUTE(1)=TIME;                          save arrival tim
QUEUE,COPIER;                                    queue for copier
ACQUIRE,COPIER;                                  use the copier
SERVE,AMOUNT=EXPONENTIAL(MEAN=4.6,STREAM=2);     service time
RELEASE,COPIER;                                 copier available
RECORD,FLOWTIME=TIME-ATTRIBUTE(1);              record total tim
EXIT;                                           customer departs
```

FIGURE 4.5
Process simulation model of The Darker Image.

will RELEASE the copier. The time required to complete the copying job is specified by the AMOUNT option of the SERVE statement. After releasing the copier, the entity will EXIT the system. (We discuss the SET and RECORD statements below.)

This description of the activities of a customer entity is quite natural, which is one reason for the popularity of simulation languages that employ a process view. However, the process view is simply a different way to describe system events and state variables. For example, the QUEUE statement automatically maintains a count of the number of waiting customers, analogous to the state variable Q_n in the discrete-event simulation. Whenever an entity enters the QUEUE, the count increases by 1, and whenever an entity leaves the QUEUE, it decreases by 1. The ACQUIRE and RELEASE statements update the status of the resource COPIER, just as we update the value of A_n in the discrete-event simulation. The ARRIVE and SERVE statements play the role of resetting the clocks C_1 and C_2, respectively, and the simulation still progresses chronologically from system event to system event.

---

*The **discrete-event view** of stochastic processes **defines** system events by describing what happens to the system as it encounters entities. The **process view** of stochastic processes **implies** system events by describing what happens to an entity as it encounters the system.*

The process view—as described here—has some advantages over the discrete-event view. The use of predefined logic such as QUEUE, ACQUIRE, RELEASE, etc., is convenient, and it allows certain statistics to be collected automatically. In this example the delay-in-queue statistics are maintained by the QUEUE statement, while the utilization of the copier is maintained by the ACQUIRE and RELEASE statements. In addition, each entity carries with it a collection of data elements called *attributes* that can be used to uniquely identify entities and collect additional statistics. The SET statement in Figure 4.5 assigns the current time in the simulation to the first ATTRIBUTE of each arriving entity; the current time is maintained in the variable TIME, which is analogous to the event epoch T_n in the discrete-event simulation. Just prior to departing the system a customer's total time in the system is recorded by taking the difference between the current time and the value stored in ATTRIBUTE(1). This occurs in the RECORD statement.

The primary disadvantage of the process view is that the restriction to predefined logic can be limiting if some new or more general logic is required. Fortunately, most of the commercial languages that employ a process view of stochastic processes also facilitate the discrete-event view and allow the two perspectives to be combined.

4.8   FINE POINTS

In many books a stochastic process is defined as a sequence of random variables defined on a common probability space (sample space and probability measure). We started with random variables and ignored the underlying probability space. But `random()`, algorithm simulation, and the system-event functions complete the description: The sample space of the experiment is $[0, 1]^\infty$, the probability measure on each dimension is the uniform distribution, and the function from an outcome to the real numbers is algorithm simulation and the system-event functions.

4.9   EXERCISES

4.1. Verify the results of the Bit Bucket Computers simulation by coding the simulation and executing it. Estimate the probability of a system failure within 5 years.

4.2. Code and execute the simulation of The Darker Image using the input distributions determined in Chapter 3, Exercise 23. Simulate the system for 100 days (replications), recording the time-average number waiting and the utilization of the copier. Compute the sample averages and empirical cdfs of these 100 values.

4.3. Formulate a stochastic-process model for the system in Exercise 6, Section 2.4, including defining the random variables and the system-event functions.

4.4. Formulate a stochastic-process model for the system in Exercise 7, Section 2.4, including defining the random variables and the system-event functions.

4.5. Formulate a stochastic-process model for the system in Exercise 9, Section 2.4, including defining the random variables and the system-event functions. Simulate 1000 replications, each of 25 manufacturing jobs, and look at the distribution of the maximum amount of buffer space used on each replication. How does this distribution compare to the class results?

4.6. Formulate a stochastic-process model for the system in Exercise 10, Section 2.4, including defining the random variables and the system-event functions. Simulate 1000 replications, each of 30 computer jobs, and look at the distribution of the maximum load imbalance from each replication. How does this distribution compare to the class results?

4.7. Formulate a stochastic-process model for the system in Exercise 11, Section 2.4, including defining the random variables and the system-event functions.

4.8. The following table gives a sample path for the simulation of The Darker Image described in Section 4.6. You are to do the following:
   (a) Plot $Y_{i,t}$ for $i = 1, 2$ for the given data.
   (b) Compute the time-average value of $Y_{i,t}$ for $i = 1, 2$ for the given data, and interpret the results in words.
   (c) Estimate the average delay using the method described at the end of Section 4.6 for the given data and interpret the result. Hint: You will need N_{19}, the total number of arrivals by time 19.

n	I	T_n	Q_n	A_n	C_1	C_2
0		0	0	0	3	∞
1	1	3	0	1	5	9
2	1	5	1	1	6	9
3	1	6	2	1	11	9
4	2	9	1	1	11	15
5	1	11	2	1	13	15
6	1	13	3	1	20	15
7	2	15	2	1	20	16
8	2	16	1	1	20	18
9	2	18	0	1	20	19
10	2	19	0	0	20	∞

4.9. A bus shuttles students from a remote parking facility to campus every 15 minutes. The capacity of the bus is 60 students. If students at the remote parking facility are unable to board the bus because it is full, then they nearly always wait for the next bus because it takes more than 15 minutes to walk to campus. Students have complained that the buses do not run frequently enough and they often have to wait for a second bus. The university would like to determine how frequently it needs to pick up students in order to keep up with demand. To do this, it would like to know something about how often the bus reaches capacity, and how many students are left waiting, if it picks up at 15, 12 and 10-minute intervals.

Students arrive sporadically. A brief time study was performed and recorded the following time gaps between the arrival of students to the bus stop (times in seconds): 18, 45, 27, 9, 11, 9, 11, 12, 17, 30, 8, 17, 14, 3, 2, 15, 1, 26, 9, 18, 3, 9, 63, 10, 30, 9, 1, 58, 31, 11. Perform a study to evaluate the impact of using 15, 12 and 10-minute intervals for the buses. Make a recommendation for the interval that should be used.

CHAPTER

5

ARRIVAL-COUNTING PROCESSES

A characteristic of many dynamic systems is that there is uncertainty due to one or more *arrival-counting processes*. An "arrival" is defined broadly to be any discrete unit that can be counted. For example, an arrival might refer to a customer (the full-service customers arriving to The Darker Image, Case 2.1, for example), a service request (the service requests for repairs arriving to Bit Bucket Computers, Case 4.1, for example), a demand for an item from inventory, an order for a product, the arrival of an electronic-mail message, or an accident in a factory. This chapter presents methods for modeling, analysis and simulation of one useful type of arrival-counting process.

5.1 THE CASE OF THE RECKLESS BEEHUNTER

Case 5.1. Citizens of Beehunter, Indiana, have complained that a busy intersection has recently become more dangerous, and they are demanding that the city council take action to make the intersection safer. There have been 103 accidents at the intersection since record keeping began. The city council agrees to undertake a study of the intersection to determine if the accident rate has actually increased above the 1 per week average that is (unfortunately) considered normal. It hires a traffic engineer from nearby Vincennes, Indiana, to perform the study.

The traffic engineer recommends that the number of accidents at the intersection be recorded for a 24-week period. If the number of accidents is significantly larger than expected, then she will declare that the intersection has indeed become more dangerous. During the study period, 36 accidents were observed.

The traffic engineer has adopted a standard approach for determining whether a system has changed: Assume that it has not changed, observe the system, and then decide whether the observed sample path is consistent with, or extremely different from, what was anticipated assuming no change. In this case the engineer assumes that the accident rate is still 1 per week, she records the number of accidents during a 24-week period, and then decides if the actual number of accidents is extremely different from the 24 accidents that were expected.

The modeling-and-analysis problem is to decide when observed behavior is "extremely different" from anticipated behavior. For instance, if 25 accidents were observed, then the traffic engineer probably would not declare the accident rate to be larger than 1 per week, since 25 is close to 24 and accidents do not occur with perfect regularity. On the other hand, if 117 accidents were observed, then the traffic engineer would surely declare the accident rate to be larger than 1 per week. The question is, How much of a difference from 24 should be considered extreme?

One way to identify extreme behavior is to propose a stochastic-process model that represents the system, and then define extreme behavior to be a collection of sample paths that have a very small probability of being generated by the model of the system *as we think it is*, but have a much larger probability of being generated by an alternative model of the system *as we think it might be*. If one of these extreme sample paths is observed, then we have some justification for concluding that the system has changed.

For example, suppose that the traffic engineer's model with an accident rate of 1 per week implies that the probability of observing more than 30 accidents in 24 weeks is 0.025. If she observes 36 accidents, then she might reasonably conclude that the intersection has become more dangerous, since greater than 30 accidents is very unlikely assuming a rate of 1 per week. Of course, this is no guarantee that the rate is larger than 1 per week since the probability of greater than 30 accidents is not zero. However, greater than 30 accidents is more likely if the accident rate is larger, which is an alternative model of the system as we think it might be. In fact, based on the observed sample path, she would estimate the accident rate to be 36 accidents/24 weeks = 1.5 accidents per week.

The preceding discussion illustrates how we use probability to make statements about the future and statistics to make statements about the past: A probability model provides statements about possible sample paths that the model might generate. In this case the statement is the probability of greater than 30 accidents in 24 weeks when the accident rate is 1 per week. After observing the system for 24 weeks and obtaining a sample path containing 36 accidents, a statistic provides a statement about the model that generated the data. In this case a statistic is the estimated accident rate of 1.5 accidents per week. Even though the 36 accidents were not actually "generated" by a model—they came from the intersection itself—we can treat the data as if they were generated by a model with unknown accident rate.

In this chapter you will learn how to formulate models of arrival-counting processes and how to derive probability statements about the sample paths of these models.

5.2 NOTATION AND REVIEW

To prepare for the study of arrival-counting processes, this section reviews a definition and two probability distributions that play an important role in their analysis.

Let $X_1, X_2, \ldots, X_n$ be n independent, time-stationary random variables with common cdf F_X. The cdf of the random variable $Z_n = X_1 + X_2 + \cdots + X_n$, denoted F_{Z_n}, is called the *n-fold convolution* of the cdf F_X. Random-variate generation for Z_n is easy: First generate $X_1, X_2, \ldots, X_n$ via any method, then set $Z_n \leftarrow X_1 + X_2 + \cdots + X_n$. Unfortunately, the cdf F_{Z_n} is usually difficult to write in a useful form. However, in the following two special cases F_{Z_n} is easy.

- If F_X is the Bernoulli distribution with parameter γ (Chapter 3, Exercise 17), then F_{Z_n} is called the *binomial* distribution with parameter γ and n trials; it has cdf

$$F_{Z_n}(a) = \sum_{j=0}^{\lfloor a \rfloor} \frac{n!}{j!(n-j)!} \gamma^j (1-\gamma)^{n-j} \tag{5.1}$$

for $0 \leq a \leq n$, and mass function

$$p_{Z_n}(b) = \frac{n!}{b!(n-b)!} \gamma^b (1-\gamma)^{n-b} \tag{5.2}$$

for $b = 0, 1, \ldots, n$. The $\mathrm{E}[Z_n] = n\gamma$ and the $\mathrm{Var}[Z_n] = n\gamma(1-\gamma)$. The binomial distribution can be interpreted as the probability distribution of the number of successes in n independent trials when each trial has success probability γ. Later in the chapter Z_n will represent the number of arrivals out of n that have a particular characteristic when each arrival has the characteristic independently with probability γ.

- If F_X is the exponential distribution with parameter λ, then F_{Z_n} is called the *Erlang* distribution with parameter λ and n phases; it has cdf

$$F_{Z_n}(a) = 1 - \sum_{j=0}^{n-1} \frac{e^{-\lambda a} (\lambda a)^j}{j!} \tag{5.3}$$

and density function

$$f_{Z_n}(a) = \frac{\lambda(\lambda a)^{n-1} e^{-\lambda a}}{(n-1)!}$$ (5.4)

for $a \geq 0$. The $E[Z_n] = n/\lambda$ and the $Var[Z_n] = n/\lambda^2$. Later in the chapter Z_n will represent the time that the nth arrival occurs when the gaps between arrivals are independent, exponentially distributed random variables.

5.3   A GENERIC ARRIVAL-COUNTING-PROCESS MODEL

The traffic engineer in Case 5.1 proposes the following model of the intersection in Beehunter, Indiana:

- There is a single system input process, which is the time between successive accidents at the intersection measured in weeks. These interarrival-time gaps are independent, time-stationary random variables with cdf F_G, where G denotes the interarrival-time gap between accidents, and $E[G] = 1$ week.
- The system logic is simply that the total accident count increases by 1 each time an accident occurs.

Let $G_1, G_2, \ldots$ denote the time gaps between successive accidents. Independence of these times is a reasonable approximation, since the occurrence of any particular accident should have little influence on the occurrence of any other accident (people typically try to avoid being in an accident). Time stationarity means that the probability distribution of the time gaps, F_G, never changes, which is a more problematic approximation. Perhaps an accident is more likely to occur during morning and evening rush hours or during bad weather. Perhaps the traffic flow is gradually increasing at the intersection, shortening the time gap between accidents. Both of these statements argue against stationarity, and both possibilities should be considered by the traffic engineer.

The single system event is the "arrival" of an accident, so there is a single clock, C, representing the time of the next arrival. Define the state of the system at event epoch n, denoted S_n, to be the total number of accidents up to and including the nth event epoch. Let time 0 be a time when the total number of accidents is defined to be 0. Then the initial system event e_0 and the accident-arrival system event e_1 below define the sample path of the stochastic process:

$e_0()$ **(initialization)**

$S_0 \leftarrow 0$ (initially no accidents)
$C \leftarrow F_G^{-1}(\texttt{random}())$ (set clock for first accident)

$e_1()$ **(accident)**

$$S_{n+1} \leftarrow S_n + 1 \qquad \text{(one more accident)}$$
$$C \leftarrow T_{n+1} + F_G^{-1}(\texttt{random}()) \qquad \text{(set clock for next accident)}$$

Because there is only a single clock, we can employ a simplified version of algorithm simulation:

algorithm simulation

1. $n \leftarrow 0$ (initialize system-event counter)
      $T_0 \leftarrow 0$ (initialize event epoch)
      $e_0()$ (execute initial system event)

2. $T_{n+1} \leftarrow C$ (update event epoch)
      $e_1()$ (update state and reset clock)
      $n \leftarrow n + 1$ (update system-event counter)

3.   repeat 2

The state-change process $\{S_n; n = 0, 1, 2, \ldots\}$ is a discrete-state process with state space $\{0, 1, 2, \ldots\}$ that represents the number of accidents. The time at which the nth accident occurs, measured in weeks, is represented by the event-epoch process $\{T_n; n = 0, 1, 2 \ldots\}$. A little thought reveals that $S_n = n$ for all n, so the important information about the arrival process is contained in the arrival times T_n.

The output process $Y_t \leftarrow S_n$ for all $T_n \leq t < T_{n+1}$ is simply the number of accidents up to and including time t, measured in weeks. Stated differently, Y_t counts the number of arrivals, and it increases by 1 at the time of the nth accident, T_n. For brevity, we refer to $\{Y_t; t \geq 0\}$ as the *arrival-counting process*, even though Y_t is defined in terms of S_n and T_n. The step function in Figure 5.1 is a typical sample path for our generic arrival-counting process.

Given a fixed point in time $a \geq 0$ weeks and an increment of $\Delta a > 0$ weeks, the number of accidents after time a and up to and including time $a + \Delta a$ is $Y_{a+\Delta a} - Y_a$. For example, the traffic engineer plans to observe the process for $\Delta a = 24$ weeks, so if she begins observing at time a weeks, then the random variable of interest is $Y_{a+24} - Y_a$. We use the convention that subscripts like t and a denote absolute times measured from time 0, while Δt and Δa denote relative-time increments without a fixed origin.

The time a when the traffic engineer begins observing the intersection may or may not coincide with an accident. In fact, unless she plans to start observing immediately after an accident, then it almost certainly will not coincide with one. Define a new random variable R_a to be the amount of time that passes until the first accident after time a; R_a is called the *forward-recurrence time* at

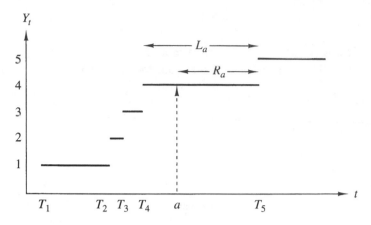

FIGURE 5.1

The forward-recurrence time R_a and gap L_a when the arrival-counting process is observed at time a.

time a. Also define L_a to be the length of the interarrival-time gap that contains time a. Figure 5.1 illustrates the relationship between R_a, L_a and the other random variables in the process. Values for R_a and L_a can be generated by adding Step 2.5 to algorithm simulation:

$$2.5 \quad \text{if } \{T_{n-1} \le a, T_n > a\} \text{ then}$$

$$R_a \leftarrow T_n - a \qquad \text{(record forward-recurrence time)}$$

$$L_a \leftarrow T_n - T_{n-1} \qquad \text{(record length of gap)}$$

$$\text{endif}$$

In order to make probability statements about the number of accidents that might occur during the 24-week observation period, the traffic engineer needs to know the distribution of the random variable $Y_{a+24} - Y_a$, specifically $\Pr\{Y_{a+24} - Y_a > 30\}$. Does this probability depend on the interarrival-time gap distribution F_G, or does it depend only on the expected value of G? Does it depend on whether the intersection is observed for *any* 24-week period or for 24 weeks beginning with an accident? What if it is not possible to define the starting time, a, because we do not know when "time 0" was? We will answer these questions in the next section by generating sample paths.

The stochastic-process model defined in this section represents the accidents that occur in Beehunter, Indiana, as described in Case 5.1. However, it is also a generic model for any arrival-counting process in which the interarrival-time gaps are independent and time stationary and arrivals occur one-at-a-time. Such an arrival-counting process is called a *renewal arrival-counting process*, or renewal process for short. The general conclusions we draw in the next section apply to any renewal process.

5.4 SIMULATING THE RECKLESS BEEHUNTER

The goal of this section is to investigate the sensitivity of the renewal arrival-counting process to the distribution of the interarrival-time gap, F_G. We begin with a thought experiment that requires no simulation.

Suppose that $\Pr\{G = 1\} = 1$. That is, suppose that the interarrival-time gap between accidents is always *exactly* 1 week. Clearly, the $E[G] = 1$ and the $\text{Var}[G] = 0$ for G defined this way. A random variable that takes a single value with probability 1 is called a *degenerate* random variable. Thinking through the logic of a stochastic process by assuming that the input random variables are degenerate is sometimes useful for developing intuition.

If G is degenerate at 1, then $\Pr\{Y_{a+24} - Y_a = 24\} = 1$; that is, we are guaranteed to observe exactly 24 accidents because 1 occurs every week. But if G is not degenerate and there is some randomness, then clearly $\Pr\{Y_{a+24} - Y_a = 24\} < 1$ since we are not certain to observe exactly 24 accidents. Therefore, we can conclude that the distribution of G—not just its expected value—may affect the distribution of Y_t.

If we begin observing at time $a = 0$, then the time until the first accident we observe—the forward-recurrence time R_a—is exactly 1 week for degenerate G. However, if we begin observing at time $a = 1/4$ week, then the forward-recurrence time is exactly $3/4$ week. Therefore, the time we begin observing an arrival-counting process may affect the distribution of R_a. *The thought experiment reveals that the distribution of the interarrival-time gap and the time we begin observing matter.*

Moving on to more realistic examples, consider the two distributions F_G described below. Both have expected value 1, but differ in other respects. The cdfs and densities of both distributions are plotted in Figure 5.2.

- The exponential distribution with parameter $\lambda = 1$. The expected value of this distribution is $1/\lambda = 1$, and the standard deviation is $\sqrt{1/\lambda} = 1$, so that the coefficient of variation is 1.
- The Weibull distribution (Section 3.6, Exercise 17) with parameters $\alpha = 2$ and $\beta = 2/\sqrt{\pi} \approx 1.128$. The expected value of this distribution is $\frac{\beta}{\alpha}\Gamma\left(\frac{1}{\alpha}\right) = 1$, the standard deviation is

$$\sqrt{\frac{\beta^2}{\alpha}\left(2\Gamma\left(\frac{2}{\alpha}\right) - \frac{1}{\alpha}\left[\Gamma\left(\frac{1}{\alpha}\right)\right]^2\right)} \approx \frac{1}{2}$$

where Γ is the gamma function, so that the coefficient of variation is $1/2$. For this choice of parameter values the Weibull distribution is only half as variable as the exponential distribution relative to its expected value.

Using the system-event functions in Section 5.3 and our algorithm simulation, we generated 1000 independent sample paths $\{Y_t; t \geq 0\}$ for each of the interarrival-time gap distributions. We observed each sample path beginning at

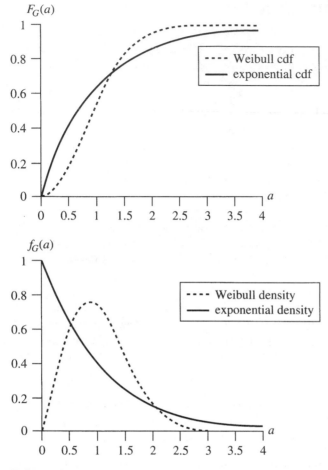

FIGURE 5.2
Densities and cdfs of the exponential and Weibull distributions.

time $a = 6$ weeks until time $a + 24 = 30$ weeks. Figure 5.3 shows one sample path that was generated when F_G is the exponential distribution.

On each replication R_6, L_6 and $Y_{30} - Y_6$ were recorded; that is, we recorded the forward-recurrence time from time 6, the length of the interarrival-time gap that contained time 6, and the number of accidents that occurred between the 6th week and the 30th week. For the sample path in Figure 5.3 the values of each of these variables are indicated. Figure 5.4 presents histograms for all 1000 values of each random variable.

Visually, the histograms for the exponential case differ from those for the Weibull case. Here are some other comparisons:

- The sample-average number of accidents between times 6 and 30, $Y_{30} - Y_6$, is about 24 in both cases. This is reasonable since the accident rate is 1 per week.

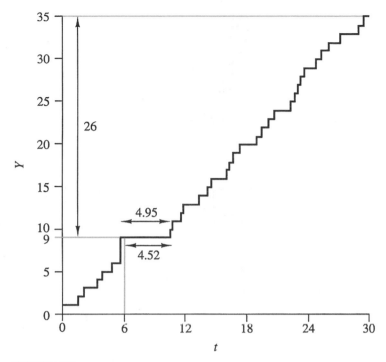

FIGURE 5.3
Sample path of the accident arrival-counting process with exponential interarrival-time gaps.

But the distribution of the number of accidents is different for the exponential and Weibull cases. The histograms show that the Weibull case includes fewer large values. The estimated $\widehat{\Pr}\{Y_{30} - Y_6 > 30\}$ is 0.10 for the exponential case, but 0.025 for the Weibull case (these probabilities come from the empirical cdf of $Y_{30} - Y_6$, which is not shown, and they are simply the number of observed values that were greater than 30, divided by 1000).

- The sample averages of the forward-recurrence times, R_6, are approximately 1 and 0.6 for the exponential and Weibull cases, respectively. The sample average for the exponential case is particularly interesting, because the expected length of the interarrival-time gaps is also 1. We might expect the forward-recurrence time to be less than 1, as it is in the Weibull case.

- In both cases the sample-average length of the interarrival-time gap that covered time $a = 6$, L_6, is greater than 1, the expected length of a gap. It is nearly 2 in the exponential case and 1.3 in the Weibull case. Figure 5.3 provides some intuition as to why this occurs: A long gap has a greater chance than a small gap to contain a particular point in time, in this case time $a = 6$.

Certainly the conclusion that the traffic engineer reaches about what is extreme behavior depends on the distribution of the time between accidents. Greater

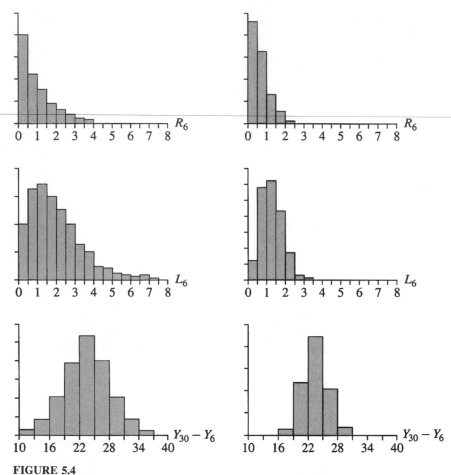

FIGURE 5.4

Summary statistics from 1000 replications of the accident arrival-counting processes, exponential results on the left, Weibull results on the right.

than 30 accidents is quite extreme if the Weibull distribution is an appropriate model (probability 0.025), but less extreme if the exponential distribution is appropriate (probability 0.10).

Given any distribution F_G and a starting-time value a, the traffic engineer can estimate the distribution of $Y_{a+24} - Y_a$ as we did here. Unfortunately, deciding what the starting time a should be is difficult. Time 0 has a precise meaning in our stochastic-process model, but it may not have any physical meaning for the system itself. One way to avoid the problem of specifying a is to start observing the system just after an arrival. Then we can define our starting time as time 0. This is a reasonable approach when arrivals occur frequently. But what do we do when we cannot wait for an arrival, or the data have already been collected, or we are specifically interested in the forward-recurrence time itself?

Look again at the histogram of R_6 values for the exponential-distribution case in Figure 5.4. It is quite similar to the density function of the exponential distribution in Figure 5.2. And recall that the sample average of the R_6 values was nearly 1, the same as the E[G]. These observations suggest that for the case of exponentially distributed interarrival-time gaps, the distribution of R_a and G might be the same. If this is true, then the time we begin observing the process does not matter if G has an exponential distribution. We investigate the structure of arrival-counting processes with exponentially distributed interarrival-time gaps in Sections 5.5 and 5.6. In Section 5.9 we return to the generic renewal process.

5.5 THE POISSON ARRIVAL PROCESS

In the previous section we studied sample paths of the renewal arrival-counting process by generating sample paths and estimating performance. In this section we derive performance measures directly from the stochastic-process model itself via mathematical analysis, rather than via statistical analysis of simulated sample paths.

There are several advantages of mathematical analysis: No decision is required regarding the number of replications (sample paths) necessary to obtain a good estimate because there is no error in a result achieved via mathematical analysis, except the numerical error (roundoff) when evaluating it on a calculator or computer. And it is possible to investigate the effect of changes in the model (such as a different arrival rate) without executing a new simulation. Unfortunately, the interarrival-time-gap distributions for which we can derive probability statements mathematically are limited. Simulation imposes no restrictions on the interarrival-time-gap distribution we employ.

5.5.1 Probability Structure of the Sample Paths

Because our renewal arrival process has a single clock, the time of the nth arrival T_n has a simple structure:

$$T_0 = 0$$

$$T_n = G_1 + G_2 + \cdots + G_n \tag{5.5}$$

for $n = 1, 2, \ldots$, where G_i is the interarrival-time gap between arrivals $i - 1$ and i. That is, T_n is the sum of the first n interarrival-time gaps, each an independent random variable with cdf F_G. Therefore, the distribution of T_n, denoted F_{T_n}, is the n-fold convolution of the distribution F_G. In many cases this knowledge is of limited use. But when F_G is the exponential distribution with parameter λ, the distribution of T_n is the Erlang distribution with parameter λ and n phases, so that

$$F_{T_n}(a) = 1 - \sum_{j=0}^{n-1} \frac{e^{-\lambda a}(\lambda a)^j}{j!} \tag{5.6}$$

The significance of knowing the distribution of T_n is that T_n completely determines the sample path of Y_t, the number of arrivals by time t, because Y_t

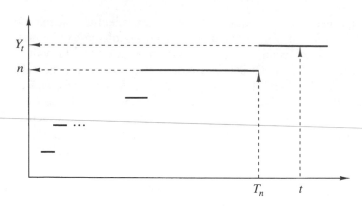

FIGURE 5.5
The relationship between T_n and Y_t.

increases by 1 at each arrival time $T_1, T_2, \ldots$. More precisely, the time of the nth arrival and the number of arrivals by time t are related in a following fundamental way:

$$\boxed{\{T_n \leq t\} \text{ if and only if } \{Y_t \geq n\}} \tag{5.7}$$

In words, the nth arrival can occur on or before time t only if the number of arrivals up to and including time t is greater than or equal to n. Conversely, there can be n arrivals by time t only if the nth arrival occurs on or before time t. Figure 5.5 illustrates this relationship. Since $\{T_n \leq t\}$ and $\{Y_t \geq n\}$ are the same event, the

$$\Pr\{T_n \leq t\} = \Pr\{Y_t \geq n\} = 1 - \sum_{j=0}^{n-1} \frac{e^{-\lambda t}(\lambda t)^j}{j!}$$

for fixed $t > 0$ and $n = 0, 1, 2, \ldots$, provided we define $\sum_{j=0}^{-1} = 0$. Written more naturally, the cdf of Y_t is

$$F_{Y_t}(n) = \Pr\{Y_t \leq n\} = 1 - \Pr\{Y_t \geq n+1\} = \sum_{j=0}^{n} \frac{e^{-\lambda t}(\lambda t)^j}{j!} \tag{5.8}$$

The associated mass function is

$$p_{Y_t}(n) = \Pr\{Y_t = n\} = F_{Y_t}(n) - F_{Y_t}(n-1) = \frac{e^{-\lambda t}(\lambda t)^n}{n!} \tag{5.9}$$

The mass function (5.9) is known as the **Poisson** distribution with parameter λt, and an arrival-counting process $\{Y_t; t \geq 0\}$ with independent, time-stationary, exponentially distributed interarrival-time gaps is called a **Poisson arrival process**.

There is substantial power in this result. Recall the intersection model for Beehunter, Indiana. If the exponential distribution is a good model of the interarrival-time gaps, then instead of generating a large number of sample paths to *estimate* a probability such as $\Pr\{Y_{24} > 30\}$ from the empirical cdf, we can *calculate* the probability directly from the actual cdf:

$$\Pr\{Y_{24} > 30\} = 1 - \Pr\{Y_{24} \le 30\} = 1 - \sum_{j=0}^{30} \frac{e^{-24}(24)^j}{j!} \approx 1 - 0.904 = 0.096$$

since $\lambda t = (1)(24)$ in this example. The only error in this result is that we have rounded the probability to three places after the decimal point.

The following properties of Poisson arrival processes provide the fundamental decompositions that allow us to derive probability statements about sample paths. All of these properties can be established using only relationship (5.7) and the Erlang distribution, and we do this in Section 5.8.

Let t be a fixed point in time, and let $\Delta t > 0$ be a time increment (for instance, t could be 6 weeks from time 0 and $\Delta t = 24$ weeks).

A Poisson arrival process has the **independent-increments property**, which means that the number of arrivals in nonoverlapping time intervals are independent random variables. Specifically,

$$\Pr\{Y_{t+\Delta t} - Y_t = m, Y_t = k\} = \Pr\{Y_{t+\Delta t} - Y_t = m\}\Pr\{Y_t = k\}$$

As a consequence

$$\Pr\{Y_{t+\Delta t} - Y_t = m | Y_t = k\} = \Pr\{Y_{t+\Delta t} - Y_t = m\}$$

regardless of the value of k.

A Poisson arrival process has the **stationary-increments property**, which means that the number of arrivals in a time increment of length Δt depends only on the length of the increment, and not when it starts. Specifically,

$$\Pr\{Y_{t+\Delta t} - Y_t = m\} = \Pr\{Y_{0+\Delta t} - Y_0 = m\} = \Pr\{Y_{\Delta t} = m\}$$

regardless of the value of t. As a consequence

$$E[Y_{t+\Delta t} - Y_t] = \lambda \Delta t \tag{5.10}$$

*for any $t \geq 0$, which shows that λ can be interpreted as the **arrival rate** of the process.*

---

*A Poisson arrival process has the **memoryless property**, which means that the forward-recurrence time R_t has the same distribution as the interarrival-time gaps. Specifically,*

$$\Pr\{R_t \leq \Delta t\} = 1 - e^{-\lambda \Delta t}$$

for any $t \geq 0$.

These properties make it possible to decompose complex probability statements about sample paths into simple probability statements about the Poisson distribution. We illustrate this for Case 5.1, again assuming that exponentially distributed gaps are appropriate.

Recall that a total of 103 accidents have occurred at the intersection up to the time the traffic engineer starts observing, the unspecified time a. The probability of interest is therefore

$$\Pr\{Y_{a+24} - Y_a > 30 | Y_a = 103\}$$

the conditional probability of greater than 30 accidents in the next 24 weeks given that 103 accidents have already occurred. The independent-increments property implies that the conditional event is not relevant, so that

$$\Pr\{Y_{a+24} - Y_a > 30 | Y_a = 103\} = \Pr\{Y_{a+24} - Y_a > 30\}$$

The stationary-increments property implies that the starting time is not relevant, either. Therefore,

$$\Pr\{Y_{a+24} - Y_a > 30\} = \Pr\{Y_{24} > 30\} \approx 0.096$$

as we determined earlier. Comparing this result to our estimate in Section 5.4, which was 0.10, we see that our estimate based on 1000 replications of the simulation was quite good. But keep in mind that the simulation provided only an estimate, while mathematical analysis provides the exact result up the numerical accuracy of our calculation. The traffic engineer can now decide if a probability of 0.096 is "extreme." If not, she can compute $\Pr\{Y_{24} > m\}$ for various values of m until an extreme probability is found.

The memoryless property confirms what we observed in the simulation study, namely that R_6, the forward-recurrence time at time 6, does indeed have an exponential distribution with parameter $\lambda = 1$. This is called the "memoryless property" because at the moment we start observing the arrival-counting process it is as if the process starts over with a new interarrival-time gap and forgets when the previous arrival occurred.

Equation (5.10) confirms something else we observed in the simulation: $E[Y_{30} - Y_6] = \lambda 24 = 24$ accidents during the 24-week study. Recall that the sample average number of accidents across the 1000 replications was nearly 24, also.

As in Chapter 4, we urge caution not to confuse simulated sample paths and probability statements about sample paths. For instance, the stationary-increments property implies that the probability of more than 30 accidents in any 24-week increment is the same, regardless of what 24-week increment we consider. But it does not imply that the same number of accidents actually occur in the time intervals $(0, 24]$ and $(16, 40]$, or even that more than 30 accidents will occur in either interval. Probability statements refer to future sample paths, not to what has happened.

The convenient decomposition of a Poisson arrival process is only of practical value if a Poisson process is frequently a useful model of real or conceptual systems. The following fundamental property justifies why it is often useful:

---

*Any arrival-counting process in which arrivals occur one-at-a-time and which has independent and stationary increments **must** be a Poisson arrival process.*

We do not prove this extraordinary result—see, for instance, Clarke and Disney (1985, Chapter 9) or Çinlar (1975, Chapter 4)—but it establishes why a Poisson arrival process often provides a good approximation for real or conceptual systems. Whenever we can argue that (a) the number of arrivals in nonoverlapping intervals of time are independent of each other and that (b) the expected number of arrivals in an interval of time is a constant rate times the length of the interval, then a Poisson process is a plausible model. Condition (a) will be reasonable when the arrival-counting process is formed by a large number of customers making individual, independent decisions about when to arrive, for instance, the arrival of telephone calls to an exchange, since people choose to make phone calls independently of who else is calling. Condition (b) is tenable when the arrival rate is steady, rather than varying with time, for instance, the arrival of human births, which occur at a relatively constant rate in a country like the United States.

5.5.2 Parameterizing Poisson Processes

To analyze the sample paths of a Poisson arrival process, we must first parameterize the model, using either the definition of λ as the arrival rate or the fact the $1/\lambda$ is the expected time between successive arrivals.

Suppose that we have observed an arrival-counting process for one or more intervals of time (not necessarily of equal length). Let c be the count of the total number of arrivals observed, and let d be the total time that the process was observed (sum of the lengths of all of the time intervals). Then an estimate of λ is $\widehat{\lambda} = c/d$, the total number of arrivals divided by the total time. The estimated standard error of $\widehat{\lambda}$ is approximately

$$\widehat{se} = \sqrt{\frac{\widehat{\lambda}}{d}} \tag{5.11}$$

For example, when the traffic engineer in Beehunter, Indiana, observed $c = 36$ accidents in $d = 24$ weeks, it implied an arrival rate of $\widehat{\lambda} = 36/24 = 1.5$ accidents per week, with a standard error of $\widehat{se} = \sqrt{1.5/24} = 0.25$ accidents per week. Notice that the $\widehat{se}$ indicates that this estimate is not very precise, since $\widehat{\lambda} \pm 2\,\widehat{se}$ gives an accident rate between 1 and 2 per week. The precision would improve with a longer observation period.

Data may also be available in the form of interarrival-time gaps, say $g_1, g_2, \ldots, g_m$. Then another estimate of λ is $\widehat{\lambda} = 1/\bar{g}$, where $\bar{g}$ is the sample average of the interarrival-time gaps. The standard error of this estimate is also approximately (5.11), where $d = \sum_{i=1}^{m} g_i$.

5.6 MORE ABOUT POISSON ARRIVAL PROCESSES

Recall Case 2.1, the design of the photocopying shop called The Darker Image. The arrival-counting process in that example is composed of two types of customers, full-service customers and self-service customers. There are two ways to view such an arrival-counting process: as a single arrival-counting process of all customers that is *decomposed* into two subprocesses, one for each type of customer; or as two arrival-counting processes for each type of customer that are *superposed* into a single, overall arrival-counting process. In the next two sections we examine the implications of each view when the arrival processes are Poisson.

The time-study data from The Darker Image were collected over a few hours one morning. Quite possibly there are times during the day when the shop is more busy than it is at other times. During these so-called rush hours the interarrival-time gaps would tend to be shorter. When the distribution of the interarrival-gaps changes over time, the process is no longer time stationary and is said to be *nonstationary*. The third section presents a model for nonstationary arrival-counting processes.

5.6.1 Decomposition of a Poisson Process

Table 2.3 contains time-study data on the interarrival-time gaps of customers arriving to The Darker Image. In summary, the sample-average time between arrivals was 6 minutes, and 8 of 20 customers were full-service customers.

Let $\{Y_t; t \geq 0\}$ be a Poisson arrival process that represents the arrival of all customers at The Darker Image. A value for the arrival rate λ is required to complete the model. Since the expected value of the time between arrivals is $1/\lambda$ for a Poisson process and the sample-average time between arrivals from the time study was 6 minutes, we can estimate the arrival rate by $\widehat{\lambda} = 1/6$ customers per minute.

We also need to model the different customer types. Associate with customer n a Bernoulli random variable, B_n, where B_n takes the value 0 to represent a self-service customer and 1 to represent a full-service customer. Suppose that these Bernoulli random variables form a stochastic process $\{B_n; n = 1, 2, \ldots\}$ of independent and time-stationary random variables that are also independent of the interarrival-time gaps. The implication of this model is that the types of successive customers are unrelated, the mix of customer types does not change throughout the day, and there is no relationship between the time that customers arrive and what type of service they require. Since 8 of 20 customers observed during the time study were full-service customers, an estimate of the "success" probability γ for the Bernoulli distribution is $\widehat{\gamma} = \widehat{\Pr}\{B_n = 1\} = 8/20 = 0.4$.

The Bernoulli random variables decompose the overall arrival-counting process into two subprocesses representing arrivals of each type of customer. Let S_n represent the total number of arrivals by event epoch n, $S_{0,n}$ represent the number of self-service customers among the first n arrivals, and $S_{1,n}$ represent the number of full-service customers among the first n arrivals. Thus, $S_n = S_{0,n} + S_{1,n}$. The system-event functions below show how to use the Bernoulli process to decompose the overall arrival-counting process into the two subprocesses. There is a single customer arrival event, e_1, but the change in system state depends on the Bernoulli process.

$e_0()$ **(start of day)**

         $S_0 \leftarrow 0$ (initially no arrivals)

         $S_{0,0} \leftarrow 0$ (initially no self-service customers)

         $S_{1,0} \leftarrow 0$ (initially no full-service customers)

         $C \leftarrow F_G^{-1}(\texttt{random}())$ (set clock for first arrival)

$e_1()$ (**customer arrives**)

$$S_{n+1} \leftarrow S_n + 1 \qquad \text{(one more arrival)}$$
$$B_{n+1} \leftarrow F_B^{-1}(\texttt{random}()) \qquad \text{(determine customer type)}$$

if $\{B_{n+1} = 1\}$ then

$\qquad\qquad$ ~~$S_{1,n+1} \leftarrow S_{1,n} + 1$~~ $\qquad\qquad$ ~~(a full-service customer)~~

else

$\qquad\qquad S_{0,n+1} \leftarrow S_{0,n} + 1 \qquad$ (a self-service customer)

endif

$$C \leftarrow T_{n+1} + F_G^{-1}(\texttt{random}()) \quad \text{(set clock for next arrival)}$$

Let the output process be

$$\mathbf{Y}_t = \begin{pmatrix} Y_t \\ Y_{0,t} \\ Y_{1,t} \end{pmatrix}$$

where Y_t is the total number of customers, and $Y_{0,t}$ and $Y_{1,t}$ are the number of self-service customers and full-service customers, respectively, that have arrived by time t minutes. We know that the overall arrival-counting process Y_t is Poisson with rate λ by the way we constructed it. The following fundamental property of Poisson processes provides a complete characterization for the subprocesses:

A Poisson process has the **decomposition property**, which means that a Poisson process with arrival rate λ that is decomposed into two subprocesses by an independent, time-stationary Bernoulli process with success probability γ yields two Poisson subprocesses with arrival rates $\lambda_0 = (1 - \gamma)\lambda$ and $\lambda_1 = \gamma\lambda$, and these subprocesses are independent of each other. The key consequence of the decomposition property is that

$$\Pr\{Y_{0,t+\Delta t} - Y_{0,t} = k, \ Y_{1,t+\Delta t} - Y_{1,t} = m | Y_{0,t} = i, Y_{1,t} = j\}$$

$$= \Pr\{Y_{0,t+\Delta t} - Y_{0,t} = k\} \Pr\{Y_{1,t+\Delta t} - Y_{1,t} = m\}$$

$$= \left(\frac{e^{-\lambda_0 \Delta t} (\lambda_0 \Delta t)^k}{k!} \right) \left(\frac{e^{-\lambda_1 \Delta t} (\lambda_1 \Delta t)^m}{m!} \right)$$

regardless of i, j and t.

For a model of The Darker Image, this property implies that the arrival process of self-service customers is Poisson with arrival rate $\lambda_0 = (1 - 0.4)(1/6) =$

1/10 customers per minute, while the arrival process of full-service customers is Poisson with arrival rate $\lambda_1 = (0.4)(1/6) = 1/15$ customers per minute, and these two processes are independent of each other. All of the properties of Poisson processes in Section 5.5.1 apply to these two arrival-counting processes individually.

For example, the probability that fewer than three self-service customers arrive during any hour the shop is open is

$$\Pr\{Y_{0,a+60} - Y_{0,a} < 3\} = \sum_{j=0}^{2} \frac{e^{-60\frac{1}{10}} \left(60\frac{1}{10}\right)^j}{j!} \approx 0.062$$

The expected number of full-service customers to arrive during any 60-minute period is

$$E[Y_{1,a+60} - Y_{1,a}] = \frac{1}{15}(60) = 4$$

A more complex question is the following: Suppose that we know that 12 customers arrived during the last hour. What is the probability that fewer than 3 of them were self-service customers? The event of interest is a conditional event, $\{Y_{0,a+60} - Y_{0,a} < 3 | Y_{a+60} - Y_a = 12\}$, the event that there are fewer than 3 self-service customers given that there were 12 customers total. To answer the question we derive $\Pr\{Y_{0,a+60} - Y_{0,a} = k | Y_{a+60} - Y_a = 12\}$, for $k = 0, 1, \ldots, 12$.

The stationary-increments property of Poisson processes implies that

$$\Pr\{Y_{0,a+60} - Y_{0,a} = k | Y_{a+60} - Y_a = 12\} = \Pr\{Y_{0,60} = k | Y_{60} = 12\}$$

By the definition of conditional probability,

$$\Pr\{Y_{0,60} = k | Y_{60} = 12\} = \frac{\Pr\{Y_{0,60} = k, Y_{60} = 12\}}{\Pr\{Y_{60} = 12\}}$$

But the event $\{Y_{0,60} = k, Y_{60} = 12\}$ is the same as the event $\{Y_{0,60} = k, Y_{1,60} = 12 - k\}$. Then using the fact that $Y_{0,t}$ and $Y_{1,t}$ are independent, we obtain

$\Pr\{Y_{0,60} = k | Y_{60} = 12\}$

$$= \frac{\Pr\{Y_{0,60} = k, Y_{1,60} = 12 - k\}}{\Pr\{Y_{60} = 12\}}$$

$$= \Pr\{Y_{0,60} = k\} \Pr\{Y_{1,60} = 12 - k\} \frac{1}{\Pr\{Y_{60} = 12\}}$$

$$= \left(\frac{e^{-60/10}(60/10)^k}{k!}\right)\left(\frac{e^{-60/15}(60/15)^{12-k}}{(12-k)!}\right)\left(\frac{12!}{e^{-60/6}(60/6)^{12}}\right)$$

$$= \frac{12!}{k!(12-k)!}\left(\frac{6}{10}\right)^k\left(\frac{6}{15}\right)^{12-k}$$

$$= \frac{12!}{k!(12-k)!}(0.6)^k(0.4)^{12-k}$$

This is just the binomial probability of k successes in 12 Bernoulli trials, where a "success" is defined to mean that an arrival is a self-service customer. This makes sense because the customer types are determined by a Bernoulli process. The $\Pr\{Y_{0,a+60} - Y_{0,a} < 3 | Y_{a+60} - Y_a = 12\}$ then comes from summing this result for $k = 0, 1, 2$, and it is 0.003 to three decimal places.

The decomposition property is derived in Section 5.8. The method of decomposition—by an independent Bernoulli process—is critical to the derivation. Other methods of decomposition do not lead to Poisson subprocesses. For instance, if all of the odd-numbered arrivals form one subprocess and the even-numbered arrivals form the other, then the subprocesses are not Poisson.

5.6.2 Superposition of Poisson Processes

The statement of Case 2.1 made it natural to model the customer arrival-counting process as one process that is decomposed into two subprocesses for each customer type. But we could have modeled it the other way, as two arrival-counting processes that are superposed to create an overall customer arrival-counting process. Table 5.1 contains the interarrival-time gaps *by customer type* that were extracted from the time study, in other words, the gaps between self-service customers (ignoring the existence of full-service customers) and the gaps between full-service customers (ignoring the existence of self-service customers). The sample-average

TABLE 5.1
Interarrival-time gaps for customers in The Darker Image time study by customer type

customer number, i	self-service gap	full-service gap
1		12
2	14	
3	3	
4	21	
5		27
6	5	
7	2	
8	7	
9		19
10		15
11	22	
12	2	
13	3	
14	9	
15		23
16	17	
17	3	
18		17
19		5
20		2

time between the arrival of self-service customers is 9 minutes, while the sample-average time between the arrival of full-service customers is 15 minutes.

Suppose we had initially modeled the self-service-customer arrival-counting process as a Poisson process with arrival rate λ_0, the full-service-customer arrival-counting process as a Poisson process with arrival rate λ_1, and the two processes as mutually independent. Independence of the processes is reasonable if we believe that individual customers act independently. Based on the time-study data, estimates of the arrival rates are $\widehat{\lambda}_0 = 1/9$ customers per minute, and $\widehat{\lambda}_1 = 1/15$ customers per minute.

Let F_{G_0} be the interarrival-time gap cdf for self-service customers (exponential with parameter λ_0) and let F_{G_1} be the interarrival-time gap cdf for full-service customers (exponential with parameter λ_1). Let S_n, $S_{0,n}$ and $S_{1,n}$ be defined as in Section 5.6.1. There are two system events, e_1 representing the arrival of a full-service customer and e_2 representing the arrival of a self-service customer. The general-purpose algorithm simulation in Chapter 4 is required to generate sample paths. The system-event functions for this arrival-counting process are as follows:

$e_0()$ **(start of day)**

$\quad S_0 \leftarrow 0$ (initially no arrivals)

$\quad S_{0,0} \leftarrow 0$ (initially no self-service customers)

$\quad S_{1,0} \leftarrow 0$ (initially no full-service customers)

$\quad C_1 \leftarrow F_{G_1}^{-1}(\texttt{random}())$ (set clock for first full-service arrival)

$\quad C_2 \leftarrow F_{G_0}^{-1}(\texttt{random}())$ (set clock for first self-service arrival)

$e_1()$ **(full-service customer arrives)**

$\quad S_{1,n+1} \leftarrow S_{1,n} + 1$ (one more full-service customer)
$\quad S_{n+1} \leftarrow S_n + 1$ (one more total arrival)

$\quad C_1 \leftarrow T_{n+1} + F_{G_1}^{-1}(\texttt{random}())$ (set clock for next f-s arrival)

$e_2()$ **(self-service customer arrives)**

$\quad S_{0,n+1} \leftarrow S_{0,n} + 1$ (one more self-service customer)
$\quad S_{n+1} \leftarrow S_n + 1$ (one more total arrival)

$\quad C_2 \leftarrow T_{n+1} + F_{G_0}^{-1}(\texttt{random}())$ (set clock for next s-s arrival)

Let Y_t be defined as in Section 5.6.1. By construction the self-service and full-service-customer arrival-counting processes are Poisson; it turns out that their superposition, $Y_t = Y_{0,t} + Y_{1,t}$, is also Poisson.

---

*A Poisson process has the **superposition property**, which means that two independent Poisson arrival processes with arrival rates λ_0 and λ_1 that are superposed form a Poisson arrival process with arrival rate $\lambda = \lambda_0 + \lambda_1$.*

For a model of The Darker Image this property implies that the overall customer arrival-counting process is Poisson with arrival rate $\lambda = 1/15 + 1/9 = 8/45$ customers per minute. Notice that this is a different model of the overall arrival-counting process than the one in Section 5.6.1 (arrival rate 8/45 rather than 1/6), which illustrates that there is no unique model of any system, and the choice of an appropriate model involves judgment.[1]

The superposition property is derived in Section 5.8. The fact that the processes that are superposed are independent is critical to the derivation. If the processes are dependent in some way, then the overall arrival-counting process may not be Poisson. For instance, if the arrivals in one process are timed to coincide with the arrivals in the other process—such as people meeting for a dinner appointment or component parts gathered to produce a final assembly—then the overall arrival-counting process is not Poisson.

5.6.3   Nonstationary Poisson Processes

The Darker Image is open each day from 9 a.m. until 3 p.m., a total of 360 minutes. The time-study data presented in Chapter 2 provided an estimated arrival rate of 1/6 customers per minute, based on the sample-average time between arrivals of 6 minutes. Suppose that further time study shows that this rate applies only until 12 noon, after which the sample-average time between arrivals drops to 5 minutes, implying an estimated arrival rate of 1/5 customers per minute from 12 until 3 p.m.

Since the arrival rate, and therefore the interarrival-time gap, depends on the time of day, the arrival-counting process is not time stationary. Nonstationary behavior occurs frequently in arrival-counting processes. In this section we show that a stationary Poisson process with arrival rate 1 can be transformed into a *nonstationary Poisson process* with any time-dependent arrival rate we desire. The nonstationary Poisson process maintains the independent-increments property of the Poisson process, but sacrifices the stationary-increments property.

To illustrate the approach, consider an arrival-rate function that has a more dramatic change than the one representing The Darker Image. Let the arrival rate

---

[1]In this example the approach in Section 5.6.1 probably makes better use of the data, since all 20 arrivals contribute to estimating both the arrival rate λ and the probability that a customer wants full service, γ.

at time τ be

$$\lambda(\tau) = \begin{cases} 2 & \text{if } 0 \le \tau < 4 \\ \frac{1}{2}, & \text{if } 4 \le \tau < 8 \end{cases}$$

where we introduce the new variable τ to represent time for reasons that will become clear later. This function implies that the arrival rate is 2 arrivals per unit time for $0 \le \tau < 4$, and $1/2$ arrival per unit time for $4 \le \tau < 8$. The expected number of arrivals by time τ is given by the *integrated-rate function*

$$\Lambda(\tau) = \int_0^\tau \lambda(a)\,da = \int_0^\tau 2\,da = 2\tau$$

for $0 \le \tau < 4$, and

$$\Lambda(\tau) = \int_0^\tau \lambda(a)\,da = \int_0^4 2\,da + \int_4^\tau \frac{1}{2}\,da = \frac{1}{2}\tau + 6$$

for $4 \le \tau < 8$. Figure 5.6 plots $\Lambda(\tau)$ as a function of τ. The integrated-rate function is continuous and increasing, so it has an inverse. The inverse is obtained by solving $t = \Lambda(\tau)$ for τ. If we do this we obtain

$$\tau = \Lambda^{-1}(t) = \begin{cases} \frac{1}{2}t, & \text{if } 0 \le \tau < 8 \\ 2(t - 6), & \text{if } 8 \le \tau < 10 \end{cases}$$

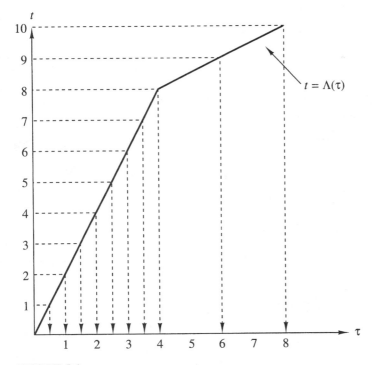

FIGURE 5.6
Transforming an arrival-counting process with rate 1 into an arrival-counting process with rate $\lambda(\tau)$.

The central idea is to let τ and t represent different time scales that are connected by the relationship $t = \Lambda(\tau)$. The t time scale will be for a stationary Poisson process with arrival rate 1, while the τ scale will be for the nonstationary process with any time-varying arrival rate we desire. To develop intuition as to why this works, we perform a thought experiment.

Suppose that on the t time scale arrivals occur at times $t = 1, 2, 3, \ldots$. In other words, they occur at rate 1 and *exactly* 1 time unit apart. Then $\tau = \Lambda^{-1}(t)$ gives the corresponding arrival times on the τ time scale. In Figure 5.6 the dashed arrows represent the arrivals. Notice that the arrivals on the vertical t scale occur at exactly rate 1, while the arrivals on the horizontal τ scale occur at a rate of 2 arrivals per unit time from time 0 to time 4, and $1/2$ arrival per unit time from time 4 to time 8, as specified by $\lambda(\tau)$. This picture provides the intuition behind our construction of nonstationary processes on the τ time scale.

Let $\{Y_t; t \geq 0\}$ be a Poisson arrival process with arrival rate 1, and let $\{T_n; n = 0, 1, 2, \ldots\}$ be the arrival times for this process. Define a new arrival-counting process $\{\mathcal{Y}_\tau; \tau \geq 0\}$ in which the arrival times are $\mathcal{T}_n = \Lambda^{-1}(T_n)$, for $n = 0, 1, 2, \ldots$. The process $\{\mathcal{Y}_\tau; \tau \geq 0\}$ is called a *nonstationary Poisson arrival process*, for which we obtain the following:

The nonstationary Poisson process $\{\mathcal{Y}_\tau; \tau \geq 0\}$ *has the property that*

$$\Pr\{\mathcal{Y}_{\tau+\Delta\tau} - \mathcal{Y}_\tau = m | \mathcal{Y}_\tau = k\} = \Pr\{\mathcal{Y}_{\tau+\Delta\tau} - \mathcal{Y}_\tau = m\}$$

$$= \frac{e^{-[\Lambda(\tau+\Delta\tau)-\Lambda(\tau)]}[\Lambda(\tau + \Delta\tau) - \Lambda(\tau)]^m}{m!}$$

for fixed time $\tau \geq 0$, *increment* $\Delta\tau > 0$, *and* $m = 0, 1, 2, \ldots$. *This is the Poisson mass function, but with a parameter that depends on* τ *and* $\Delta\tau$.

This property establishes that the increments of $\mathcal{Y}_\tau$ are independent, but the probability distribution of the number of arrivals in the interval $(\tau, \tau + \Delta\tau]$ depends on both $\Delta\tau$ and τ. A direct consequence of this property is that the

$$E[\mathcal{Y}_{\tau+\Delta\tau} - \mathcal{Y}_\tau] = \Lambda(\tau + \Delta\tau) - \Lambda(\tau)$$

As a second example, the proposed arrival-rate function for The Darker Image is

$$\lambda(\tau) = \begin{cases} \frac{1}{6} & \text{if } 0 \leq \tau < 180 \\ \frac{1}{5}, & \text{if } 180 \leq \tau < 360 \end{cases}$$

where τ is measured in minutes. The integrated rate function is

$$\Lambda(\tau) = \begin{cases} \frac{1}{6}t & \text{if } 0 \leq \tau < 180 \\ \frac{1}{5}t - 6, & \text{if } 180 \leq \tau < 360 \end{cases}$$

Therefore, the expected number of customers arriving during a day is

$$E[\mathcal{Y}_{360}] = \Lambda(360) = \frac{1}{5}(360) - 6 = 66$$

The expected number of arrivals between 11 a.m. and 1 p.m. (times 120 to 240 minutes) is $\Lambda(240) - \Lambda(120) = 42 - 20 = 22$. The probability distribution of the number of customers that arrive between these times is

$$\Pr\{\mathcal{Y}_{240} - \mathcal{Y}_{120} = m\} = \frac{e^{-22}(22)^m}{m!}$$

for $m = 0, 1, 2, \ldots$. As a consequence of the independent-increments property this distribution does not depend on how many arrivals have occurred before 11 a.m., but it does depend on the specific 2-hour period between 11 a.m. and 1 p.m.

5.7 THE CASE OF THE MEANDERING MESSAGE

Case 5.2. A computer at The Ohio State University receives electronic-mail messages (e-mail) from outside the university. This computer is called osu.edu, and it distributes the mail it receives to other computers on campus, including those in The College of Engineering.[2] The College of Engineering's central computer is called eng.ohio-state.edu, and it distributes e-mail received from osu.edu to other computers and workstations. However, eng.ohio-state.edu also receives mail directly without passing through osu.edu. The College of Engineering plans to replace eng.ohio-state.edu with a newer machine and therefore needs to do some capacity planning.

Data are maintained on e-mail traffic through osu.edu. Records from two randomly selected days show that 88,322 e-mail messages were received on one day and 84,478 on the other, during normal business hours (7:30 a.m.–7:30 p.m.). Historically 20% of all incoming mail goes to the College of Engineering. System managers say that the average message size is about 12K bytes. The College of Engineering does not keep such detailed records, but it is willing to say that the direct traffic to eng.ohio-state.edu is about two-and-a-half times the traffic it receives from osu.edu. The College of Engineering wants its new machine to have enough capacity to handle even extreme bursts of traffic.

The e-mail traffic is naturally viewed as an arrival-counting process with the e-mail messages as the arrivals. The overall system should probably be viewed as a queueing process (see Exercise 21, Chapter 8), but here we concentrate only on the arrivals.

---

[2]The computer systems in this case are real. However, the description of the e-mail system is a simplified version of what really happens. And while the numbers are fabricated, they are consistent with actual data.

Provided that the rate of message arrivals is reasonably steady throughout the business day, a Poisson arrival process is a plausible model for the e-mail directly to `osu.edu` and directly to `eng.ohio-state.edu`, because it is the result of a large number of senders acting independently. However, it is not a perfect model, since it does not represent "bulk mail" that occurs when a single message is simultaneously sent to multiple users on a list. And a stationary arrival process is not appropriate for all time, because e-mail traffic is significantly lighter at night, so our conclusions are restricted to business hours.

The total of $88{,}322 + 84{,}478 = 172{,}800$ arrivals to `osu.edu` occurred over $(2 \text{ days}) \times (12 \text{ hours/day}) \times (60 \text{ minutes/hour}) \times (60 \text{ seconds/minute}) = 86{,}400$ seconds, giving an estimated arrival rate of

$$\widehat{\lambda}_{\text{osu}} = \frac{172{,}800}{86{,}400} = 2 \text{ arrivals per second}$$

with standard error of only

$$\widehat{\text{se}} = \sqrt{\frac{\widehat{\lambda}_{\text{osu}}}{86{,}400}} \approx 0.005 \text{ arrivals per second.}$$

We can conclude that our estimate is quite precise.

What can we say about the overall arrival-counting process to `eng.ohio-state.edu`? It is composed of the direct arrivals and those arrivals that are distributed by `osu.edu`. If we say that each arrival to `osu.edu` has a 0.2 probability of being routed to Engineering and that routing is instantaneous, then the arrivals to `osu.edu` are decomposed into arrivals routed to `eng.ohio-state.edu` and arrivals routed somewhere else. The arrivals routed to `eng.ohio-state.edu` therefore form a Poisson arrival process with rate

$$\widehat{\lambda}_{\text{routed}} = (0.2)\widehat{\lambda}_{\text{osu}} = 0.4 \text{ arrivals per second}$$

The routing is not instantaneous, of course, but if `osu.edu` processes messages much faster than they are received, then this is a reasonable approximation.

Based on speculation, the direct arrival rate to `eng.ohio-state.edu` is taken to be $\widehat{\lambda}_{\text{direct}} = (2.5)\widehat{\lambda}_{\text{routed}} = 1$ per second. We can assign no standard error to this estimate since it is not based on data. Thus, we may want to do a sensitivity analysis over a range of values for $\widehat{\lambda}_{\text{direct}}$.

The overall arrival process to `eng.ohio-state.edu` is the superposition of two Poisson arrival processes, and is therefore a Poisson process $\{Y_t; t \geq 0\}$ with rate

$$\widehat{\lambda}_{\text{eng}} = \widehat{\lambda}_{\text{direct}} + \widehat{\lambda}_{\text{routed}} = 1.4 \text{ arrivals per second}$$

Given this model, some rough capacity planning can be done by looking at the probability of extreme bursts. For example, if the new `eng.ohio-state.edu` can easily handle 3 messages per second, we might be interested in the probability of more than 3 arrivals in any 1-second interval:

$$\Pr\{Y_{t+1} - Y_t > 3\} = 1 - \Pr\{Y_{t+1} - Y_t \leq 3\}$$

$$= 1 - \sum_{j=0}^{3} \frac{e^{-\widehat{\lambda}_{\text{eng}}} \left(\widehat{\lambda}_{\text{eng}}\right)^j}{j!}$$

$$\approx 0.05$$

If processing of e-mail also depends on the size of the message, then we can say something about this based on an average message size of 12K bytes. Provided the size of the message is independent of the time it arrives, the average number of bytes `eng.ohio-state.edu` receives every second is (12K bytes/message)$\widehat{\lambda}_{\text{eng}} = 16.8K$ bytes per second. Unfortunately, we cannot make probability statements about the number of bytes received without knowing the distribution of message size or treating all messages as exactly 12K bytes.

5.8   DERIVATIONS

This section can be skipped by readers interested only in applications.

Let $\{Y_t; t \geq 0\}$ be a Poisson arrival process with arrival rate $0 < \lambda < \infty$. In this section we derive the properties that were stated in the previous sections. The notation is consistent with the earlier sections.

5.8.1   Memoryless Property

Here we derive the memoryless property of Y_t by showing that the forward-recurrence time R_t is exponentially distributed with parameter λ.

Suppose that the event $\{Y_t = k\}$ occurs, meaning that there are k arrivals by time t. Then we know that T_k, the time of the kth arrival, is also the time of the last arrival on or before time t. By similar reasoning we know that $\{T_k + G_{k+1} > t\}$ because $T_{k+1} = T_k + G_{k+1}$ and T_{k+1} is the time of the first arrival after time t. Figure 5.7 illustrates the situation. Expressing what we know in terms of a probability statement, we have

$$\Pr\{R_t > \Delta t\} = \Pr\{T_{k+1} - t > \Delta t | T_{k+1} > t\}$$

$$= \Pr\{T_k + G_{k+1} - t > \Delta t | T_k + G_{k+1} > t\}$$

$$= \Pr\{G_{k+1} > t + \Delta t - T_k | G_{k+1} > t - T_k\}$$

Applying the definition of conditional probability and recalling that G_{k+1} is

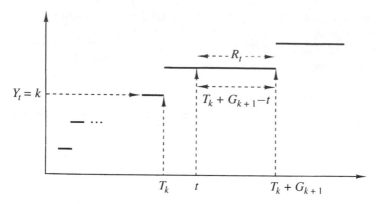

FIGURE 5.7
The forward-recurrence time R_t when $\{Y_t = k\}$.

exponentially distributed with parameter λ, we obtain

$$
\begin{aligned}
\Pr\{R_t > \Delta t\} &= \Pr\{G_{k+1} > t + \Delta t - T_k | G_{k+1} > t - T_k\} \\
&= \frac{\Pr\{G_{k+1} > t + \Delta t - T_k,\ G_{k+1} > t - T_k\}}{\Pr\{G_{k+1} > t - T_k\}} \\
&= \frac{\Pr\{G_{k+1} > t + \Delta t - T_k\}}{\Pr\{G_{k+1} > t - T_k\}} \\
&= \frac{e^{-\lambda(t+\Delta t - T_k)}}{e^{-\lambda(t - T_k)}} \\
&= e^{-\lambda \Delta t}
\end{aligned}
\tag{5.12}
$$

Therefore $\Pr\{R_t \leq \Delta t\} = 1 - e^{-\lambda \Delta t}$. And since the result did not depend on the value of k, we have also established that R_t and Y_t are independent, a property we will need below.

Our derivation of the memoryless property is somewhat informal, because we did not explicitly include the condition $\{Y_t = k\}$ in our probability statements and we treated T_k as a constant in (5.12). The additional work required to make it rigorous is tedious, but not substantial.

5.8.2   Independent-Increments and Stationary-Increments Properties

The independent-increments property follows immediately from the memoryless property by noticing that

$$
\Pr\{Y_{t+\Delta t} - Y_t \geq m, Y_t = k\}
$$
$$
= \Pr\{Y_{t+\Delta t} - Y_t \geq m | Y_t = k\} \Pr\{Y_t = k\}
$$

$$= \Pr\{R_t + G_{k+2} + G_{k+3} + \cdots + G_{k+m} \leq \Delta t | Y_t = k\} \Pr\{Y_t = k\}$$

$$= \Pr\{R_t + G_{k+2} + G_{k+3} + \cdots + G_{k+m} \leq \Delta t\} \Pr\{Y_t = k\}$$

$$= \Pr\{Y_{t+\Delta t} - Y_t \geq m\} \Pr\{Y_t = k\}$$

where the second-to-last equality follows because R_t is independent of Y_t. This also establishes the stationary-increments property, because $R_t + G_{k+2} + G_{k+3} + \cdots + G_{k+m}$ is the sum of m exponentially distributed random variables with parameter λ and, therefore, has an Erlang distribution with parameter λ and m phases. Thus, $\Pr\{Y_{t+\Delta t} - Y_t \geq m\} = \Pr\{Y_{\Delta t} \geq m\}$.

5.8.3 Decomposition Property

To derive the decomposition property, let B_n be an independent Bernoulli process with success probability γ, and let $Z_n = \sum_{j=0}^{n} B_j$, a binomial random variable with success probability γ and n trials. If $\{B_n = 0\}$ then the nth arrival is of type 0 and increments arrival-counting process $Y_{0,t}$; otherwise it is of type 1 and increments arrival-counting process $Y_{1,t}$. Therefore, Z_n is the number of arrivals out of the first n that are of type 1.

The approach we take is to convert probability statements about the subprocesses $Y_{0,t}$ and $Y_{1,t}$ into statements about the overall arrival-counting process Y_t and the number of arrivals of type 1, Z_n. Remember that these two processes are independent by assumption. Therefore,

$$\Pr\{Y_{0,t+\Delta t} - Y_{0,t} = k, Y_{1,t+\Delta t} - Y_{1,t} = m | Y_{0,t} = i, Y_{1,t} = j\}$$

$$= \Pr\{Y_{t+\Delta t} - Y_t = k + m, Z_{i+j+k+m} - Z_{i+j} = m | Y_t = i + j, Z_{i+j} = j\}$$

$$= \Pr\{Y_{t+\Delta t} - Y_t = k + m, Z_{i+j+k+m} - Z_{i+j} = m\}$$

$$= \Pr\{Y_{t+\Delta t} - Y_t = k + m\} \Pr\{Z_{i+j+k+m} - Z_{i+j} = m\}$$

$$= \left(\frac{e^{-\lambda \Delta t} (\lambda \Delta t)^{k+m}}{(k+m)!} \right) \left(\frac{(k+m)!}{k!m!} \gamma^m (1-\gamma)^k \right)$$

$$= \left(\frac{e^{-(1-\gamma)\lambda \Delta t} [(1-\gamma)\lambda \Delta t]^k}{k!} \right) \left(\frac{e^{-\gamma \lambda \Delta t} (\gamma \lambda \Delta t)^m}{m!} \right)$$

$$= \left(\frac{e^{-\lambda_0 \Delta t} (\lambda_0 \Delta t)^k}{k!} \right) \left(\frac{e^{-\lambda_1 \Delta t} (\lambda_1 \Delta t)^m}{m!} \right)$$

where $\lambda_0 = (1-\gamma)\lambda$, $\lambda_1 = \gamma\lambda$, and we make use of the independent-increments and stationary-increments properties of Y_t.

5.8.4 Superposition Property

To derive the superposition property, let $Y_{0,t}$ and $Y_{1,t}$ be independent Poisson processes with rates λ_0 and λ_1, respectively, and let $Y_t = Y_{0,t} + Y_{1,t}$ be the superposition process. Define $\lambda = \lambda_0 + \lambda_1$. Then

$$
\Pr\{Y_t = m\} = \sum_{j=0}^{m} \Pr\{Y_{0,t} = j, Y_{1,t} = m - j\}
$$

$$
= \sum_{j=0}^{m} \Pr\{Y_{0,t} = j\} \Pr\{Y_{1,t} = m - j\}
$$

$$
= \sum_{j=0}^{m} \left(\frac{e^{-\lambda_0 t}(\lambda_0 t)^j}{j!} \right) \left(\frac{e^{-\lambda_1 t}(\lambda_1 t)^{m-j}}{(m-j)!} \right)
$$

$$
= \frac{e^{-(\lambda_0 + \lambda_1)t}[(\lambda_0 + \lambda_1)t]^m}{m!} \sum_{j=0}^{m} \frac{m!}{j!(m-j)!} \left(\frac{\lambda_0}{\lambda_0 + \lambda_1} \right)^j
$$

$$
\times \left(\frac{\lambda_1}{\lambda_0 + \lambda_1} \right)^{m-j}
$$

$$
= \frac{e^{-\lambda t}(\lambda t)^m}{m!} \times 1
$$

where we make use of the fact that $Y_{0,t}$ and $Y_{1,t}$ are independent, and that

$$
\frac{m!}{j!(m-j)!} \left(\frac{\lambda_0}{\lambda_0 + \lambda_1} \right)^j \left(\frac{\lambda_1}{\lambda_0 + \lambda_1} \right)^{m-j}
$$

is a binomial mass function with success probability $\lambda_0/(\lambda_0 + \lambda_1)$. Therefore

$$
\sum_{j=0}^{m} \frac{m!}{j!(m-j)!} \left(\frac{\lambda_0}{\lambda_0 + \lambda_1} \right)^j \left(\frac{\lambda_1}{\lambda_0 + \lambda_1} \right)^{m-j} = 1
$$

5.8.5   Nonstationary Poisson Process

We establish that

$$
\Pr\{\mathcal{Y}_{\tau + \Delta\tau} - \mathcal{Y}_\tau = m | \mathcal{Y}_\tau = k\} = \frac{e^{-[\Lambda(\tau + \Delta\tau) - \Lambda(\tau)]}[\Lambda(\tau + \Delta\tau) - \Lambda(\tau)]^m}{m!}
$$

for a nonstationary Poisson arrival process with integrated-rate function $\Lambda(\tau)$. The derivation proceeds by transforming the probability statement about the nonstationary process into a probability statement about the underlying Poisson process Y_t with rate 1:

$$
\Pr\{\mathcal{Y}_{\tau + \Delta\tau} - \mathcal{Y}_\tau = m | \mathcal{Y}_\tau = k\} = \Pr\{Y_{\Lambda(\tau + \Delta\tau)} - Y_{\Lambda(\tau)} = m | Y_{\Lambda(\tau)} = k\}
$$

$$
= \frac{e^{-\Delta t}(\Delta t)^m}{m!}
$$

where $\Delta t = \Lambda(\tau + \Delta\tau) - \Lambda(\tau)$, since Y_t is a Poisson process with rate 1 and $\Lambda(\tau)$ is nondecreasing.

5.9 RESULTS FOR THE RENEWAL ARRIVAL-COUNTING PROCESS

The generic arrival-counting process $\{Y_t; t \geq 0\}$ described in Section 5.3 is called a *renewal arrival-counting process*, and the time of the nth arrival is called the nth renewal. Recall that the central characteristics of this process are that Y_t counts the number of arrivals by time t, and the interarrival-time gaps between arrivals, $G_1, G_2, \ldots$, are independent, time-stationary random variables with common cdf F_G. The term "renewal" comes from the fact that the process effectively starts fresh at the moment of each arrival.

The simulations in Section 5.4 established that F_G matters when we make probability statements about Y_t. Unfortunately, making these statements necessitates working with the n-fold convolution of F_G. In the special case when F_G is the exponential distribution, the renewal process is a Poisson process and the convolution is easy. In this section we describe some useful long-run performance measures for the general renewal process, results that do not depend on the n-fold convolution of F_G. The only restrictions we require are that:

- $F_G(a) = 0$ for $a < 0$, which implies that the interarrival-time gaps are not negative
- $F_G(0) < 1$, which implies that the interarrival-time gaps are not degenerate at 0

Let $\delta \equiv \mathrm{E}[G_n]$ be the expected time between renewals. The following result confirms what we observed in the simulations: The long-run arrival rate depends only on δ, even though probability statements about the number of arrivals by a particular time depends on the entire distribution.

For a renewal process $\{Y_t; t \geq 0\}$ with expected time between renewals δ, the observed arrival rate Y_t/t has the property that

$$\frac{Y_t}{t} \to \frac{1}{\delta} \; as \; t \to \infty \tag{5.13}$$

with probability 1. In other words, $1/\delta$ can be interpreted as the long-run **arrival rate**.

Thus, the long-run arrival rate is 1 accident per week for both the exponential and Weibull models in Section 5.4, because both distributions have expected value $\delta = 1$.

This result, and another that depends on it, can be put to good use, as illustrated in the following section.

5.9.1   The Case of the Perpetual Payoff

Case 5.3. Mr. Conservative plans to invest $300,000$, spending only the interest earned and keeping the principal (original $300,000$ investment) intact. Because Mr. Conservative is cautious, he will invest only in certificates of deposit (CDs) that mature within 1 to 5 years. Market conditions change, so Mr. Conservative cannot be sure of attaining a constant rate of return on his investment, but he is willing to settle for a good long-run rate of return.

Investment problems often involve considerable uncertainty. The uncertainty in Mr. Conservative's investment problem could include the interest rate attainable, the manner in which interest is calculated and paid, and the time to maturity of a CD. For the purpose of illustration, suppose that Mr. Conservative can always obtain an interest rate of 6%, compounded continuously, with the interest paid at maturity. However, the time to maturity changes depending on market conditions. Accounting for uncertainty in the other factors, such as the interest rate, does not alter the approach presented here, but it does make the calculations more complex.

First we need some background concerning interest calculations. If d dollars are invested in a CD for G years at interest rate $i100\%$ per year, compounded continuously, then the net return to the investor is

$$R = \text{value of investment after } G \text{ years} - \text{principal} = de^{iG} - d = d(e^{iG} - 1)$$

For example, investing $1000 for 5 years at an interest rate of 6% per year gives a net return of $1000(e^{0.06(5)} - 1) \approx \$349.86$.

Each time Mr. Conservative's CD matures, he deducts the net return and immediately reinvests the principal in another CD. Since the time to maturity of the new CD is not known until the day he is ready to invest, we can model it as a random variable G with cdf F_G. Mr. Conservative believes that a reasonable approximation is the uniform distribution with minimum $\alpha = 1$ year and maximum $\beta = 5$ years. That is,

$$F_G(a) = \frac{a - \alpha}{\beta - \alpha}$$

$$f_G(a) = \frac{1}{\beta - \alpha}$$

for $\alpha \le a \le \beta$. Since CDs typically have maturities measured in months, a more precise approximation might be a discrete distribution on 1 to 60 months.

Let G_n be the time to maturity for the nth CD in which Mr. Conservative invests. If $G_1, G_2, \ldots$ are independent and time stationary, then the number of times he has reinvested by time t years is a renewal arrival-counting process. Notice that the assumption of independence implies that the time to maturity of the current CD gives no indication of what the time to maturity will be for the next CD purchased; and time stationarity implies that there will be no fundamental changes in the way that maturity dates are assigned to CDs. The validity of both assumptions should be investigated.

For the nth reinvestment, Mr. Conservative receives a net return of $R_n = d(e^{iG_n} - 1)$ dollars, with $d = 300,000$ and $i = 0.06$. The returns $R_1, R_2, \ldots$ also form an independent, time-stationary sequence, although R_n and G_n are clearly dependent. Let $\eta \equiv E[R_n]$ be the expected net return from each CD.

Since $\{Y_t; t \geq 0\}$ is a renewal process, the long-run rate at which reinvesting occurs is $1/\delta$ investments per year, where $\delta \equiv E[G] = (\alpha + \beta)/2 = 3$ years, the expected time until a CD matures. Of more direct concern to Mr. Conservative is the quantity

$$\frac{\sum_{n=1}^{Y_t} R_n}{t}$$

which is the total net returns by time t divided by t, or the rate of return in dollars per year. To understand the long-run return from his investment, Mr. Conservative needs to know this quantity as t becomes large. The following result, which is a consequence of (5.13), provides his answer.

For a renewal process $\{Y_t; t \geq 0\}$ with expected time between renewals $\delta < \infty$, if the nth renewal brings a reward R_n with expected value $\eta < \infty$, and $R_1, R_2, \ldots$ are independent and time stationary, then

$$\frac{\sum_{n=1}^{Y_t} R_n}{t} \to \frac{\eta}{\delta} \quad as \; t \to \infty$$

with probability 1. In other words, η/δ can be interpreted as the long-run **reward rate.**

The result above is called the *renewal-reward theorem.* It establishes that the long-run rate of reward is the ratio of the expected reward per renewal divided by the expected time per renewal.

To compute the long-run rate of return for Mr. Conservative, we first find

$$\eta = E[R]$$
$$= E\left[d(e^{iG} - 1)\right]$$
$$= \int_{\alpha}^{\beta} d(e^{ia} - 1)\frac{1}{\beta - \alpha}\, da$$
$$= d\left(\frac{e^{i\beta} - e^{i\alpha}}{i(\beta - \alpha)} - 1\right)$$

For the specific case $\alpha = 1, \beta = 5, i = 0.06$ and $d = 300,000$, we obtain $\eta \approx 300,000(0.20) = 60,000$. Therefore, the long-run rate of return is $\eta/\delta = 60,000/3 = \$20,000$ per year. This result does not *guarantee* Mr. Conservative $\$20,000$ a year every year, but rather tells him what his long-run average return will be.

Compare this to the rate of return Mr. Conservative would obtain if at each renewal, he could find a guaranteed 5-year CD paying 6%, which is

$$\frac{\text{net return}}{\text{number of years}} = \frac{300,000(e^{0.06(5)} - 1)}{5} \approx \$20,991.53 \text{ per year}$$

or nearly \$1000 more. In other words, the uncertainty in the maturity date costs Mr. Conservative nearly \$1000 a year compared to his ideal investment, not to mention some sleepless nights fretting about how well he will do on his next investment.

5.9.2   Derivations

This section can be skipped by readers interested only in applications.

Recall that Y_t is the number of arrivals by time t and T_n is the time of the nth arrival. Thus, T_{Y_t} is the time of the last arrival before time t, while T_{Y_t+1} is the time of the next arrival after time t. Therefore,

$$T_{Y_t} \le t < T_{Y_t+1}$$

from which it follows that

$$\frac{T_{Y_t}}{Y_t} \le \frac{t}{Y_t} < \frac{T_{Y_t+1}}{Y_t} \tag{5.14}$$

We will prove (5.13) by showing that as $t \to \infty$ the left-hand and right-hand sides of (5.9.2) both converge to δ, with probability 1. Therefore, the middle term t/Y_t must also converge to δ, since it is squeezed between the other two.

Recall that $G_1, G_2, \ldots$ are independent and time stationary with common expectation δ. Therefore, by the Strong Law of Large Numbers

$$\frac{T_{Y_t}}{Y_t} = \frac{\sum_{n=1}^{Y_t} G_n}{Y_t} \to \delta$$

as $Y_t \to \infty$. But the number of renewals $Y_t \to \infty$ as $t \to \infty$, so the limit holds as $t \to \infty$ also.

Similarly,

$$\frac{T_{Y_t+1}}{Y_t} = \left(\frac{\sum_{n=1}^{Y_t+1} G_n}{Y_t + 1}\right)\left(\frac{Y_t + 1}{Y_t}\right) \to (\delta)(1) = \delta$$

as $t \to \infty$ by applying the previous result. Therefore, the inequality (5.9.2) implies that $t/Y_t \to \delta$, from which it follows that $Y_t/t \to 1/\delta$.

To prove the renewal-reward theorem, we can write

$$\frac{\sum_{n=1}^{Y_t} R_n}{t} = \left(\frac{\sum_{n=1}^{Y_t} R_n}{Y_t}\right)\left(\frac{Y_t}{t}\right)$$

As $t \to \infty$ the first term converges to η by the Strong Law of Large Numbers, while the second term converges to $1/\delta$ from (5.13). This establishes the result.

5.9.3 Other Arrival-Counting Processes

Certainly not all arrival-counting processes are Poisson arrival processes, but not all arrival-counting processes can be represented as renewal processes either. Three arrival-counting processes that fall outside the scope of this chapter are the following:

- Arrival processes with scheduled arrivals, such as patients arriving at a dentist's office, in which arrivals may deviate from the schedule by being early or late. In this situation it is natural to model the deviation from the schedule, rather than the interarrival-time gap.

- Arrival processes in which the arrival rate depends on the state of the system, such as the arrival of machines to a repair facility when there are only a small number of machines serviced by the facility. If at any time all of the machines are under repair, then the arrival rate to the facility is 0 because no arrivals are possible. In this situation it is natural to model the arrival of each machine individually; see Chapter 8.

- Arrival processes in which the length of an interarrival-time gap depends on the lengths of previous interarrival-time gaps, such as the arrival of orders to buy or sell a particular stock on the stock market where buying and selling by some investors triggers buying and selling by other investors. In this case it is natural to model the dependence among interarrival-time gaps as a time series.

5.10 FINE POINTS

The approach that the traffic engineer chose to solve Case 5.1 is akin to hypothesis testing in statistics. To see the correspondence, notice that the traffic engineer's null hypothesis was $H_0: \lambda = 1$, and her alternative hypothesis was $H_1: \lambda > 1$. The only difference relative to elementary statistics is that the Poisson distribution was used to determine what was extreme, rather than the normal distribution that arises in many other applications. The probability of observing an outcome as extreme, or more extreme, than what was actually observed is sometimes called a "p-value."

To be concise, the decomposition and superposition properties of Poisson arrival processes were stated in terms of decomposing one process into two subprocesses, or superposing two processes into one overall process. These properties readily extend to decomposition into m processes or superposition of m processes.

Let $\gamma_1, \gamma_2, \ldots, \gamma_m$ be probabilities such that $0 < \gamma_i < 1$ and $\sum_{i=1}^{m} \gamma_i = 1$. If a Poisson arrival process with arrival rate λ is decomposed into m subprocesses in such a way that each arrival has probability γ_i of joining process i, then the m subprocesses are independent Poisson processes with arrival rates $\gamma_1 \lambda, \gamma_2 \lambda, \ldots, \gamma_m \lambda$.

On the other hand, the superposition of m independent Poisson arrival processes with arrival rates $\lambda_1, \lambda_2, \ldots, \lambda_m$ is a Poisson process with arrival rate $\lambda = \lambda_1 + \lambda_2 + \cdots + \lambda_m$.

Our presentation of the nonstationary Poisson process was limited to piecewise-constant arrival-rate functions, $\lambda(\tau)$. More general rate functions are permitted, and the fundamental property applies as long as the integrated rate function $\Lambda(\tau)$ is strictly increasing and continuous. Bratley, Fox and Schrage (1987) and Law and Kelton (1991) provide excellent discussions. In addition, the superposition of m independent, nonstationary Poisson processes with integrated rate functions $\Lambda_1(\tau), \Lambda_2(\tau), \ldots, \Lambda_m(\tau)$ is a nonstationary Poisson arrival process with integrated rate function $\Lambda(\tau) = \Lambda_1(\tau) + \Lambda_2(\tau) + \cdots + \Lambda_m(\tau)$.

5.11   EXERCISES

5.1. For a Poisson arrival process $\{Y_t; t \geq 0\}$ with arrival rate $\lambda = 2$ per hour, compute the following:
(a) $\Pr\{Y_2 = 5\}$
(b) $\Pr\{Y_4 - Y_3 = 1\}$
(c) $\Pr\{Y_6 - Y_3 = 4 | Y_3 = 2\}$
(d) $\Pr\{Y_5 = 4 | Y_4 = 2\}$

5.2. Let X be a random variable having a Poisson distribution with parameter η (typically in this chapter $\eta = \lambda t$). When η is large (greater than about 20), the Poisson distribution can be approximated by the standard normal distribution. Specifically,

$$\Pr\{X \leq c\} \approx \Phi\left(\frac{c + 1/2 - \eta}{\sqrt{\eta}}\right)$$

where Φ is the cdf of the standard normal distribution.

Evaluate the accuracy of this approximation for $\eta = 10$, 20 and 30 and $c = \eta + \lfloor\sqrt{\eta}\rfloor$, $\eta + 2\lfloor\sqrt{\eta}\rfloor$ and $\eta + 3\lfloor\sqrt{\eta}\rfloor$ (notice that $\sqrt{\eta}$ is the standard deviation of X). Do this by writing a computer program to compute $\Pr\{X \leq c\}$ directly, and compare the result to the approximation. This approximation may be helpful in other exercises.

5.3. Patients arrive to a hospital emergency room at a rate of 2 per hour. A doctor works a 12-hour shift from 6 a.m. until 6 p.m. Answer the following questions by approximating the arrival-counting process as a Poisson process:
(a) If the doctor has seen six patients by 8 a.m., what is the probability that the doctor will see a total of nine patients by 10 a.m.?
(b) What is the expected time between the arrival of successive patients? What is the probability that the time between the arrival of successive patients will be more than 1 hour?
(c) What is the expected time after coming on duty until the doctor sees her first patient? What is the probability that she will see her first patient in 15 minutes or less after coming on duty?
(d) What is the probability that the doctor will see her thirteenth patient before 1 p.m.?
(e) Of patients admitted to the emergency room, 14% are classified as "urgent." What is the probability that the doctor will see more than six urgent patients during her shift?
(f) The hospital also has a walk-in clinic to handle minor problems. Patients arrive at this clinic at a rate of 4 per hour. What is the probability that the

total number of patients arriving at both the emergency room and clinic from 6 a.m. until 12 noon will be greater than 30?

5.4. Suppose that the traffic engineer from Vincennes, Indiana, thinks that an event with probability 0.05 or less is extreme. Find the smallest value of m such that $\Pr\{Y_{30} - Y_6 > m\} \leq 0.05$ for (a) the Poisson arrival process, by using mathematical analysis, and (b) the Weibull arrival process, by simulating 1000 sample paths and looking at the empirical cdf or histogram. You will need to add steps to our algorithm simulation to generate values for $Y_{30} - Y_6$.

5.5. Suppose that during the observation period in Beehunter, Indiana, the traffic engineer had recorded 20 accidents. Should she conclude that the intersection has become safer? Compute appropriate probabilities to support your conclusion by approximating the arrival-counting process as a Poisson process.

5.6. An automated optical scanner looks for defects on a continuous sheet of metal. If the metal is being produced according to specifications (the process is "in control"), then defects should occur at a rate of 1 defect per 50 square meters of metal. Answer the following questions assuming that the occurrence of defects can be modeled as a Poisson arrival process. Notice that "time" corresponds to square meters of metal in this situation.

  (a) Assuming that the process is in control, what is the probability of finding 7 or more defects in any 200 meters^2 of metal?

  (b) The company wants to set up an inspection plan that detects when the process is "out of control." Inspection will be performed as follows: The first 200 meters^2 of metal produced is inspected. If more than a critical number c of defects is found, then the process is declared out of control. Specifically, if the true defect rate is as high as 4 defects per 50 meters^2, then the probability of declaring the process out of control should be at least 0.95. Determine the critical number c to achieve this goal. Using this value of c, what is the probability of declaring an "in control" process "out of control"?

5.7. Rework Exercise 9, Chapter 4, using the tools you learned in this chapter.

5.8. Several arrival processes are described below. State whether or not it is reasonable to approximate them as Poisson arrival processes, and why or why not.

  (a) The arrival of customers to the purchase doughnuts from a campus doughnut shop.

  (b) The arrival of students at a university football game or concert.

  (c) The arrival of patients at a doctor's office.

  (d) The discovery of "bugs" in a new software product.

  (e) The arrival of emergency calls to a fire station.

5.9. Students in the local IIE Chapter at A Buckeye State University sell doughnuts daily in the lobby of their building. They are curious if there is a day-of-the-week effect on donut sales. The number of customers that arrived between 7 a.m. and 10 a.m. on three Mondays was observed to be 60, 72 and 68 customers. Suppose that the arrival of customers is well modeled as a Poisson arrival process.

  (a) Estimate the arrival rate of customers on Mondays in terms of customers per hour. Give a standard error for your estimate.

(b) Suppose that this Friday only 56 customers arrived. Would you conclude that Fridays are different from Mondays? Justify your conclusion with appropriate analysis. Be sure to account for the sensitivity of your analysis to the error in your estimate of the arrival rate.

5.10. When a power surge occurs on local electric lines, it can damage a computer plugged into the line if the computer does not have a "surge protector." There are various types of surges: "Tiny" surges occur at the rate of 8 per hour, but they cannot damage a computer. "Small" surges occur at the rate of 1 every 18 hours; a small surge will damage an unprotected computer with probability 0.005. "Moderate" surges occur at a rate of 1 every 46 hours; a moderate surge will damage an unprotected computer with probability 0.08. Suppose that the arrival of each type of surge can be modeled as an independent Poisson process.

(a) What is the expected number of surges of *any* type during 8 hours of computer work?

(b) What is the expected number of *computer-damaging surges* during 8 hours of computer work?

(c) What is the probability that there will be *no* computer-damaging surge during 8 hours of computer work?

5.11. Investigate how the distribution of the forward-recurrence time R_t is affected by the interarrival-time-gap distribution, F_G, and the time of interest, t, when $E[G] = 1$. Use the Erlang distribution with different numbers of phases n as F_G. To obtain $E[G] = 1$, set $\lambda = n$. Notice that the coefficient of variation of the Erlang distribution with parameter $\lambda = n$ and n phases is $CV[G] = 1/\sqrt{n}$, so the relative variability decreases as n increases. When $n = 1$, the Erlang distribution is the exponential distribution.

For each value of n that you try, generate 1000 sample paths and look at the histogram and empirical cdf of R_6 and R_{30}. Propose a theory of how the $CV[G]$ and t affect the distribution of R_t.

5.12. ON-LINE INFO is a company that provides computerized data-base searches to customers over telephone lines. The requests from customers can be classified into those that require printing a document and those that do not require printing. Requests that require printing arrive at a rate of 400 per hour; those that do not require printing arrive at a rate of 1000 per hour. Suppose that both arrival processes are well modeled as independent Poisson processes.

(a) All incoming requests are initially handled by a single front-end computer. What is the probability that this computer receives more than 2000 requests between 1 p.m. and 2:30 p.m.?

(b) Suppose that the front-end computer distributes the requests randomly between two processing computers (called A and B) that fulfill the requests, in such a way that each of the processing computers is equally likely to receive a new request. Assuming that the time for the front-end computer to do this is essentially 0, what is the probability that **both** of the processing computers receive more than 1000 requests between 1 p.m. and 2:30 p.m.?

5.13. Automobiles pass a point on the highway at a rate of 1 per minute. If 5% of all automobiles are trucks and the arrival process is well approximated as a Poisson process, then answer the following questions:

(a) What is the probability that at least one truck passes by during an hour?

(b) Given that 10 trucks have passed by in an hour, what is the expected total number of automobiles that have passed by in that time?

(c) If 50 automobiles have passed by in an hour, what is the probability that 5 of them were trucks?

5.14. Suppose that the occurrence of typographical errors in the first draft of a book is well approximated as a Poisson process with a rate of 1 error per 1000 words. Each time the author proofreads the book, the error rate is reduced by 50%. For example, after the first proofreading, the error rate is 0.5 errors per 1000 words, or 1 error per 2000 words. Notice that we are saying that the "error rate" goes down by 50%, not that the actual number of errors is reduced by that amount since the author can (and did) introduce new errors while correcting other errors.

The author wants to know how many times he needs to proofread the book so that the probability of there being no errors is at least 0.98. Assume that the book contains 200,000 words. Compute this number.

5.15. This problem concerns capacity planning for a manufacturing company. The company has two salespersons (John and Louise) who each cover one half of the United States. At the end of each week the salespersons report their sales to the company, which then manufactures the products that have been ordered.

The company has three products, which it calls A, B and C for simplicity. Each salesperson obtains 10 orders per week, on average, of which approximately 20% are for A, 70% are for B, and 10% are for C. In terms of capacity, it takes 25 person-hours to produce one A, 15 person-hours to produce one B, and 40 person-hours to produce one C.

Help the company do its capacity planning by answering the following questions. You may assume that the arrival of orders to each salesperson can be well approximated as a Poisson process.

(a) A Poisson process is time stationary. What is the physical interpretation of "time stationary" in this situation?

(b) What is the probability that the total sales for 1 week will be more than 30 products?

(c) Capacity can only be changed on a monthly basis. What is the expected number of person-hours the company will need over a 1-month (4-week) period?

(d) What is the probability that Louise will sell more than 5 product Bs on *each* of 2 consecutive weeks?

5.16. Suppose that the National Weather Service believes that the arrival of hurricanes to the Florida coast may be well modeled as a Poisson process. The hurricane season is 3 months long, and during the last 10 hurricane seasons there have been 8 hurricanes.

(a) Estimate the arrival rate for the National Weather Service's Poisson process model, and compute the standard error for this estimate.

(b) Using the estimated value from above, what is the probability that no hurricane will hit the Florida coast during the next hurricane season?

5.17. This problem concerns customer traffic on a local telephone system. The local telephone company in Bicknell, Indiana, handles telephone calls that arrive at rate of 1000 per hour during business hours (7 a.m. to 6 p.m.). Most calls are local (87%); the remainder are long-distance calls that require a connection to a long-distance carrier. The Bicknell phone company charges customers $0.08 for each local call, and for each long-distance call it charges customers $0.13 for the connection to the outside carrier.

Help the phone company do its budgeting and capacity planning by answering the following questions. You may assume that the arrival of calls during business hours is well approximated as a Poisson arrival-counting process.

(a) A Poisson process has independent increments. What is the physical interpretation of "independent increments" in this situation?

(b) The record for long-distance calls placed between 2–3 p.m. is 412 calls. What is the probability that this record will be broken on any given day?

(c) What is the phone company's expected revenue per day from local-call and long-distance-connect charges?

(d) If 1200 calls are placed during an hour, what is the probability that more than 300 of them were long-distance calls?

5.18. A small restaurant on a highway serves passengers in both trucks and cars. Trucks come down the highway at the rate of 10/hour, while cars come at the rate of 20/hour. These arrival processes are independent and are well represented as Poisson processes. Suppose that 10% of the trucks and 10% of the cars stop at the restaurant.

(a) What is the expected number of cars and trucks (total) arriving at the restaurant during a given hour?

(b) What is the probability that there are no arrivals during an hour?

(c) Trucks always carry just one passenger. However, only 30% of the cars carry just one passenger; 50% carry two passengers, and the remainder carry three passengers. What is the expected number of passengers arriving at the restaurant during an hour?

5.19. Suppose that traffic engineers claim that the number of accidents at the intersection of Olentangy River Road and Henderson Road in Columbus, Ohio, is well modeled as a Poisson process with a rate of 2 per week. A city council person, who thinks that the intersection has become more dangerous, challenges that model with some recent data: Last year (52 weeks) there were 100 accidents, but in the last month of the year (4 weeks) there were 20 accidents. Do these statistics indicate a change? Answer the questions below *assuming the traffic engineers' model is correct.*

(a) What is the probability of 20 or more accidents in a 2-week period?

(b) Given 100 total accidents in the year, what is the probability that they will be distributed in the way that happened last year (80 accidents in the first 11 months, and 20 accidents in the last month of the year)? Does this probability support or refute the council person's claim, or neither?

(c) The traffic engineers have said that in the most recent 2 weeks there have been no accidents. What is the probability of this event, assuming their model is correct?

5.20. A more refined model for the arrival-counting process to the emergency room described in Exercise 3 is as a nonstationary Poisson arrival process with arrival-rate function

$$\lambda(t) = \begin{cases} 1 & \text{if } 0 \le t < 6 \\ 2 & \text{if } 6 \le t < 13 \\ \frac{1}{2} & \text{if } 13 \le t < 24 \end{cases}$$

where time is measured in hours and time 0 is 6 a.m.

(a) Derive the integrated-rate function for this model.

(b) What is the probability that the doctor will see more than 12 patients between 8 a.m. and 2 p.m. for this model? What is the expected number of patients the doctor will see during that time?

(c) If the doctor has seen six patients by 8 a.m., what is the probability that the doctor will see a total of nine patients by 10 a.m.?

(d) What is the probability that she will see her first patient in 15 minutes or less after coming on duty?

(e) What is the probability that the doctor will see her thirteenth patient before 1 p.m.?

5.21. One method for estimating the arrival-rate function $\lambda(t)$ of a nonstationary Poisson arrival process is as follows: Break up the time period of interest into intervals $[t_0, t_1), [t_1, t_2), \ldots$, not necessarily of equal length. Record the number of arrivals in each interval; then estimate $\lambda(t)$ as a piecewise-constant function over each time interval, in other words, $\lambda(t) = \lambda_i$ for $t_{i-1} \le t < t_i$.

For example, suppose that traffic engineers place a counter along one side of a road and obtain the following counts of number of cars passing the counter for the same 4-hour period over 5 different days:

hour	1	2	day 3	4	5
1	123	170	171	110	147
2	212	250	217	226	242
3	412	374	365	380	382
4	97	100	88	103	92

(a) Estimate $\lambda(t)$ for $0 \le t \le 4$, and give a standard error for each piece of the piecewise-constant rate function.

(b) Derive the integrated-rate function $\Lambda(t)$ for your model.

(c) What is the expected number of cars passing this location between times 1.75 and 3.4 hours for your model?

(d) What is the probability of more than 700 cars passing this location between times 1.75 and 3.4 hours for your model?

5.22. Write random-variate-generation algorithms to generate values from the following distributions: (a) the Erlang distribution with parameters λ and n phases; (b) the binomial distribution with success probability γ and n trials; (c) the number of arrivals by time t in a Poisson process with arrival rate λ; and (d) the number of arrivals by time t in the nonstationary Poisson process given in Exercise 20. Code these algorithms, generate values, and examine the empirical cdfs and histograms as you vary λ, n, γ and t.

5.23. Show that $E[Y_{t+\Delta t} - Y_t] = \lambda \Delta t$ for a Poisson process with arrival rate λ by deriving the expected value.

5.24. For a Poisson arrival process with rate λ, derive the distribution of the *backward-recurrence time* at time t. The backward-recurrence time, H_t, is the time gap between the last arrival before time t and t. If no arrivals have occurred by time t, then we define H_t to be t. (Hint: The distribution of H_t is a mixed discrete-continuous distribution, since $\Pr\{H_t = t\} > 0$. Start by deriving $\Pr\{H_t = t\}$.) What is the limiting distribution of H_t as $t \to \infty$?

5.25. The simulation study in Section 5.4 suggests that $E[L_t] > E[G]$. That is, the expected length of the interval that contains a particular point in time, t, is longer than the expected length of an interarrival-time gap. Prove this for a Poisson arrival process. (Hint: Notice that $L_t = R_t + H_t$.) This phenomenon, known as *length-biased sampling*, has implications for work-sampling studies.

5.26. Rework Case 5.3 in Section 5.9.1 with the distribution of the time to maturity being the discrete uniform distribution on 1 to 60 months. How much does this change affect the long-run rate of return?

5.27. A high-intensity spotlight is replaced immediately when it burns out. The times until burnout for the last 10 lights were 227, 312, 300, 198, 260, 203, 330, 266, 179, and 212 hours. Estimate the long-run replacement rate of this type of light. Carefully justify your response by stating your modeling approximations.

5.28. A florist shop receives orders every 15 minutes, on average. The charge for each order varies depending on what is ordered, but averages $27. What is the long-run revenue per minute for the florist shop? Carefully justify your response by stating your modeling approximations.

5.29. A machine tool wears over time and may eventually fail. According to company rules the tool must be replaced after no more than 30 days, even if it has not failed. The number of days until failure or required removal has been modeled as a random variable X (days) with a right-triangular distribution and density function

$$f_X(a) = \begin{cases} 0, & a < 0 \\ \frac{2a}{900}, & 0 \le a \le 30 \\ 0, & 30 < a \end{cases}$$

If the tool fails, it must be replaced immediately at a cost of $1000. If it is replaced prior to failure, the cost is only $300. Does it makes sense, over the long run, to schedule replacements for less than 30 days? When should replacements be scheduled to minimize long-run cost? (Hint: Use the renewal-reward theorem to solve this problem, with the replacement cost being the reward. Let the scheduled time until replacement be c. Since the tool is replaced when it fails or at time c, whichever is earlier, the expected time between replacements is $\int_0^c a f_X(a)\, da + c \int_c^{30} f_X(a)\, da$.)

5.30. Boilermaker Land theme park conducts free tours of the engineering department that designs its rides. The tour times are not scheduled; they begin whenever there are a sufficient number of patrons wanting to take a tour. The arrival of patrons wanting a tour is well modeled as a renewal-arrival process with expected time between arrivals of 1 minute. It costs Boilermaker Land $10 each time it conducts a tour, regardless of how many people are in the tour group. But there is also a cost to Boilermaker Land of having patrons waiting for a tour, because if they are waiting, then they are not spending money in the park. Accountants have estimated that patrons in the park spend money at the rate of $0.50 per minute.

(*a*) What should the size of each tour group be to minimize long-run cost to Boilermaker Land? (Hint: Use the renewal-reward theorem to solve this problem, with waiting and tour costs being the rewards and the beginning of each tour being the renewal process. Let n be the size of each tour group, and notice that the nth patron to arrive for a tour incurs no waiting cost, the $(n-1)$st incurs an expected waiting cost of (1 minute)($0.50 per minute), the $(n-2)$d incurs an expected waiting cost of (2 minutes)($0.50 per minute), and so on.)

(*b*) What is the expected time between the departure of tours for your optimal-size tour group?

5.31. Prove the three decomposition and superposition properties described in Section 5.10.

5.32. Derive the integrated-rate functions for the nonstationary Poisson arrival processes described in Brennan, J. E., B. L. Golden and H. K. Rappoport. 1992. Go with the flow: Improving Red Cross bloodmobiles using simulation analysis. *Interfaces* **22**, 1–13.

5.33. Read the article by Farina, R., G. A. Kochenberger and T. Obremski. 1989. The computer runs the Bolder Boulder: A Simulation of a Major Running Race. *Interfaces* **19**, 48–55. The authors used a simulation model to solve their problem. See if you can come up with a solution mathematically, using the idea of a nonstationary Poisson arrival process.

5.34. Read the article by Youngren, M. A. 1991. Modeling relocatable nuclear targets. *Interfaces* **21**, 132–140. Describe how the model extends the idea of a renewal process.

CHAPTER

6

DISCRETE-TIME
PROCESSES

We have characterized the sample path of a stochastic process as a sequence of state changes occurring at randomly spaced points in time. To solve some modeling and analysis problems, it is sufficient to analyze the state changes and ignore the times at which they occur. For example, consider a model of the number of defective items produced by a manufacturing system in which the state of the system changes only when an item is produced. For the purpose of quality control it may be sufficient to derive probability statements about the number of defective items produced and ignore the time it takes to produce the items. This chapter presents methods for modeling, analysis and simulation of one useful type of discrete-time state-change process, a process that underlies the continuous-time stochastic processes described in Chapters 7–8.

6.1 THE CASE OF THE RANDOM BEHAVIOR

Case 6.1. Data Driven, Inc., provides on-line data-base services to thousands of customers who connect to Data Driven's computer network via their personal computers and modems. In order to evaluate changes to Data Driven's computer architecture and operating system prior to installing the changes, Data Driven's performance-modeling group needs a model that describes customer behavior. Customer behavior consists of the times at which customers log on to the system, the sequence of transactions that customers request (for example, fetch a document, and read a page of a document), and the "think times" that customers spend determining their next request. The model must be suitable for generating sample customer behavior as input to a simulation of Data Driven's computer network. Fortunately, Data Driven, Inc., has collected substantial data on customer behavior.[1]

Since Data Driven, Inc., has abundant data on customer behavior, the peformance-modeling group could use it directly as input to its simulation (see Chapter 2). However, a stochastic-process model has several advantages over data: A model provides a concise description of customer behavior that is easier to manipulate than (perhaps millions of) customer records. Developing a model can provide useful insights about how customers behave. And a model can be modified more easily to reflect anticipated changes in customer behavior.

The different aspects of customer behavior can perhaps be represented by separate models. The customer log-on times can be viewed as an arrival process, for which the models in Chapter 5 are appropriate. After a customer "arrives," the transaction requests and think times can be represented by a second model. In this chapter we focus on the sequence of transaction requests and defer incorporating the think times into the model until Chapter 7.

6.2   NOTATION AND REVIEW

This chapter makes extensive use of matrix algebra, specifically the multiplication of matrices, the inverse of square matrices, and systems of linear equations represented in matrix form. You should review these topics before continuing if they are not comfortable.

Boldface type is used to denote vectors or matrices. To conform to established conventions, certain matrices are denoted by boldface *capital* letters even though they are *not* random variables. These include $\mathbf{F}, \mathbf{I}, \mathbf{P}$ and $\mathbf{R}$. An $\mathbf{I}$ denotes an identity matrix, and the symbols $\mathbf{1}$ and $\mathbf{0}$ denote column vectors of all 1s and 0s, respectively.

Let $\mathbf{P}$ be an $m \times m$ matrix with elements p_{ij}, for $i, j = 1, 2, \ldots, m$. For instance, if $m = 5$, then

$$\mathbf{P} = \begin{pmatrix} p_{11} & p_{12} & p_{13} & p_{14} & p_{15} \\ p_{21} & p_{22} & p_{23} & p_{24} & p_{25} \\ p_{31} & p_{32} & p_{33} & p_{34} & p_{35} \\ p_{41} & p_{42} & p_{43} & p_{44} & p_{45} \\ p_{51} & p_{52} & p_{53} & p_{54} & p_{55} \end{pmatrix}$$

Let $\mathcal{A}$ and $\mathcal{B}$ be ordered subsets of the elements $\{1, 2, \ldots, m\}$. For instance, when $m = 5$, we could have $\mathcal{A} = \{2, 5\}$ and $\mathcal{B} = \{1, 3, 4\}$. Then $\mathbf{P}_{\mathcal{A}\mathcal{B}}$ denotes the

---

[1] This case is based on W. S. Keezer, A. P. Fenic and B. L. Nelson. 1992. Representation of user transaction processing behavior with a state transition matrix. *1992 Winter Simulation Conference Proceedings* (J. J. Swain, D. Goldsman, R. C. Crain and J. R. Wilson, eds.), 1223–1231.

submatrix whose elements are p_{ij} with $i \in \mathcal{A}$ and $j \in \mathcal{B}$. In the example

$$\mathbf{P}_{\mathcal{AB}} = \begin{pmatrix} p_{21} & p_{23} & p_{24} \\ p_{51} & p_{53} & p_{54} \end{pmatrix}$$

If $\mathbf{F}$ is an $h \times k$ matrix with elements f_{ij} and $\mathbf{R}$ is a $k \times m$ matrix with elements r_{ij}, then $\mathbf{FR}$ denotes their matrix product, an $h \times m$ matrix with i, jth element $\sum_{n=1}^{k} f_{in} r_{nj}$. For an $m \times m$ matrix $\mathbf{P}$, $\mathbf{P}^k$ denotes the product of $\mathbf{P}$ with itself k times. We define $\mathbf{P}^0 \equiv \mathbf{I}$. The matrix $\mathbf{P}^{-1}$ is the inverse of $\mathbf{P}$, when it exists, such that $\mathbf{PP}^{-1} = \mathbf{P}^{-1}\mathbf{P} = \mathbf{I}$.

6.3 SIMULATING THE RANDOM BEHAVIOR

Data Driven's performance-modeling group, Case 6.1, has identified four types of customer transactions and assigned them identification numbers: log on (1), fetch a document (2), read a page of a document (3) and log off (4). A sequence of transaction requests from a customer constitutes a *session*. For example, the session 1, 2, 3, 3, 2, 4 occurs when a customer first logs on, then fetches a document, reads a page, reads another page, fetches another document, and finally logs off. Each session begins with a log on and ends with a log off.

The performance-modeling group believes that the next transaction that a customer requests is strongly influenced by the last transaction requested and insignificantly influenced by anything else.[2] It has defined N to be a random variable representing the Next transaction a customer requests, L to be a random variable representing the Last transaction requested, and $F_{N|L=i}$ to be the conditional cdf of N given that the event $\{L = i\}$ has occurred. For example, $F_{N|L=2}$ is the cdf of the next transaction given that the last transaction was a fetch, and is

$$F_{N|L=2}(a) \equiv \Pr\{N \le a | L = 2\} = \begin{cases} 0, & a < 2 \\ 0.27, & 2 \le a < 3 \\ 0.90, & 3 \le a < 4 \\ 1, & 4 \le a \end{cases} \tag{6.1}$$

with associated mass function

$$p_{N|L=2}(a) \equiv \Pr\{N = a | L = 2\} = \begin{cases} 0.27, & a = 2 \\ 0.63, & a = 3 \\ 0.10, & a = 4 \\ 0, & \text{otherwise} \end{cases} \tag{6.2}$$

Thus, the most likely transaction to follow a "fetch" ($\{L = 2\}$) is a "read" ($\{N = 3\}$), which has conditional probability 0.63. In Section 6.4.2 we present

---

[2]We use the word "last" to mean the most recent transaction requested, not the final transaction of the session.

the conditional distributions for other values of L and describe how they were estimated from data.

Let $\{S_n; n = 0, 1, 2, \ldots\}$ be a stochastic process representing the sequence of transactions a customer requests; it is a discrete-state process with state space $\{1, 2, 3, 4\}$ representing the transaction types, and a discrete-time process with the time index $n = 0, 1, 2, \ldots$ representing the number of transaction requests. Given the characterization above, we can generate sample paths of S_n via simulation. The single system event, e_1, corresponds to the customer choosing the next transaction. Because we are ignoring the actual time at which the state changes (customer choices) occur and because there is only one system event, we can employ a simplified version of the algorithm simulation.

$e_0()$ (**log on**)

$S_0 \leftarrow 1$ (initial state is log on)

$e_1()$ (**next transaction**)

$$S_{n+1} \leftarrow F^{-1}_{N|L=S_n}(\text{random}())$$ (next state depends on current state)

algorithm simulation

1. $n \leftarrow 0$ (initialize system-event counter)
     $e_0()$ (execute initial system event)

2. $e_1()$ (update state of the system)
     $n \leftarrow n + 1$ (update system-event counter)

3.   repeat 2

Figure 6.1 shows two sample paths generated by this simulation after specifying seeds or streams for `random()`. Both sample paths begin with a log on (state 1), but one of them reaches a log off (state 4) after 10 transactions ($n = 9$), while the other reaches a log off after 16 transactions ($n = 15$). Notice that when the sample paths reach state 4, they remain in state 4. In other words, all sequences of transactions ultimately end with $\ldots, 4, 4, 4, \ldots$. This is a convenient representation, and it is consistent with our convention that all stochastic processes are infinitely long.

Let X be a random variable representing the length of a session, *excluding log on* ($n = 0$). A value of X can be generated by adding Step 2.5 to the algorithm simulation:

2.5   if $\{S_n = 4\}$ then   (check if log off)
         $X \leftarrow n$ (record session length)
         stop   (terminate simulation)
      endif

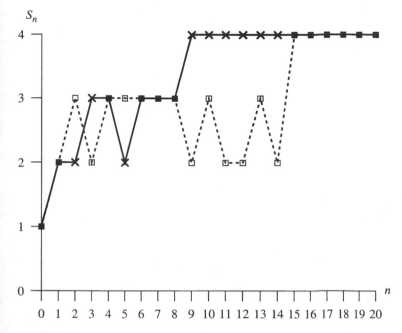

FIGURE 6.1
Two sample paths of customer transaction requests (sessions), where ∗ represents the states of one path and □ represents the states of the other.

Figure 6.2 shows a histogram of the session lengths $X_1, X_2, \ldots, X_{500}$ from 500 simulated sessions. The sample-mean session length was 6.8 transactions, which is an estimate of $E[X]$, the expected session length. If the model is a faithful representation of customer behavior, then this distribution should be similar to the session lengths observed in the actual customer behavior data, and comparing them provides a check on the fidelity of the model.

The stochastic-process model defined in this section represents the behavior of Data Driven's customers, as described in Case 6.1. However, it is also a generic model for any discrete-state, discrete-time stochastic process in which the probability distribution of the next state of the process depends only on the last state of the process. Such a stochastic process is called a *Markov chain*. In the remainder of this chapter we study Markov chains in detail, including one generalization: We allow the initial state of the process, S_0, to be a random variable with cdf F_{S_0}. The generic event e_0 is therefore

$e_0()$ **(generate initial state)**

$$S_0 \leftarrow F_{S_0}^{-1}(\texttt{random}())   \text{(initial state of the process)}$$

In the model of Data Driven's customers, S_0 is a degenerate random variable with $\Pr\{S_0 = 1\} = 1$ to represent the fact that a session always begins with a log on.

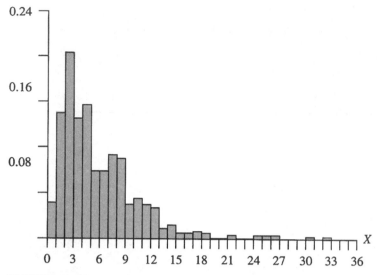

FIGURE 6.2
Histogram of 500 observed session lengths, X, excluding log on.

6.4 MARKOV CHAINS

In this section we investigate the probability structure of sample paths of Markov chains and show how to parameterize Markov chain models given data. The generic Markov chain is a discrete-time, discrete-state stochastic process $\{S_n; n = 0, 1, 2, \ldots\}$ with state space $\mathcal{M} = \{1, 2, \ldots, m\}$, where m can go to ∞. We force the state space to be numbered $1, 2, \ldots, m$ so that formulas involving summations—such as $\sum_{i=1}^{m}$—can be written simply. In applications the states may be numbered in any convenient manner, and summations are over all, or a subset of, the elements of $\mathcal{M}$. For example, $\sum_{i \in \mathcal{M}}$ also means to sum over all states. We continue to use the customer behavior model of Case 6.1 as an illustration.

6.4.1 Probability Structure of the Sample Paths

Again let S_n represent the nth transaction that a customer requests. Customer behavior is characterized by the probability of possible sessions, such as

$$\Pr\{S_0 = 1, S_1 = 2, S_2 = 3, S_3 = 3, S_4 = 2, S_5 = 4\} \tag{6.3}$$

In principle, the performance-modeling group needs to know the joint probability of all such sample paths to completely characterize customer behavior. This is a formidable task, but it is somewhat easier if we use the definition of conditional probability in Section 3.1.4 to rewrite (6.3) as

$$\Pr\{S_5 = 4, S_4 = 2, S_3 = 3, S_2 = 3, S_1 = 2, S_0 = 1\}$$
$$= \Pr\{S_5 = 4 | S_4 = 2, S_3 = 3, S_2 = 3, S_1 = 2, S_0 = 1\}$$
$$\times \Pr\{S_4 = 2, S_3 = 3, S_2 = 3, S_1 = 2, S_0 = 1\}$$

$$= \Pr\{S_5 = 4|S_4 = 2, S_3 = 3, S_2 = 3, S_1 = 2, S_0 = 1\}$$
$$\times \Pr\{S_4 = 2|S_3 = 3, S_2 = 3, S_1 = 2, S_0 = 1\}$$
$$\times \Pr\{S_3 = 3, S_2 = 3, S_1 = 2, S_0 = 1\}$$

$$\vdots$$

$$= \Pr\{S_5 = 4|S_4 = 2, S_3 = 3, S_2 = 3, S_1 = 2, S_0 = 1\}$$
$$\times \Pr\{S_4 = 2|S_3 = 3, S_2 = 3, S_1 = 2, S_0 = 1\}$$
$$\times \Pr\{S_3 = 3|S_2 = 3, S_1 = 2, S_0 = 1\}$$
$$\times \Pr\{S_2 = 3|S_1 = 2, S_0 = 1\}$$
$$\times \Pr\{S_1 = 2|S_0 = 1\}$$
$$\times \Pr\{S_0 = 1\} \tag{6.4}$$

For instance, $\Pr\{S_5 = 4|S_4 = 2, S_3 = 3, S_2 = 3, S_1 = 2, S_0 = 1\}$ is the conditional probability that the fifth transaction is a log off, given that the previous transactions (in chronological order) were log on, fetch, read, read, and fetch. From a modeling perspective, the advantage of expression (6.4) is that we can concentrate on determining the probability of the *next* transaction given a history of previous transactions, rather than on the entire joint distribution of a sequence of transactions.

Fortunately, the performance-modeling group is comfortable with the approximation that the next transaction a customer requests depends on the last one requested and no others. Therefore, $\Pr\{S_5 = 4|S_4 = 2, S_3 = 3, S_2 = 3, S_1 = 2, S_0 = 1\} = \Pr\{S_5 = 4|S_4 = 2\}$, for instance, and (6.4) simplifies to

$$\Pr\{S_5 = 4, S_4 = 2, S_3 = 3, S_2 = 3, S_1 = 2, S_0 = 1\}$$
$$= \Pr\{S_5 = 4|S_4 = 2\} \Pr\{S_4 = 2|S_3 = 3\}$$
$$\times \Pr\{S_3 = 3|S_2 = 3\} \Pr\{S_2 = 3|S_1 = 2\}$$
$$\times \Pr\{S_1 = 2|S_0 = 1\} \Pr\{S_0 = 1\}$$

This simplification is the defining characteristic of a Markov chain.

A process for which

$$\Pr\{S_{n+1} = j|S_n = i, S_{n-1} = a, \ldots, S_0 = z\} = \Pr\{S_{n+1} = j|S_n = i\}$$

*has the **Markov property** and is called a **Markov chain**. A consequence of the Markov property is the more general statement*

$$\Pr\{S_{n+k} = j|S_n = i, S_{n-1} = a, \ldots, S_0 = z\} = \Pr\{S_{n+k} = j|S_n = i\} \tag{6.5}$$

for $k = 1, 2, \ldots$.

The performance-modeling group also believes that the next transaction requested depends *solely* on the last transaction and therefore does not depend on how many transactions have been requested. For example, the conditional probability that the next transaction requested is a fetch given that the last one was a read does not depend on whether the last transaction was the 3d, the 47th, or the 1079th transaction. This important simplification applies to all Markov chains studied in this book:

A Markov chain for which

$$\Pr\{S_{n+1} = j \mid S_n = i\} \text{ is the same for all } n \geq 0$$

has the **time-stationarity property**. A consequence of the time-stationarity property is the more general statement

$$\Pr\{S_{n+k} = j \mid S_n = i\} \text{ is the same for all } n \geq 0 \tag{6.6}$$

Let $p_{ij} \equiv \Pr\{S_{n+1} = j \mid S_n = i\}$. The p_{ij} are called *one-step transition probabilities* and are interpreted as the probability that the state of the process changes (has a "transition") from state i to state j in one time step. Applying time stationarity, we can further simplify (6.4) to

$$\Pr\{S_5 = 4, S_4 = 2, S_3 = 3, S_2 = 3, S_1 = 2, S_0 = 1\}$$

$$= \Pr\{S_5 = 4 \mid S_4 = 2\} \Pr\{S_4 = 2 \mid S_3 = 3\}$$

$$\times \Pr\{S_3 = 3 \mid S_2 = 3\} \Pr\{S_2 = 3 \mid S_1 = 2\}$$

$$\times \Pr\{S_1 = 2 \mid S_0 = 1\} \Pr\{S_0 = 1\}$$

$$= p_{24}\, p_{32}\, p_{33}\, p_{23}\, p_{12} \Pr\{S_0 = 1\}$$

$$= \Pr\{S_0 = 1\} p_{12}\, p_{23}\, p_{33}\, p_{32}\, p_{24}$$

In this example $\Pr\{S_0 = 1\} = 1$, but in general we let $p_i \equiv \Pr\{S_0 = i\}$ denote the *initial-state probability*. Therefore,

$$\Pr\{S_0 = 1, S_1 = 2, S_2 = 3, S_3 = 3, S_4 = 2, S_5 = 4\} = p_1\, p_{12}\, p_{23}\, p_{33}\, p_{32}\, p_{24}$$

This last expression is quite intuitive, since we can think of the probability of the sample path $1, 2, 3, 3, 2, 4$ being decomposed into the probability of starting with transaction type 1, multiplied by the transition probability of following transaction type 1 with type 2, multiplied by the transition probability of following 2 with 3, and so on. The same decomposition can be applied to the sample paths observed in Figure 6.1 to derive the probability that the simulation generates those sample paths. The fact that the process is trapped forever in state 4 does not affect these probabilities if we let $p_{44} = 1$, since $(1)(1)(1)\cdots = 1$.

The example illustrates our fundamental decomposition of the sample paths of time-stationary Markov chains:

The sample paths of a Markov chain are completely characterized by the one-step transition probabilities, p_{ij}, $i, j = 1, 2, \ldots, m$, and initial-state probabilities, p_i, $i = 1, 2, \ldots, m$. Given values for these probabilities, the joint probability of any finite sample path can be decomposed into a product of an initial-state probability and the appropriate sequence of transition probabilities.

For convenience of exposition and calculation we organize the one-step transition probabilities into a *one-step transition matrix*

$$\mathbf{P} \equiv \begin{pmatrix} p_{11} & p_{12} & \cdots & p_{1m} \\ p_{21} & p_{22} & \cdots & p_{2m} \\ \vdots & \vdots & \vdots & \vdots \\ p_{m1} & p_{m2} & \cdots & p_{mm} \end{pmatrix} \tag{6.7}$$

The rows of this matrix represent the last state of the process, while the columns represent the next state of the process. Each row of $\mathbf{P}$ corresponds to a conditional mass function and therefore sums to 1. For instance, row i corresponds to $p_{N|L=i}(j) = \Pr\{N = j | L = i\}$, for $j = 1, 2, \ldots, m$. For the purpose of simulation the matrix can be rewritten so that each row corresponds to a conditional cdf:

$$\Sigma\mathbf{P} \equiv \begin{pmatrix} p_{11} & p_{11} + p_{12} & p_{11} + p_{12} + p_{13} & \cdots & 1 \\ p_{21} & p_{21} + p_{22} & p_{21} + p_{22} + p_{23} & \cdots & 1 \\ \vdots & \vdots & \vdots & & \vdots & \vdots \\ p_{m1} & p_{m1} + p_{m2} & p_{m1} + p_{m2} + p_{m3} & \cdots & 1 \end{pmatrix} \tag{6.8}$$

Row i of $\Sigma\mathbf{P}$ corresponds to $F_{N|L=i}(j) = \Pr\{N \le j | L = i\}$, for $j = 1, 2, \ldots, m$.

Similarly, we organize the initial-state probabilities into an *initial-state vector*

$$\mathbf{p} \equiv \begin{pmatrix} p_1 \\ p_2 \\ \vdots \\ p_m \end{pmatrix} \tag{6.9}$$

For the purpose of simulation the vector can be rewritten so that it corresponds to the cdf F_{S_0}

$$\Sigma\mathbf{p} \equiv \begin{pmatrix} p_1 \\ p_1 + p_2 \\ p_1 + p_2 + p_3 \\ \vdots \\ 1 \end{pmatrix}$$

For example, the one-step transition matrix derived by Data Driven's performance-modeling group—see Section 6.4.2 below for how it was determined—is

$$\mathbf{P} = \begin{pmatrix} 0 & 0.95 & 0.01 & 0.04 \\ 0 & 0.27 & 0.63 & 0.10 \\ 0 & 0.36 & 0.40 & 0.24 \\ 0 & 0 & 0 & 1 \end{pmatrix} \tag{6.10}$$

and the initial-state vector is $\mathbf{p}' = (1, 0, 0, 0)$. Notice that $p_{i1} = 0$ for all i, indicating that a log on cannot follow any other transaction. And also notice that $p_{44} = 1$, indicating that when a log off occurs, the process remains in that state. A state i for which $p_{ii} = 1$ is called an *absorbing state*. The initial-state probability $p_1 = 1$ indicates that a session always begins with a log on. Written as a matrix of conditional cdfs, the transition matrix becomes

$$\Sigma\mathbf{P} = \begin{pmatrix} 0 & 0.95 & 0.96 & 1 \\ 0 & 0.27 & 0.90 & 1 \\ 0 & 0.36 & 0.76 & 1 \\ 0 & 0 & 0 & 1 \end{pmatrix}$$

For instance, the second row of $\Sigma\mathbf{P}$ corresponds to $F_{N|L=2}$. Therefore, if the last state was 2, then $F_{N|L=2}^{-1}(\text{random}())$ returns the next state as 2 when $0 \le \text{random}() \le 0.27$; as 3 when $0.27 < \text{random}() \le 0.90$; and as 4 when $0.90 < \text{random}() \le 1$.

A special case of a Markov chain occurs when the successive states of the process are independent random variables; in other words when $\Pr\{S_{n+1} = j | S_n = i\} = \Pr\{S_{n+1} = j\}$ regardless of i. Time stationarity then implies that all the S_n have the same marginal distribution. The one-step transition matrix of an independent, time-stationary Markov chain has all rows identical

$$\mathbf{P} = \begin{pmatrix} p_1 & p_2 & \cdots & p_m \\ p_1 & p_2 & \cdots & p_m \\ \vdots & \vdots & \vdots & \vdots \\ p_1 & p_2 & \cdots & p_m \end{pmatrix}$$

Viewed in this way, the Markov property can be considered the simplest sort of dependence structure beyond independence.

6.4.2 Parameterizing Markov Chains

To analyze the sample paths of a Markov chain, we must first parameterize $\mathbf{P}$ and $\mathbf{p}$. In many applications the analyst has to "invent" values based on knowledge of the system of interest or expert opinion. In some applications the transition probabilities are derived from other random variables (see, for instance, Chapter 7). However, when the system can be observed and individual transitions identified, as in Case 6.1, then the transition probabilities and initial-state probabilities can be estimated. The following method is valid if either a single sample path is observed for many transitions or many sample paths are observed.

Organize the available data into a table

	1	2	$\cdots$	m	
1	c_{11}	c_{12}	$\cdots$	c_{1m}	$c_1 = \sum_{j=1}^{m} c_{1j}$
2	c_{21}	c_{22}	$\cdots$	c_{2m}	$c_2 = \sum_{j=1}^{m} c_{2j}$
$\vdots$	$\vdots$	$\vdots$	$\vdots$	$\vdots$	$\vdots$
m	c_{m1}	c_{m2}	$\cdots$	c_{mm}	$c_m = \sum_{j=1}^{m} c_{mj}$

in which c_{ij} is the count of observed transitions from state i to state j, and c_i is the total count of transitions that began in state i. If state i is known to be an absorbing state, then set $p_{ii} = 1$ and $p_{ij} = 0$, for $j \neq i$. Otherwise, estimate p_{ij} by the fraction of transitions from state i to state j,

$$\widehat{p}_{ij} = \begin{cases} \frac{c_{ij}}{c_i}, & c_i > 0 \\ 0, & c_i = 0 \end{cases}$$

When $c_{ij} > 0$, the standard error is of $\widehat{p}_{ij}$ is approximately

$$\widehat{se} = \sqrt{\frac{\widehat{p}_{ij}(1 - \widehat{p}_{ij})}{c_i}}$$

In order to estimate $\mathbf{p}$, multiple sample paths must be observed so that multiple initial states can be observed. Let d_i be the number of sample paths that started in state i, and let $d = \sum_{i=1}^{m} d_i$ be the number of sample paths observed. Then an estimate of p_i is

$$\widehat{p}_i = \frac{d_i}{d}$$

The standard error of $\widehat{p}_i$ is approximately

$$\widehat{se} = \sqrt{\frac{\widehat{p}_i(1 - \widehat{p}_i)}{d}}$$

For example, the transition matrix for Data Driven's customer behavior model was based on the following observed data for 2600 sessions:

	1	2	3	4	
1	0	2470	26	104	2600
2	0	2457	5733	910	9100
3	0	3204	3560	2136	8900
4	0	0	0	0	0

The entry $c_{12} = 2470$ means that of the 2600 observed customer sessions there were 2470 cases in which a log on was immediately followed by a fetch. Therefore, $\widehat{p}_{12} = 2470/2600 = 0.950$, with a standard error of $\widehat{se} = \sqrt{(0.95)(0.05)/2600} \approx$ 0.004. The standard error indicates that this probability is correct to about two decimal places since $\widehat{p}_{12} \pm 2\,\widehat{se}$ is 0.950 ± 0.008.

Notice also that $c_{44} = 0$, since we cannot observe a transition from log off to log off, but our knowledge of the system prescribes that $p_{44} = 1$.

6.4.3   Transition Diagrams

A Markov chain transition matrix can be represented graphically as a *transition-probability diagram*. Figure 6.3 displays the transition-probability diagram for Data Driven's customer behavior model. Each node ◯ represents a state of the system and is numbered accordingly. A directed arc ⟶ connects state i to state j if a one-step transition from i to j is possible. The one-step transition probability p_{ij} is written next to the arc. Notice that a transition from a state to itself is represented by a loop.

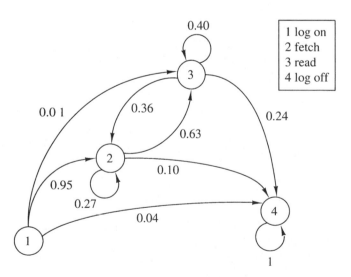

FIGURE 6.3
Transition-probability diagram for Data Driven's customer behavior model.

The transition diagram is particularly useful as a formulation tool: First draw a node for each physical state of the system. During formulation it may be helpful to label each node with a descriptive name, such as "log on," rather than a number. For each ordered pair of states, decide whether a direct transition is possible in one time step; if it is, add a directed arc from one node to the other. Finally, assign appropriate transition probabilities to the arcs. Each arc corresponds to a nonzero element of the transition matrix **P**.

The transition diagram is also useful for describing the model to others, since it is easy to visualize a sample path evolving by following arcs ("chaining") from node to node. The same intuition is difficult to attain from looking at the transition matrix. And notice that the probability of any sample path through a finite number of nodes is just the product of the probabilities on the arcs connecting them.

6.5   THE CASE OF THE DEFECTIVE DETECTIVE

In Section 6.4.1 we showed that, at least in principle, we can compute the probability of any finite sample path of a Markov chain given values for **P** and **p**. In the remainder of the chapter we derive formulas for the probabilities of certain *collections* of sample paths, formulas that do not require us to explicitly calculate the probability of each sample path individually. We also consider probability statements about sample paths as the time index n becomes large. Before presenting these results, we introduce a new modeling-and-analysis problem in this section.

Case 6.2. Quality control engineers at KRN Corporation use acceptance sampling to monitor the ongoing performance of a manufacturing system that produces an electronic component. To detect a change in the performance of the system, they take a sample of 15 components produced in sequence, inspect them, and count the number of defective components. If no more than one defective component is found, then the system is regarded as operating in control; otherwise it is shut down and adjusted. This sampling plan—inspect 15, accept if less than or equal to 1 defective—comes from a standard table of acceptance-sampling plans that assumes that the quality of successive components can be modeled as independent, time-stationary random variables with a common probability of being defective. The engineers are concerned that the approximation of independence is not appropriate when components are inspected in the order that they are produced, and they would like to know if dependence between successive components affects the ability of their sampling plan to detect poor quality.

Standard acceptance-sampling plans assume that the number of defective components in a sample of components has a binomial distribution (Section 5.2) with parameter $\gamma = \Pr\{\text{defective}\}$. Therefore, if Z_{15} represents the number of defective components in a sample of size 15, then

$$\Pr\{Z_{15} \leq 1\} = \sum_{j=0}^{1} \frac{15!}{j!(15-j)!}\gamma^{j}(1-\gamma)^{15-j}$$

Acceptance-sampling plans are designed so that this probability will be large when γ is small—implying that we are likely to find one or fewer defective components when quality is good—and small when γ is large—implying that we are unlikely to find as few as one defective component when quality is poor. For example, $\Pr\{Z_{15} \leq 1\} \approx 0.99$ when $\gamma = 0.01$, but $\Pr\{Z_{15} \leq 1\} \approx 0.08$ when $\gamma = 0.25$. The engineers want to know how the probabilities change if there is dependence between successive components.

A Markov chain has been suggested as one way to model the dependence. Let the state space of the process be $\mathcal{M} = \{1, 2\}$, where state 1 represents an acceptable component and state 2 represents a defective component.[3] The time index n counts the number of components produced in sequence. The one-step transition matrix is

$$\mathbf{P} = \begin{pmatrix} p_{11} & p_{12} \\ p_{21} & p_{22} \end{pmatrix}$$

for which a specific example is

$$\mathbf{P} = \begin{pmatrix} 0.995 & 0.005 \\ 0.495 & 0.505 \end{pmatrix} \tag{6.11}$$

Notice that the probability of a defective component following an acceptable component is quite small (0.005), but the probability of a defective component following another defective component is much larger (0.505).

Standard acceptance-sampling plans assume that this matrix has the form

$$\mathbf{P} = \begin{pmatrix} 1 - \gamma & \gamma \\ 1 - \gamma & \gamma \end{pmatrix}$$

implying that successive components are independent. By allowing the rows of the transition matrix to be different, the engineers can cause the probability that the next component is defective to depend on whether or not the last component was defective.

6.6 TIME-DEPENDENT PERFORMANCE MEASURES

In this section we present formulas for time-dependent performance measures of Markov chain sample paths, performance measures that are functions of the time index, n. These performance measures are probability statements about the state of a Markov chain at time n.

We present formulas for the following probabilities:

$$\boxed{p_{ij}^{(n)} \equiv \Pr\{S_n = j \mid S_0 = i\}}$$

---

[3]It is perfectly valid, and perhaps more natural, to let the state space be $\{0, 1\}$, but in this chapter we retain our convention of numbering the state space beginning with 1.

This is the probability that the process is in state j after n time steps, given that it started in state i. These quantities are called the *n-step transition probabilities*. Because of time stationarity, $p_{ij}^{(n)} = \Pr\{S_{k+n} = j | S_k = i\}$ for any $k \geq 0$.

For example, in the quality control model of Case 6.2, $p_{12}^{(3)}$ is the probability that the fourth component produced is defective, given that the first component was not defective (recall that the first component is numbered $n = 0$). Notice that this is the probability of a *collection* of sample paths, since the states of the second and third components (time steps $n = 1, 2$) are not restricted.

We also present a formula for

$$p_j^{(n)} \equiv \Pr\{S_n = j\}$$

This is the unconditional probability that the process is in state j after n time steps without knowledge of the initial state. These quantities are called the *n-step state probabilities*.

For example, in the quality control model $p_2^{(3)}$ is the unconditional probability that the fourth component is defective, without restriction on the states of the first, second and third components ($n = 0, 1, 2$).

Let $\mathcal{A}$ and $\mathcal{B}$ be two disjoint subsets of the state space $\mathcal{M}$. The term *disjoint* means that they have no states in common. Let the number of states in $\mathcal{A}$ and $\mathcal{B}$ be denoted by $m_\mathcal{A}$ and $m_\mathcal{B}$, respectively. For initial state $i \in \mathcal{A}$ and final state $j \in \mathcal{B}$, we present a formula for

$$f_{ij}^{(n)} \equiv \Pr\{S_n = j, S_{n-1} \in \mathcal{A}, S_{n-2} \in \mathcal{A}, \ldots, S_1 \in \mathcal{A} | S_0 = i\}$$

This is the probability that the process is confined to states in $\mathcal{A}$ for the first $n - 1$ time steps, but at time step n it enters state $j \in \mathcal{B}$. These quantities are called *first-passage probabilities*. Because of time stationarity, $f_{ij}^{(n)} = \Pr\{S_{k+n} = j, S_{k+n-1} \in \mathcal{A}, S_{k+n-2} \in \mathcal{A}, \ldots, S_{k+1} \in \mathcal{A} | S_k = i\}$ for any $k \geq 0$.

For example, let $\mathcal{A} = \{1, 2, 3\}$ and $\mathcal{B} = \{4\}$ in the customer behavior model (Case 6.1). Then $f_{14}^{(9)}$ is the probability that the customer's tenth transaction is a log off and all earlier transactions were log ons, reads or fetches, given the initial transaction was a log on. Stated differently, $f_{14}^{(9)}$ is the probability that the session ends at precisely 10 transactions including log on, which is the same as nine transactions excluding log on.

All three of these performance measures can be computed via matrix calculations. Let $\mathbf{P}^{(n)}$ be the $m \times m$ matrix with elements $p_{ij}^{(n)}$. Then

$$\mathbf{P}^{(n)} = \mathbf{P}^n, \ n = 1, 2, \ldots \tag{6.12}$$

In words, the matrix of n-step transition probabilities is obtained by raising the one-step transition matrix to the nth power.

Now let $\mathbf{p}^{(n)}$ be the $m \times 1$ vector with elements $p_j^{(n)}$. Then

$$\mathbf{p}^{(n)'} = \mathbf{p}'\mathbf{P}^n, \; n = 1, 2, \ldots \tag{6.13}$$

In words, the vector of n-step state probabilities is obtained by taking the product of the initial-state vector and the n-step transition matrix.

Finally, let $\mathbf{F}_{AB}^{(n)}$ be the $m_A \times m_B$ matrix whose elements are $f_{ij}^{(n)}$ with $i \in A$ and $j \in B$. Then

$$\mathbf{F}_{AB}^{(n)} = \mathbf{P}_{AA}^{n-1}\mathbf{P}_{AB}, \; n = 1, 2, \ldots \tag{6.14}$$

(Recall that $\mathbf{P}_{AA}^0 = \mathbf{I}$.) In words, the matrix of first-passage probabilities is obtained by taking the one-step transition matrix restricted to states in A and raising it to the $(n - 1)$st power, then multiplying the result by the matrix of transition probabilities from states in A to states in B.

We first give an example of each of these calculations and then derive the formulas.

6.6.1 Examples

To illustrate the n-step transition and state probabilities, consider again the quality control model with one-step transition matrix (6.11). To obtain $p_{12}^{(3)}$, the probability that the fourth component is defective given that the first component is not, compute

$$\mathbf{P}^{(3)} = \begin{pmatrix} p_{11}^{(3)} & p_{12}^{(3)} \\ p_{21}^{(3)} & p_{22}^{(3)} \end{pmatrix}$$

$$= \mathbf{P}^3$$

$$= \begin{pmatrix} 0.995 & 0.005 \\ 0.495 & 0.505 \end{pmatrix}^3$$

$$= \begin{pmatrix} 0.99125 & 0.00875 \\ 0.86625 & 0.13375 \end{pmatrix}$$

Therefore, $p_{12}^{(3)} = 0.00875$. The calculation requires a *matrix multiplication*, not a component-by-component multiplication, because $p_{ij}^{(n)} \neq p_{ij}^n$ in general. Notice that each row of $\mathbf{P}^3$ sums to 1, and this is true of $\mathbf{P}^n$ for all n.

If we do not know the state of the first component but we have a probability distribution for the initial state given by the vector $\mathbf{p}' = (0.96, 0.04)$, then the unconditional probability that the fourth component is defective is $p_2^{(3)}$, which is obtained from

$$\mathbf{p}^{(3)'} = \left(p_1^{(3)}, p_2^{(3)}\right)$$

$$= \mathbf{p}'\mathbf{P}^3$$

$$= (p_1, p_2) \begin{pmatrix} p_{11}^{(3)} & p_{12}^{(3)} \\ p_{21}^{(3)} & p_{22}^{(3)} \end{pmatrix}$$

$$= (0.96, 0.04) \begin{pmatrix} 0.99125 & 0.00875 \\ 0.86625 & 0.13375 \end{pmatrix} = (0.98625, 0.01375)$$

Therefore, $p_2^{(3)} = 0.01375$.

In the calculations above all digits were retained for the purpose of illustration. They are certainly not all meaningful since they were derived from transition probabilities p_{ij} that were correct to only three decimal places. Thus, $p_2^{(3)} \approx 0.014$ is a more appropriate statement.

To illustrate first-passage probabilities, consider again the customer behavior model with one-step transition matrix (6.10). The length of a session, represented by the random variable X, is the number of transactions requested until a log off, excluding log on. Let $\mathcal{A} = \{1, 2, 3\}$ and $\mathcal{B} = \{4\}$. Then $f_{14}^{(n)}$ is the probability that the first n transactions are confined to log on, fetch and read, and a log off is the $(n + 1)$st transaction (given the customer begins with a log on). Notice that since the log on is transaction 0, the time step n corresponds to the length of the session, which excludes log on.

We compute the $f_{ij}^{(n)}$ as follows:

$$\mathbf{F}_{AB}^{(n)} = \begin{pmatrix} f_{14}^{(n)} \\ f_{24}^{(n)} \\ f_{34}^{(n)} \end{pmatrix}$$

$$= \mathbf{P}_{AA}^{n-1}\mathbf{P}_{AB}$$

$$= \begin{pmatrix} 0 & 0.95 & 0.01 \\ 0 & 0.27 & 0.63 \\ 0 & 0.36 & 0.40 \end{pmatrix}^{n-1} \begin{pmatrix} 0.04 \\ 0.10 \\ 0.24 \end{pmatrix}$$

For instance, rounding to two decimal places

$$\mathbf{F}_{AB}^{(9)} \approx \begin{pmatrix} 0.05 \\ 0.04 \\ 0.03 \end{pmatrix}$$

so that $f_{14}^{(9)} \approx 0.05$.

Contrast the first-passage probability $f_{14}^{(9)}$ to the n-step transition probability $p_{14}^{(9)}$, which is the probability that at time step 9 the process is in state 4, without regard to when the process first entered state 4. In this case $p_{14}^{(9)} \approx 0.80$. The probability of finding the process in state 4 at time $n = 9$ is much larger than the probability that the process enters state 4 *for the first time* at $n = 9$. In general $f_{ij}^{(n)} \leq p_{ij}^{(n)}$.

Figure 6.4 shows a plot of $f_{14}^{(n)}$, for $n = 1, 2, \ldots, 20$, which is the probability mass function of X truncated at $n = 20$ transactions. In other words, $\Pr\{X = n\} = f_{14}^{(n)}$. Figure 6.2 is an *estimate* of this distribution based on simulation, while Figure 6.4 is the *true* distribution without any sampling error.

The expected session length is

$$E[X] = \sum_{n=1}^{\infty} n \Pr\{X = n\} = \sum_{n=1}^{\infty} n f_{14}^{(n)}$$

In the exercises we derive a formula for $E[X]$ that does not involve an infinite summation. As a practical matter it is often sufficient to approximate an infinite summation by a truncated version if the terms in the summation are getting smaller

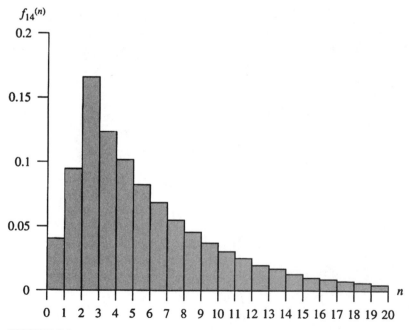

FIGURE 6.4
Probability distribution of session length for Data Driven's customers based on the Markov chain model.

and smaller. For example,

$$E[X] \approx \sum_{n=1}^{n^*} n f_{14}^{(n)}$$

where n^* is chosen so that the remaining terms add a negligible amount to the summation. For example, at $n = 31$ we have $f_{14}^{(31)} \approx 0.0005$, implying that $31 \times f_{14}^{(31)} \approx 0.016$, which is pretty small. If we truncate at $n^* = 30$, then $E[X] \approx 6.5$ transactions per session. Recall that the simulation estimate was 6.8. The true value to one decimal place is 6.6.

6.6.2 Derivations

This section can be skipped by readers interested only in applications.

We first derive the formula $\mathbf{P}^{(n)} = \mathbf{P}^n$ using a method called *induction*. Induction begins by establishing that the formula is correct for some n, and then shows that if it is correct for any n, it must be also be correct for $n + 1$.

Express $p_{ij}^{(2)}$ as

$$p_{ij}^{(2)} = \Pr\{S_2 = j | S_0 = i\} = \sum_{h=1}^{m} \Pr\{S_2 = j, S_1 = h | S_0 = i\} \qquad (6.15)$$

by exploiting that fact that to reach state j at time 2, the process must be in some state h at time 1. Since the probability we actually start in state i is not our interest, we assume that $\Pr\{S_0 = i\} > 0$, but nothing more. Then using the definition of conditional probability

$$\Pr\{S_2 = j, S_1 = h | S_0 = i\} = \frac{\Pr\{S_2 = j, S_1 = h, S_0 = i\}}{\Pr\{S_0 = i\}}$$

But since S_n is a Markov chain, we can decompose the joint probability in the numerator into a product of an initial-state probability and transition probabilities; thus

$$\frac{\Pr\{S_2 = j, S_1 = h, S_0 = i\}}{\Pr\{S_0 = i\}} = \frac{p_i \, p_{ih} \, p_{hj}}{p_i} = p_{ih} \, p_{hj}$$

Substituting this expression into (6.15) gives

$$p_{ij}^{(2)} = \sum_{h=1}^{m} p_{ih} \, p_{hj}$$

which is the definition of the i, jth element of $\mathbf{P}^2$. Therefore, the formula is correct for $n = 2$. Now suppose that the formula is correct for all $n \leq k$. This is the induction hypothesis. We next show that it must also be correct for $n = k + 1$.

By reasoning similar to the $n = 2$ case,

$$p_{ij}^{(k+1)} = \Pr\{S_{k+1} = j \mid S_0 = i\}$$

$$= \sum_{h=1}^{m} \Pr\{S_{k+1} = j, S_k = h \mid S_0 = i\}$$

$$= \sum_{h=1}^{m} \frac{\Pr\{S_{k+1} = j, S_k = h, S_0 = i\}}{\Pr\{S_0 = i\}}$$

$$= \sum_{h=1}^{m} \frac{\Pr\{S_{k+1} = j \mid S_k = h, S_0 = i\} \Pr\{S_k = h \mid S_0 = i\} \Pr\{S_0 = i\}}{\Pr\{S_0 = i\}}$$

$$= \sum_{h=1}^{m} \Pr\{S_{k+1} = j \mid S_k = h\} \Pr\{S_k = h \mid S_0 = i\} \tag{6.16}$$

$$= \sum_{h=1}^{m} p_{hj}\, p_{ih}^{(k)} \tag{6.17}$$

$$= \sum_{h=1}^{m} p_{ih}^{(k)}\, p_{hj}$$

which is the definition of the i, jth element of $\mathbf{P}^{(k)}\mathbf{P}$. But our induction hypothesis is that the formula is correct for $n = k$, so $\mathbf{P}^{(k)}\mathbf{P} = \mathbf{P}^k\mathbf{P} = \mathbf{P}^{k+1}$. Therefore, if it is correct for $n = k$, it is also correct for $n = k + 1$.

This completes the derivation: We have established that the result is correct for $n = 2$, so it must also be correct for $n = 3$. But if it is correct for $n = 3$, then it must also be correct for $n = 4$, and so on. By induction we have shown that it is correct for all n.

Notice that steps (6.16) and (6.17) in the derivation make use of the Markov property and time-stationarity properties, respectively. A very similar derivation can be used to establish a slightly more general result, the *Chapman-Kolmogorov equation*,

$$\boxed{p_{ij}^{(n)} = \sum_{h=1}^{m} p_{ih}^{(k)}\, p_{hj}^{(n-k)}} \tag{6.18}$$

for any $k = 1, 2, \ldots, n - 1$. The Chapman-Kolmogorov equation shows that the probability of the process being in state j after n time steps can be decomposed into the probability of being in state h after k time steps, then moving from state h to j in the remaining $n - k$ time steps, provided we account for all possible intermediate states, h.

The formula for $\mathbf{p}^{(n)}$ follows directly from the law of total probability:

$$p_j^{(n)} = \Pr\{S_n = j\} = \sum_{i=1}^m \Pr\{S_n = j | S_0 = i\} \Pr\{S_0 = i\} = \sum_{i=1}^m p_{ij}^{(n)} p_i$$

which is the definition of the jth element of $\mathbf{p}'\mathbf{P}^{(n)} = \mathbf{p}'\mathbf{P}^n$.

We conclude by describing the intuition behind the formula for first-passage probabilities. The matrix $\mathbf{P}_{\mathcal{AA}}$ contains the one-step transition probabilities from states in $\mathcal{A}$ to other states in $\mathcal{A}$. Therefore, $\mathbf{P}_{\mathcal{AA}}^{n-1}$ contains the $(n-1)$-step transition probabilities from states in $\mathcal{A}$ to other states in $\mathcal{A}$ (and restricted only to states in $\mathcal{A}$), using the same reasoning we used to derive $\mathbf{P}^{(n)} = \mathbf{P}^n$. The matrix $\mathbf{P}_{\mathcal{AB}}$ contains the one-step transition probabilities from states in $\mathcal{A}$ to $\mathcal{B}$. Therefore the product $\mathbf{P}_{\mathcal{AA}}^{n-1}\mathbf{P}_{\mathcal{AB}}$ contains the probabilities of remaining in states in $\mathcal{A}$ for $n-1$ steps, then moving to a state in $\mathcal{B}$ on the nth step.

6.7 TIME-INDEPENDENT (LONG-RUN) PERFORMANCE MEASURES

In this section we present formulas for time-independent performance measures of Markov chain sample paths, performance measures that do not depend on the time index, n. Such performance measures tell us about the long-run behavior of a system without having to select a particular time of interest.

Specifically, we consider the *limiting probability*

$$\bar{p}_{ij} = \lim_{n \to \infty} p_{ij}^{(n)} \tag{6.19}$$

Let $\vec{\mathbf{P}}$ be the $m \times m$ matrix with elements $\bar{p}_{ij}$. For the quality control model with transition matrix (6.11) we will show that

$$\vec{\mathbf{P}} = \begin{pmatrix} 0.99 & 0.01 \\ 0.99 & 0.01 \end{pmatrix}$$

This indicates that as the number of components produced goes to infinity, the probability that a component is defective becomes 0.01 regardless of the state of the first component. For the customer behavior example with transition matrix (6.10) we can show that

$$\vec{\mathbf{P}} = \begin{pmatrix} 0 & 0 & 0 & 1 \\ 0 & 0 & 0 & 1 \\ 0 & 0 & 0 & 1 \\ 0 & 0 & 0 & 1 \end{pmatrix}$$

This indicates that every customer ultimately logs off.

These two examples exhibit quite different types of limiting behavior. There are still other cases in which $\bar{p}_{ij}$ does not exist. To compute these probabilities, and to know how to interpret them, we must first be able to classify the different cases, which requires classifying the states of a Markov chain. The time-dependent performance measures in Section 6.6 apply to *all* Markov chains, regardless of the classification of their states.

Obtaining formulas for time-independent performance measures is particularly important because these performance measures are often difficult to estimate via simulation. We can let time go to infinity mathematically, but we cannot simulate an infinitely long sample path, even if we are very patient!

6.7.1 Classification of States

To develop some intuition about the classification of states, we concentrate on a two-state Markov chain with transition matrix

$$\mathbf{P} = \begin{pmatrix} p_{11} & p_{12} \\ p_{21} & p_{22} \end{pmatrix}$$

and consider the limiting probability $\vec{p}_{11}$. The quality control model is a special case. *The results in this section depend on the Markov chain having a finite number of states ($m < \infty$), which is often the case. We elaborate on the modifications that must be made for the infinite state-space case in Section 6.9.*

First suppose that

$$\mathbf{P} = \begin{pmatrix} 0 & 1 \\ 1 & 0 \end{pmatrix}$$

Since the process alternates between states 1 and 2, $p_{11}^{(n)} = 0$ for $n = 1, 3, 5, \ldots$, while $p_{11}^{(n)} = 1$ for $n = 2, 4, 6, \ldots$. Therefore $p_{11}^{(n)}$ does not converge to a limit, and $\vec{p}_{11}$ does not exist for this Markov chain.

---

*If $p_{ii}^{(n)} > 0$ when $n = \delta, 2\delta, 3\delta, \ldots$ for some integer $\delta > 1$, and $p_{ii}^{(n)} = 0$ otherwise, then state i is **periodic** with period δ and $\vec{p}_{ii}$ does not exist. A state that is not periodic is **aperiodic**.*

In this book we do not compute $\vec{p}_{ij}$ for periodic states j. However, most of the analysis for aperiodic states applies to periodic states provided we define $\vec{p}_{ij} \equiv \lim_{n \to \infty} p_{ij}^{(n\delta)}$ and consider time steps of size δ rather than 1.

Now suppose that

$$\mathbf{P} = \begin{pmatrix} 0.6 & 0.4 \\ 0 & 1 \end{pmatrix}$$

When the process starts in state 1, the probability it remains there for n transitions is $p_{11} p_{11} \cdots p_{11} = p_{11}^n$. Once the process leaves state 1, it never returns. Therefore the probability that the process is in state 1 at time n is also the probability that it has never departed state 1, and $p_{11}^{(n)} = p_{11}^n$ for this example (although not in general). Thus,

$$\vec{p}_{11} = \lim_{n \to \infty} (0.6)^n = 0$$

implying that the process eventually leaves state 1, never to return.

---

*If the limiting probability $\vec{p}_{ii} = 0$, then state i is **transient** and the process will eventually leave state i never to return. If the limiting probability $\vec{p}_{ii} > 0$ then state i is **recurrent** and the process is guaranteed to return to state i again and again, if it ever reaches state i for the first time.*

To compute $\vec{p}_{ij}$, we need to classify the states of a Markov chain as recurrent or transient. We address this matter by looking at another example.

Consider the following one-step transition matrix for a Markov chain with state space $\mathcal{M} = \{1, 2, 3, 4, 5\}$:

$$P = \begin{pmatrix} 0.1 & 0.3 & 0.6 & 0 & 0 \\ 0 & 1 & 0 & 0 & 0 \\ 0 & 0 & 0.8 & 0.2 & 0 \\ 0 & 0 & 0.7 & 0.3 & 0 \\ 0.5 & 0 & 0 & 0.5 & 0 \end{pmatrix} \qquad (6.20)$$

A transition diagram is displayed in Figure 6.5. Let $\mathcal{T} = \{1, 5\}$, $\mathcal{R}_1 = \{2\}$ and $\mathcal{R}_2 = \{3, 4\}$ be subsets of states. Then notice that

$$\mathbf{P}_{\mathcal{R}_1 \mathcal{R}_1} = \begin{pmatrix} 1 \end{pmatrix}$$

and

$$\mathbf{P}_{\mathcal{R}_2 \mathcal{R}_2} = \begin{pmatrix} 0.8 & 0.2 \\ 0.7 & 0.3 \end{pmatrix}$$

are themselves self-contained Markov chain transition matrices, since their rows sum to 1. If the process ever enters a state in $\mathcal{R}_1$ or in $\mathcal{R}_2$, then it will remain in that set of states and never return to states in $\mathcal{T}$ again. This is easy to see from the transition diagram, since there are only arcs into $\mathcal{R}_1$ and $\mathcal{R}_2$ and none coming out.

---

*A Markov chain in which no proper subset of states also forms a Markov chain is **irreducible**; otherwise the Markov chain is **reducible** into one or more subsets of states that form irreducible Markov chains.*

The Markov chain defined by (6.20) is reducible, while the Markov chains defined by $\mathbf{P}_{\mathcal{R}_1 \mathcal{R}_1}$ and $\mathbf{P}_{\mathcal{R}_2 \mathcal{R}_2}$ are irreducible. Once the concept of reducibility is

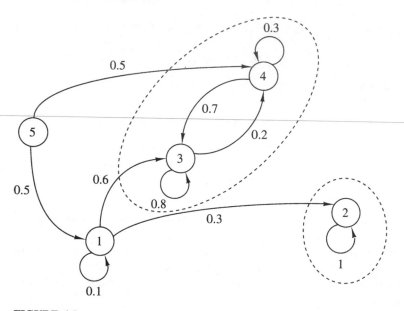

FIGURE 6.5
Transition-probability diagram for the classification example with the irreducible subsets of states highlighted.

understood, the classification of states when $\mathcal{M}$ is finite is accomplished by the following method:

---

1. *Find all irreducible proper subsets of the state space $\mathcal{M}$ (there may be no such subsets, in which case the entire state space is irreducible).*
2. *All states that are members of an irreducible set are recurrent.*
3. *All states that are not members of an irreducible set are transient.*

In the example the states in $\mathcal{R}_1$ and $\mathcal{R}_2$ are recurrent, while the states in $\mathcal{T}$ are transient. Notice that

$$\mathbf{P}_{\mathcal{T}\mathcal{T}} = \begin{pmatrix} 0.1 & 0.0 \\ 0.5 & 0.0 \end{pmatrix}$$

is not a Markov chain transition matrix since the rows do not sum to 1, implying that the process will eventually escape the states in $\mathcal{T}$, never to return.

6.7.2 Performance Measures

Let $\mathbf{P}$ be the transition matrix of a Markov chain with state space $\mathcal{M}$, and suppose that it is reducible into a set of transient states, $\mathcal{T}$, and irreducible, recurrent sets of states $\mathcal{R}_1, \mathcal{R}_2, \mathcal{R}_3, \ldots$. Our goal is to compute the limiting probabilities $\vec{p}_{ij}$. Two cases are easy:

- $\vec{p}_{ij} = 0$ if $j \in \mathcal{T}$. The limiting probability of being in state j is 0 if j is a transient state, regardless of where the process starts. This makes sense, because even if the process reaches state j, it will eventually leave state j and never return, since $\vec{p}_{jj} = 0$ for a transient state.
- $\vec{p}_{ij} = 0$ if $i \in \mathcal{R}_1$ and $j \in \mathcal{R}_2$. The limiting probability of being in state j given the process starts in state i is 0 if i and j are in different irreducible sets of states. This makes sense because a process that is initially in one irreducible set of states will never leave that set of states.

From here on we focus on a single set of irreducible, recurrent states, denoted $\mathcal{R}$ to keep the notation simple. The more interesting cases are when $i, j \in \mathcal{R}$ (i and j are both in the same irreducible set of states) and when $i \in \mathcal{T}$ and $j \in \mathcal{R}$ (i is transient and j is recurrent). The following results are derived after providing examples. For convenience, suppose that $\mathcal{R} = \{1, 2, \ldots, m_{\mathcal{R}}\}$.

- For $i, j \in \mathcal{R}$, $\vec{p}_{ij} = \pi_j > 0$, independent i. The vector $\pi'_{\mathcal{R}} = (\pi_1, \pi_2, \ldots, \pi_{m_{\mathcal{R}}})$ is obtained by solving the system of linear equations

$$
\boxed{
\begin{aligned}
\pi'_{\mathcal{R}} &= \pi'_{\mathcal{R}} \mathbf{P}_{\mathcal{R}\mathcal{R}} \\
1 &= \pi'_{\mathcal{R}} \mathbf{1}
\end{aligned}
}
\tag{6.21}
$$

This system of equations will always have $m_{\mathcal{R}}+1$ equations and $m_{\mathcal{R}}$ unknowns, and any one of the equations other than $1 = \pi'_{\mathcal{R}} \mathbf{1}$ can always be eliminated. For actually solving the system of equations, it is convenient to rewrite $\pi'_{\mathcal{R}} = \pi'_{\mathcal{R}} \mathbf{P}_{\mathcal{R}\mathcal{R}}$ as

$$0 = \pi'_{\mathcal{R}}(\mathbf{I} - \mathbf{P}_{\mathcal{R}\mathcal{R}})$$

The π_j are called *steady-state probabilities*. They can be interpreted as the probability of finding the process in state j after a very long time, or as the long-run fraction of time that the process spends in state j. The vector $\pi_{\mathcal{R}}$ is called the *steady-state distribution*, and it is a true probability distribution since $\pi'_{\mathcal{R}} \mathbf{1} = \sum_{j=1}^{m_{\mathcal{R}}} \pi_j = 1$.

- For $i \in \mathcal{T}$, and $\mathcal{R} = \{j\}$ an absorbing state, $\vec{p}_{ij} = \alpha_{ij} \geq 0$. The vector $\alpha'_{T\mathcal{R}} = (\alpha_{ij}; i \in \mathcal{T})$ is obtained from

$$
\boxed{\alpha_{T\mathcal{R}} = (\mathbf{I} - \mathbf{P}_{TT})^{-1} \mathbf{P}_{T\mathcal{R}}}
\tag{6.22}
$$

The α_{ij} are called *absorption probabilities*. They can be interpreted as the probability that the process is ultimately "absorbed" into state j.

6.7.3   Examples

To illustrate the steady-state probabilities, consider again the quality control model (Case 6.2) with one-step transition matrix (6.11). This Markov chain is irreducible, so $\mathcal{R} = \{1, 2\}$, $\mathcal{T}$ is empty, and we can compute steady-state probabilities. To obtain π_2, the steady-state probability of a defective component, solve

$$(\pi_1, \pi_2) = \pi'\mathbf{P}$$

$$= (\pi_1, \pi_2) \begin{pmatrix} 0.995 & 0.005 \\ 0.495 & 0.505 \end{pmatrix}$$

and

$$1 = \pi'\mathbf{1}$$

$$= (\pi_1, \pi_2) \begin{pmatrix} 1 \\ 1 \end{pmatrix}$$

The complete system of equations is therefore

$$\begin{aligned}
\pi_1 &= 0.995\pi_1 + 0.495\pi_2 \\
\pi_2 &= 0.005\pi_1 + 0.505\pi_2 \\
1 &= \pi_1 + \pi_2
\end{aligned}$$

which can be rewritten as

$$\begin{aligned}
0 &= 0.005\pi_1 - 0.495\pi_2 \\
0 &= -0.005\pi_1 + 0.495\pi_2 \\
1 &= \pi_1 + \pi_2
\end{aligned}$$

and either one of the first two equations can be eliminated. The solution is $\pi_1 = 0.99$ and $\pi_2 = 0.01$. The quantity π_2 can be interpreted as the long-run fraction of components that are defective, or as the limiting probability that a component is defective. If KRN makes a profit of $4 for each acceptable component and no profit for a defective component, then the expected profit per component, over the long run, is

$$\$4\pi_1 + \$0\pi_2 = \$3.96$$

To illustrate the absorption probabilities, consider a modification of the customer behavior model with one-step transition matrix (6.10). Suppose that a customer may have a session terminated by spending too much time determining the next transaction. Let the new state space be $\mathcal{M} = \{1, 2, 3, 4, 5\}$, with state 5 representing session termination, and let the revised one-step transition matrix be

$$\mathbf{P} = \begin{pmatrix}
0 & 0.95 & 0.01 & 0.03 & 0.01 \\
0 & 0.27 & 0.63 & 0.09 & 0.01 \\
0 & 0.36 & 0.40 & 0.23 & 0.01 \\
0 & 0 & 0 & 1 & 0 \\
0 & 0 & 0 & 0 & 1
\end{pmatrix}$$

The subset of states $\mathcal{T} = \{1, 2, 3\}$ are transient states, and the subsets $\mathcal{R}_1 = \{4\}$ and $\mathcal{R}_2 = \{5\}$ are both irreducible, recurrent (absorbing) states. To obtain α_{14}, the probability that a session ends with a log off, compute

$$\boldsymbol{\alpha_{\mathcal{T}\mathcal{R}_1}} = \begin{pmatrix} \alpha_{14} \\ \alpha_{24} \\ \alpha_{34} \end{pmatrix}$$

$$= (\mathbf{I} - \mathbf{P}_{\mathcal{T}\mathcal{T}})^{-1} \mathbf{P}_{\mathcal{T}\mathcal{R}_1}$$

$$= \left[\begin{pmatrix} 1\ 0\ 0 \\ 0\ 1\ 0 \\ 0\ 0\ 1 \end{pmatrix} - \begin{pmatrix} 0 & 0.95 & 0.01 \\ 0 & 0.27 & 0.63 \\ 0 & 0.36 & 0.40 \end{pmatrix} \right]^{-1} \begin{pmatrix} 0.03 \\ 0.09 \\ 0.23 \end{pmatrix}$$

$$\approx \begin{pmatrix} 0.93 \\ 0.94 \\ 0.95 \end{pmatrix}$$

Therefore, the probability that a session ends with a log off, rather than a termination, is $\alpha_{14} \approx 0.93$. Since a session must end one way or the other, $\alpha_{15} = 1 - \alpha_{14} \approx 0.07$.

6.7.4 Derivations

This section can be skipped by readers interested only in applications.

We first establish that the steady-state probabilities must satisfy $\boldsymbol{\pi}'_{\mathcal{R}} = \boldsymbol{\pi}'_{\mathcal{R}} \mathbf{P}_{\mathcal{R}\mathcal{R}}$ and $\boldsymbol{\pi}'_{\mathcal{R}} \mathbf{1} = 1$, if they exist. We leave unproven that $\boldsymbol{\pi}_{\mathcal{R}}$ does exist for an irreducible, aperiodic Markov chain with one-step transition matrix $\mathbf{P}_{\mathcal{R}}$. For ease of presentation we assume that the entire state space $\mathcal{M}$ is irreducible, so that $\mathcal{R} = \mathcal{M}$.

Suppose $\lim_{n \to \infty} p_{ij}^{(n)} = \pi_j > 0$ for all j, independent of the initial state i. Then if we take the limit of both sides of the Chapman-Kolmogorov equation (6.18) with $k = n - 1$, we obtain

$$\pi_j = \lim_{n \to \infty} p_{ij}^{(n)}$$

$$= \lim_{n \to \infty} \sum_{h=1}^{m} p_{ih}^{(n-1)} p_{hj}$$

$$= \sum_{h=1}^{m} \left(\lim_{n \to \infty} p_{ih}^{(n-1)} \right) p_{hj}$$

$$= \sum_{h=1}^{m} \pi_h p_{hj}$$

But $\sum_{h=1}^{m} \pi_h p_{hj}$ is the definition of the jth element of $\boldsymbol{\pi}'\mathbf{P}$. Therefore we have established that $\boldsymbol{\pi}' = \boldsymbol{\pi}'\mathbf{P}$, which is the same as $\boldsymbol{\pi}'(\mathbf{I} - \mathbf{P}) = \mathbf{0}$. However, since the rows of $\mathbf{P}$ all sum to 1, the matrix $\mathbf{I} - \mathbf{P}$ is not of full rank and $\boldsymbol{\pi}'(\mathbf{I} - \mathbf{P}) = \mathbf{0}$ has an infinite number of solutions. The particular solution that makes $\boldsymbol{\pi}$ a probability distribution must satisfy the additional equation $\boldsymbol{\pi}'\mathbf{1} = 1$.

To establish the absorption probabilities, let $\mathcal{T}$ be the set of transient states and let $\mathcal{R} = \{j\}$ be an absorbing state. Then for $i \in \mathcal{T}$,

$$\alpha_{ij} = \sum_{n=1}^{\infty} f_{ij}^{(n)}$$

where $f_{ij}^{(n)}$ is the first-passage probability from state i to state j. In other words, the probability that the process is absorbed in state j is equal to the probability that the process reaches state j for the first time at *some* time step n. Since j is an absorbing state, once the process reaches j for the first time, it never leaves. In matrix notation

$$\alpha_{\mathcal{T}\mathcal{R}} = \sum_{n=1}^{\infty} \mathbf{F}_{\mathcal{T}\mathcal{R}}^{(n)}$$

$$= \mathbf{P}_{\mathcal{T}\mathcal{R}} + \sum_{n=2}^{\infty} \mathbf{P}_{\mathcal{T}\mathcal{T}}^{n-1} \mathbf{P}_{\mathcal{T}\mathcal{R}}$$

$$= \left(\mathbf{I} + \sum_{n=1}^{\infty} \mathbf{P}_{\mathcal{T}\mathcal{T}}^{n} \right) \mathbf{P}_{\mathcal{T}\mathcal{R}}$$

$$= (\mathbf{I} - \mathbf{P}_{\mathcal{T}\mathcal{T}})^{-1} \mathbf{P}_{\mathcal{T}\mathcal{R}}$$

The last equality requires a result from linear algebra about the convergence of an infinite sum of square matrices: Provided $\lim_{n \to \infty} \mathbf{P}_{\mathcal{T}\mathcal{T}}^{n} = \mathbf{0}$, the matrix of all zeros, then

$$\sum_{n=0}^{\infty} \mathbf{P}_{\mathcal{T}\mathcal{T}}^{n} = \mathbf{I} + \sum_{n=1}^{\infty} \mathbf{P}_{\mathcal{T}\mathcal{T}}^{n} = (\mathbf{I} - \mathbf{P}_{\mathcal{T}\mathcal{T}})^{-1}$$

A special case of this result is the well-known sum of a geometric series, $\sum_{n=0}^{\infty} p^{n} = (1 - p)^{-1}$ for $|p| < 1$. Since $\mathbf{P}_{\mathcal{T}\mathcal{T}}$ is the matrix of transition probabilities among transient states, $\lim_{n \to \infty} \mathbf{P}_{\mathcal{T}\mathcal{T}}^{n} = \mathbf{0}$ because the process eventually leaves the transient states, never to return.

6.8 THE MARKOV AND STATIONARITY PROPERTIES REVISITED

Whether or not a Markov chain is an appropriate model for a particular modeling-and-analysis problem depends on whether or not the Markov property is a reasonable approximation for the dependence in the process. The tools presented in this chapter also rely on the Markov chain being time stationary. This section presents a collection of examples and discussion to help develop intuition about these properties.

Case 6.3. A model of the movement of a taxi defines the state of the system to be the region of the city that is the destination of the current rider, and the time index to be the number of riders the taxi has transported. When the taxi delivers a rider, it stays in the destination region until it picks up another rider.

The Markov property seems plausible here, because the destination of a rider does not depend on the destination of previous riders (it makes no sense for a rider to ask the driver, "Where have you been today?" and then decide where to go). Time stationarity may be violated because traffic patterns change throughout the day in most cities.

Case 6.4. A model of computer keyboard use defines the state of the system to be the key that a person is currently typing, and the time index to be the number of keys typed.

Applicability of the Markov property depends on the material being typed. If the material is predominantly data (numbers and symbols), then perhaps the next key typed depends only on the last key typed, if it depends on any previous keys at all. On the other hand, if the material is text (such as this book), then the requirement to spell words influences the next letter typed. For instance, consider the probability that the next letter typed is "y" when the last letter typed is "r" compared to the same probability if the previous four letters typed are known to be "Larr." Nevertheless, a Markov chain model of typing may be a useful approximation. Time stationarity will often be satisfied, since the sequence of keys typed should not change much throughout the material.

Case 6.5. A model of the preferences of consumers for brands of toothpaste defines the state of the system to be the brand of toothpaste the consumer currently uses, and the time index to be the number of tubes of toothpaste purchased.

The Markov property implies that consumers' choices depend only on the brand that they currently use, and not on past experience with other brands (that is, consumers have short memories or they think brand characteristics change so rapidly that past experience with a brand is not relevant). Time stationarity can apply only over a brief time period, since advertising campaigns, discounts, etc., influence consumers' choices.

Case 6.6. A model of the weather in Columbus, Ohio, defines the state of the system to be the high temperature, in whole degrees, and the time index to be number of days.

Since weather patterns move quickly through Columbus, today's temperature may be a good indicator of tomorrow's temperature, while the value of information about previous days may be negligible; therefore the Markov property is plausible. But seasonal fluctuations in temperature make the time-stationarity property a poor approximation over any period of time spanning more than one season.

Case 6.7. A model of an industrial robot defines the state of the system to be the task that the robot is performing, and the time index to be the number of tasks performed. The robot works on two different kinds of assemblies, each with its own distinct collection of tasks, and it requires a different tool for each kind of assembly. Changing the tool is one of the robot's tasks.

The Markov property may be reasonable if the different kinds of assemblies the robot encounters are randomly mixed. However, if each kind of assembly is produced in a batch of, say, 50 assemblies, then the robot will change tools every fiftieth assembly and the next state of the system will depend on more than just the current state; it also depends on how many assemblies of a particular kind have been produced. Time stationarity may be plausible if the mix of different assemblies does not change over time. Periodicity may arise because the tool-change state can only be entered after completing an assembly.

Case 6.8. A model of an auto insurance policy defines the state of the system to be whether or not the policyholder had an accident last year, and the time index to be the number of years. The insurance company believes that after having an accident, a policyholder is a more careful driver for the next 2 years.

The validity of the Markov property depends on how we define the state space. Since a policyholder is more careful for 2 years following an accident, the Markov property does not apply if the state space is simply whether or not the policyholder had an accident this year—$\mathcal{M} = \{1, 2\} \equiv \{$accident, no accident$\}$. However, if we expand the state space to be the accident history for the last 2 years—$\mathcal{M} = \{1, 2, 3, 4\} \equiv \{$(accident, accident), (accident, no accident), (no accident, accident), (no accident, no accident)$\}$—then the Markov property may apply. Clearly insurance companies do not believe that time stationarity holds generally since they charge premiums based on the age of the policyholder.

The Markov and time-stationarity properties are always approximations and therefore better for some modeling-and-analysis problems than others. These examples demonstrate two modeling principles: Conformance to the Markov property can sometimes be made better by redefining the state space of the process; and time stationarity is usually more appropriate over a restricted time period than over all time.

6.9   FINE POINTS

This chapter focused on Markov chain models in which the state space $\mathcal{M}$ is finite; that is, $m < \infty$. There are modeling and analysis problems in which there is no natural upper bound on the number of states (see Chapter 8 in which the state of the system is the number of customers waiting to be served in a queueing system). An infinite state space does not change any of the time-dependent performance measures in Section 6.6, but does introduce complications in the determination and interpretation of time-independent performance measures, such as $\vec{p}_{ii} = \lim_{n \to \infty} p_{ii}^{(n)}$ defined in Section 6.7. Although these complications rarely arise in practical modeling-and-analysis problems, we describe them in this section. For a thorough discussion see, for instance, Bhat (1984, Chapter 6) or Çinlar (1975, Chapter 6).

Suppose that $\mathcal{R}$ is an infinite but irreducible set of states with corresponding one-step transition matrix $\mathbf{P}_{\mathcal{R}\mathcal{R}}$. In other words, $\mathbf{P}_{\mathcal{R}\mathcal{R}}$ is a Markov chain, but no proper subset of the states in $\mathcal{R}$ forms a Markov chain. Then all states in $\mathcal{R}$ will have the same classification, either *positive recurrent*, *null recurrent* or *transient*. "Positive recurrent" matches our previous definition of recurrent: State $i \in \mathcal{R}$ is positive recurrent if $\vec{p}_{ii} > 0$. Both null recurrent and transient classifications of state i imply that $\vec{p}_{ii} = 0$. The difference between these two classifications is subtle: If j is transient, then the process will eventually leave state j never to return; if j is null recurrent, then the process will always return to state j, but the expected number of transitions between returns is infinite!

We determine which classification applies as follows:

1. If Equation (6.21) has a solution, then that solution will have all $\pi_j > 0$, the π_j can be interpreted as the steady-state probabilities, and all of the states in $\mathcal{R}$ are positive recurrent.
2. If Step 1 fails, then drop any one state, h, out of $\mathcal{R}$ to form a subset of states $\mathcal{R}^*$ and solve Equation (6.21) with $\mathbf{P}_{\mathcal{R}^*\mathcal{R}^*}$. If the only solution is $\pi_{\mathcal{R}^*} = \mathbf{0}$, then all states in $\mathcal{R}$ are null recurrent.
3. If Steps 1 and 2 fail, then all states in $\mathcal{R}$ are transient.

The intuition behind Step 2 is that $\pi_{\mathcal{R}^*} = \mathbf{0}$ implies that state h, which was dropped from $\mathcal{R}$, must be recurrent. Since all of the states in an irreducible set have the same classification, all of the states in $\mathcal{R}$ must also be recurrent (but null, since Step 1 failed).

Understanding the system being modeled should indicate to the analyst whether states are intended to be recurrent or transient. Although not a perfect rule, the existence of null-recurrent states usually signifies that the model has been incorrectly formulated, since it is difficult to conceive of a physical system that returns to a particular state with certainty, but at intervals that are expected to be infinitely long.

In a process with transient states $\mathcal{T}$ and two or more subsets of irreducible, recurrent states, $\mathcal{R}_1, \mathcal{R}_2, \ldots$, it may be of interest to know which set of recurrent states ultimately traps the process. Recall that the absorption probabilities α_{ij} are only defined if j is an absorbing state. To adapt the α_{ij} for this purpose, we first replace the states in $\mathcal{R}_1, \mathcal{R}_2, \ldots$ with a single absorbing state, say $r_1, r_2, \ldots$, that represents the entire subset and appropriately modify the transition probabilities in $\mathbf{P}$. The $\alpha_{ir_1}, \alpha_{ir_2}, \ldots$ for the modified process then apply to all states represented by the single absorbing state.

6.10   EXERCISES

6.1. Compute the probabilities of the customer sessions shown in Figure 6.1.

6.2. Determine the customer session generated by transition matrix (6.10) if `random()` returns the following sequence of values. Assume that the process starts in state 1 (log on).
(a) $0.18, 0.22, 0.87, 0.08, 0.93, \ldots$
(b) $0.96, 0.38, 0.02, 0.54, 0.15, 0.66, 0.85, \ldots$

6.3. Determine the sequence of parts generated by transition matrix (6.11) if `random()` returns the following sequence of values. Assume that the initial-state probabilities are $\mathbf{p'} = (0.990, 0.010)$.
(a) $0.187, 0.224, 0.870, 0.089, 0.937, \ldots$
(b) $0.992, 0.718, 0.026, 0.548, 0.152, 0.663, 0.999, \ldots$

6.4. For a Markov chain $\{S_n; n = 0, 1, \ldots\}$ with state space $\mathcal{M} = \{1, 2\}$ and one-step transition matrix

$$\mathbf{P} = \begin{pmatrix} 0.6 & 0.4 \\ 0.8 & 0.2 \end{pmatrix}$$

compute the following:
(a) $\Pr\{S_1 = 2 | S_0 = 1\}$
(b) $\Pr\{S_2 = 1 | S_0 = 1\}$
(c) $\Pr\{S_2 = 1, S_1 = 1 | S_0 = 2\}$

6.5. For the Markov chain with state space $\mathcal{M} = \{1, 2, 3, 4\}$ and one-step transition matrix

$$\mathbf{P} = \begin{pmatrix} 0.2 & 0.8 & 0 & 0 \\ 0 & 0 & 1 & 0 \\ 0 & 0.5 & 0.4 & 0.1 \\ 0 & 1 & 0 & 0 \end{pmatrix}$$

draw the transition diagram and classify the states as recurrent or transient.

6.6. Classify as recurrent or transient the states of the Markov chains with state space $\{1, 2, 3, 4, 5\}$ and the one-step transition matrices below by first finding all of the irreducible subsets of states. Draw a transition diagram to assist you.
(a)

$$\mathbf{P} = \begin{pmatrix} 0.1 & 0.3 & 0.4 & 0.0 & 0.2 \\ 0.5 & 0.1 & 0.1 & 0.0 & 0.3 \\ 0.8 & 0.0 & 0.0 & 0.2 & 0.0 \\ 0.0 & 0.1 & 0.0 & 0.9 & 0.0 \\ 0.3 & 0.1 & 0.1 & 0.0 & 0.5 \end{pmatrix}$$

(b)

$$\mathbf{P} = \begin{pmatrix} 0.1 & 0.5 & 0.1 & 0.1 & 0.2 \\ 0.0 & 0.8 & 0.0 & 0.0 & 0.2 \\ 0.0 & 0.0 & 0.3 & 0.0 & 0.7 \\ 0.0 & 0.0 & 1.0 & 0.0 & 0.0 \\ 0.0 & 0.5 & 0.0 & 0.0 & 0.5 \end{pmatrix}$$

(c)

$$P = \begin{pmatrix} 0.1 & 0.3 & 0.1 & 0.3 & 0.2 \\ 0.0 & 0.8 & 0.0 & 0.0 & 0.2 \\ 0.0 & 0.0 & 0.3 & 0.4 & 0.3 \\ 0.0 & 1.0 & 0.0 & 0.0 & 0.0 \\ 0.1 & 0.5 & 0.0 & 0.4 & 0.0 \end{pmatrix}$$

6.7. For the Markov chains defined in Exercise 6, do the following:
   (a) Compute the n-step transition matrix $\mathbf{P}^{(n)}$ for $n = 2, 5, 20$.
   (b) Compute the steady-state probabilities for each set of irreducible, recurrent states. Compare π_i to $p_{ii}^{(n)}$ for $n = 2, 5, 20$ for each recurrent state i.

6.8. Consider the Markov chain with state space $\mathcal{M} = \{1, 2, 3, 4, 5, 6\}$ and one-step transition matrix

$$P = \begin{pmatrix} 0 & 0.5 & 0 & 0.4 & 0.1 & 0 \\ 0 & 1 & 0 & 0 & 0 & 0 \\ 0.2 & 0 & 0.4 & 0.3 & 0 & 0.1 \\ 0 & 0.9 & 0 & 0 & 0.1 & 0 \\ 0 & 0 & 0 & 0 & 1 & 0 \\ 0 & 0 & 0 & 0 & 0 & 1 \end{pmatrix}$$

If the process starts in state 3, what is the probability it will be absorbed in state 2? In state 5? In state 6?

6.9. Consider the Markov chain with state space $\mathcal{M} = \{1, 2, 3, 4, 5\}$ and one-step transition matrix

$$P = \begin{pmatrix} 0 & 1 & 0 & 0 & 0 \\ 0 & 0 & 0.5 & 0.5 & 0 \\ 1 & 0 & 0 & 0 & 0 \\ 0 & 0 & 0 & 0 & 1 \\ 0 & 1 & 0 & 0 & 0 \end{pmatrix}$$

The states of this Markov chain are periodic with period 3. Draw the transition diagram. Then simulate the chain for 30 transitions with initial state 1, and plot the sequence of states to see the effect of periodicity.

6.10. On February 23, 1994, All-American basketball player Glenn Robinson of Purdue University scored 40 points in a game against Ohio State University. He shot the ball 45 times, scoring on 30 of them. The sequence of made and missed shots was as follows: M, 1, 0, 0, 0, 0, 2, 1, 1, 2, 0, 2, 1, 1, 1, 1, 0, 0, 2, 1, 1, 1, 0, 2, 2, 2, 0, 2, 1, 0, 1, 1, 2, 2, 1, 1, 0, 0, 1, 1, 1, M, 0, 1, 1 (a "2" corresponds to a score on a shot from the floor; a "1" corresponds to a score on a free throw; a "0" corresponds to a missed shot from the floor; and an "M" corresponds to a missed free throw).

There is a long-standing debate in basketball about the existence of the so-called hot hand. Proponents of the "hot hand" claim that players experience periods of time when they "just can't miss." There is a similar theory of the "cold hand" which has the obvious meaning. Others argue that there is no "hot" or "cold hand," just the natural runs of made and missed shots that would be expected in a random process.

(a) Develop a model of Glenn Robinson's shooting that could be used to predict his sequence of made and missed shots in n attempts if there is *no* such thing as a "hot hand." Use the data given above to parameterize your model, and include a standard error for any parameter you estimate.

(b) Develop a Markov chain model of Glenn Robinson's shooting that could be used to predict his sequence of made and missed shots in n attempts if there *is* such thing as a "hot hand," where the next shot depends on the last shot. Use the data given above to parameterize your model, and include a standard error for any parameter you estimate.

(c) Develop a Markov chain model where the next shot depends on the last *two* shots. Use the data given above to parameterize your model, and include a standard error for any parameter you estimate.

(d) In the Purdue vs. Ohio State game the longest run of shots Glenn Robinson missed was 4. How likely is a run of this length under each of your proposed models? Does this make you favor one model over the other? How might you test which model is best?

6.11. A food manufacturer plans to introduce a new potato chip, Box O' Spuds, into a local market that already has three strong competitors. Because the manufacturer makes a superior product, the marketing analysts believe that brand loyalty to Box O' Spuds will be high after people try it. However, it will be difficult to persuade people to switch from the three established brands. The marketing analysts would like to forecast the long-term market share for Box O' Spuds to determine whether it is worth entering the market.

Suppose the marketing analysts formulate a Markov chain model of customer brand switching in which the state space $\mathcal{M} = \{1, 2, 3, 4\}$ corresponds to which of the three established brands or Box O' Spuds, respectively, that a customer currently purchases. The time index is the number of bags of chips purchased. Based on market research and experience with other products, the one-step transition matrix the marketing analysts anticipate is

$$\mathbf{P} = \begin{pmatrix} 0.70 & 0.14 & 0.14 & 0.02 \\ 0.14 & 0.70 & 0.14 & 0.02 \\ 0.14 & 0.14 & 0.70 & 0.02 \\ 0.05 & 0.05 & 0.05 & 0.85 \end{pmatrix}$$

What is the long-term market share for Box O' Spuds?

6.12. Todd and Diane plan to take out a mortgage from Humorous Money Company to purchase a home. The current fixed-interest rate on a 30-year loan is 8%. For no additional service charge Humorous Money will allow Todd and Diane to "lock in" the current rate of 8% (option 1). For a service charge of $50 they can take whatever the current rate happens to be in 4 weeks (option 2). For a service charge of $100 Humorous Money will let the rate drift for 4 weeks but lock it in if it hits 8.2% or 7.8% (option 3). The sales agent says that option 3 protects Todd and Diane from getting stuck with a very high rate while giving them a chance for a lower rate.

Interest rates can change weekly, and Diane believes that interest rates are more likely to go down than up. For the next 4 weeks she believes that the following model is a good representation of how the interest rate will change each week: The rate will go up 0.1% with probability 0.3, go down 0.1% with probability 0.4, and stay the same with probability 0.3. Under this model, help Todd and Diane decide which option is best for them.

6.13. Find a building with at least five floors that has an elevator with a floor indicator so that you can observe the elevator's movement from floor to floor. Define the state of your system to be the floor on which the elevator is currently stopped, and the time index to be the number of stops. Observe this system for a substantial period of time, and record the sample path (sequence of floors on which it stops).

(a) Suppose that the movement from floor to floor is well modeled as a Markov chain, so that p_{ij} is the probability that an elevator currently stopped on floor i next stops on floor j. Estimate the one-step transition matrix, and provide a standard error for each estimate. Do you think that the Markov property is appropriate? Why or why not?

(b) To test the Markov property, select two intermediate floors j and k (for example floors $j = 2$ and $k = 3$ in a five-story building). Let $p_{ijk} \equiv \Pr\{S_{n+1} = k | S_n = j, S_{n-1} = i\}$. Let n_{ijk} be the number of times you observed a stop on i followed by a stop on j followed by a stop on floor k, for all possible floors i. Estimate p_{ijk} by

$$\widehat{p}_{ijk} = \frac{n_{ijk}}{\sum_k n_{ijk}}$$

If the Markov property applies, then $p_{ijk} = p_{jk}$ for all i. See if your estimates approximately satisfy this property. If they do not, then the Markov property may not be appropriate. The same test could be applied to each combination j, k.

(c) Suppose that the movement of the elevator depends only on the last two floors visited. Formulate a Markov chain model and estimate the transition probabilities from your data.

6.14. Define Markov chain models for Cases 6.3–6.8 by specifying the state space in detail and providing a one-step transition matrix using appropriate notation to stand in for the unknown probabilities.

6.15. (Continuation of Case 6.4.) The standard typewriter or computer keyboard, known as the QWERTY keyboard, was originally designed to slow typing so that early mechanical typewriter keys would not jam. Of course, there is no such problem today. Thus, new keyboard layouts, such as the Dvorak keyboard, have been designed to increase typing speed. These keyboards are designed to reduce the number of times your fingers must leave the home row by putting frequently used letters on the home row. The home row is ASDFGHJKL; on the QWERTY keyboard. The Dvorak keyboard also considers the sequence of keys that will be typed, so that keys are grouped together that often follow each other.

Develop a Markov chain model of the QWERTY keyboard. Let the state of the system be the current key being typed, and the time index the number of keys typed. To simplify this exercise, aggregate the keys into six sets: (ZXCVB), (NM, . /), (ASDFG), (HJKL; '), (QWERT), (YUIOP). Ignore other symbols, and count the upper and lowercase (or shift, not shift) as the same key.

Assuming that a Markov chain model is reasonable, pick a nontechnical book (since you are ignoring numbers and special symbols) and estimate the one-step transition matrix P for this model. Collect a long enough sequence of letters so that you get at least five transitions from each state to every other state (that is, all $c_{ij} > 5$). Include a standard error for each estimate.

6.16. The quality control engineers at KRN Corporation have developed a Markov chain model of their manufacturing system when it is performing improperly and should be adjusted (see Case 6.2). The one-step transition matrix is

$$P = \begin{pmatrix} 0.80 & 0.20 \\ 0.60 & 0.40 \end{pmatrix}$$

Recall that the state space $\mathcal{M} = \{1, 2\}$ corresponds to good and defective components.

(a) Compute the long-run fraction of defective components for this model, π_2.

(b) Write a simulation of this process, and record the fraction of defective components out of the first n components produced, when the initial component is acceptable. Experiment with letting $n = 10, 100, 1000$, and observe how your estimate approaches π_2.

(c) For this transition matrix and the one given in (6.11), compute the probability that no more than 1 defective component is discovered in the first 15 components. Assume that $p = \pi$ for each model (that is, the initial-state distribution is the same as the steady-state distribution). Compare these probabilities to the corresponding probabilities for the binomial-distribution model. (Hint: No single formula presented in this chapter will solve the problem. Think in terms of all the possible sample paths that satisfy the condition.)

6.17. An automated guided vehicle (AGV) transports parts between four locations: a release station A, machining station B, machining station C, and an output buffer D. The movement of the AGV can be described as making trips from location to location based on requests to move parts. More specifically:

- If the AGV is at the output buffer, it is equally likely to move next to any of the other three locations, A, B, C.
- If the AGV is at the release station, it is equally likely to move next to machining station B or C.
- If the AGV is waiting at either of the machining stations, it is equally likely to move to any of the other three locations.

(a) Develop a model of this system that is capable of answering the questions below. Be sure to define the states, time index, and one-step transition matrix.

(*b*) If the AGV is currently at machining station B, what is the probability that it will be at the output buffer after five trips?

(*c*) What is the long-run fraction of time the AGV spends traveling to the output buffer?

6.18. In a study of the effect of advertising on brand shifting in the U.S. brewing industry,[4] a Markov chain model was estimated to represent the chance of consumers shifting preferences among Anheuser-Busch (state 1), Miller (state 2) and "other" (state 3) beers from 1978 to 1979. The one-step transition matrix is

$$\mathbf{P} = \begin{pmatrix} 0.9950 & 0.0046 & 0.0004 \\ 0.0000 & 0.9971 & 0.0029 \\ 0.0084 & 0.0522 & 0.9394 \end{pmatrix}$$

Clearly brand loyalty is quite strong since the diagonal elements are all near 1.

Suppose that there were 6 million beer drinkers in 1978, divided equally among the three brands. Answer the following questions:

(*a*) How many consumers are expected to prefer Miller products in 1979?

(*b*) If the transition matrix does not change over time (typically it does) and the number of beer drinkers is unchanged (certainly not true), how many consumers are expected to prefer Anheuser-Busch products today?

6.19. Suppose that there is a prison that will be closed in m months, or as soon as all of the prisoners currently there are granted parole, whichever comes first. In each month the probability that each prisoner is paroled is p, independent of the other prisoners and independent of how long the prisoner has been there. Since the prison is to be closed, no new prisoners are admitted. Suppose that there are currently k prisoners. We want to determine how long it will take the prison to become empty.

(*a*) Derive a Markov chain model capable of answering the questions below. Be sure to define your state space, time index, and one-step transition matrix.

(*b*) What would you compute to determine the probability that the prison is closed before m months? Set up the calculation in general; then compute a numerical result for the case $m = 12$, $p = 0.1$ and $k = 6$.

(*c*) What would you compute to determine the expected number of prisoners in the prison at the end of m months? Set up the calculation in general; then compute a numerical result for the case $m = 12$, $p = 0.1$ and $k = 6$.

(*d*) Suppose that k is large, say $k = 2000$. A very large transition matrix will be required to represent this case, either for analysis or simulation. Find a way either to calculate the probability that the prison is closed before m months or to estimate it via simulation that does not require explicitly working with the entire transition matrix.

---

[4]C. M. L. Kelton and W. D. Kelton. 1982. Advertising and intraindustry brand shift in the U.S. brewing industry. *The Journal of Industrial Economics* **30**, 293–303.

6.20. Suppose that the quality of successive items is described by the two-state Markov chain described in Case 6.2 with transition matrix given by (6.11) and that each item is inspected and defective items are removed. However, there is a chance that a defective item will be missclassified; that is, an item will be declared good when it is actually defective. Suppose that the probability that a defective item will be missclassified as good is 0.06, independent of what happened to previous items.

    (a) Develop another Markov chain model that is capable of answering the question below. Be sure to define your states, time index, and one-step transition matrix.

    (b) What is the long-run fraction of items that pass inspection but are actually defective?

6.21. An auto insurance company charges customers a yearly premium based on their accident history. A customer who has had no accident during the past 2 years is charged $500 per year. A customer who has had an accident in only one of the past 2 years is charged $625. A customer who has had accidents in both of the past 2 years is charged $800.

    Actuarial records show that a customer who had an accident during the past year has a 0.03 chance of having an accident during the current year, while a customer who did not have an accident during the past year has a 0.08 chance of having an accident during the current year.

    (a) Derive a Markov chain model capable of answering the questions below. Be sure to define your states, time index, and one-step transition matrix.

    (b) Over the long run, what fraction of the company's customers are accident free during the previous 2 years?

    (c) If the company has 2000 customers, what is its long-run expected revenue from premiums each year?

6.22. Formally derive the result $\mathbf{F}_{AB}^{(n)} = \mathbf{P}_{AA}^{n-1}\mathbf{P}_{AB}$ for $n = 1, 2, \ldots$.

6.23. For a Markov chain with finite state space $\mathcal{M}$ and one-step transition matrix $\mathbf{P}$, let $\mathcal{T}$ be the set of transient states (assume that there are $m_{\mathcal{T}} > 0$ transient states). Let $\mathbf{M}$ be the $m_{\mathcal{T}} \times m_{\mathcal{T}}$ matrix with elements μ_{ij} denoting the expected number of times the process is in transient state j given $\{S_0 = i\}$. Show that

$$\mathbf{M} = (\mathbf{I} - \mathbf{P}_{\mathcal{T}\mathcal{T}})^{-1}$$

The matrix $\mathbf{M}$ is sometimes called the *fundamental matrix*. Hint: First establish that

$$\mu_{ij} = \mathcal{I}(i = j) + \sum_{n=1}^{\infty} p_{ij}^{(n)}$$

where $\mathcal{I}(i = j)$ is the indicator function taking the value 1 if $i = j$ and 0 otherwise.

6.24. Use the result in Exercise 23 to calculate the expected session length for the customer behavior model, Case 6.1.

6.25. For a Markov chain that consists of a single, irreducible set of m recurrent states and one-step transition matrix $\mathbf{P}$, let V_{ij} be the number of transitions for the process to move from state i to state j for the first time. Let $v_{ij} = E[V_{ij}]$ be the expected number of transitions to move from state i to j. Define the $m \times m$ matrix $\mathbf{V}$ with elements v_{ij} and the $m \times m$ diagonal matrix $\mathbf{D}$ with diagonal elements v_{ii}. Finally, let $\mathbf{1}$ be an $m \times 1$ vector of all 1s.

(a) Show that

$$\mathbf{V} = \mathbf{P}(\mathbf{V} - \mathbf{D}) + \mathbf{11}'$$

Hint: Start by showing that

$$v_{ij} = p_{ij} + \sum_{h \neq j}(1 + v_{hj})p_{ih}$$

(b) Use the previous result to show that

$$v_{ii} = \frac{1}{\pi_i}$$

In other words, the expected time between returns to recurrent state i is the reciprocal of the steady-state probability of being in state i. Hint: Multiply the previous result by π'.

6.26. A children's game called HI HO! CHERRY-O by Golden is played by two to four players.[5] Each player initially has 10 cherries, and the goal of the game is to be the first player to get all 10 cherries into a basket. Players take turns spinning a spinner that determines how many cherries they add or subtract from their basket. The actual spinner has pictures, but it is equivalent to Figure 6.6. The outcome $-\infty$ means remove all cherries currently in your basket; the outcome -2 means remove min{2, all} cherries from your basket.

Suppose that Golden wants to design a version of the game that takes less time to play. Your job is to analyze the distribution of the playing time and redesign the spinner.

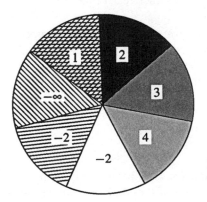

FIGURE 6.6
Spinner for HI HO! CHERRY-O. All sectors are equal sized.

Compute the distribution of the number of spins required for a single player to fill a basket and its expected value (see Exercise 23 for some help). Revise the spinner to reduce this expected value by 10%.

Advanced exercise: Compute the distribution of the total number of spins to complete the game when there are two players and also its expected value. Is this expected value smaller, larger or the same as two times the expected number of spins for a single player to fill a basket? Can you give an intuitive explanation for the result?

6.27. In this problem the goal is to model the use of automated teller machines (ATMs) at a bank.

After inserting a card into an ATM a customer may perform three types of transactions: deposit, withdrawal, and obtain account information. The bank believes that 50% of all customers start with a withdrawal, 40% start with a deposit, and the remainder start by requesting account information.

After completing a transaction, 90% of the customers complete their business (obtain their card and leave); those who do not complete their business are equally likely to select one of the other two types of transactions (for instance, if they just made a withdrawal and they do not complete their business, then they are equally likely next to select a deposit or request account information). This pattern continues until their business finally is completed.

(a) Derive a Markov chain model capable of answering the questions below. Be sure to define your state space, time index, and one-step transition matrix.

(b) Evaluate the Markov and stationarity properties for this situation. Do you think they are appropriate? Why or why not?

(c) Including inserting the card as a transaction, what is the probability distribution of the number of transactions that a customer performs on an ATM? Carry the calculation out to $n = 20$ transactions.

(d) If customers withdraw \$100 on each withdrawal they make, what is the expected amount of money withdrawn by customers each time they use an ATM? See Exercise 23 for some help.

6.28. Section 3.6, Exercise 17, introduced the *geometric distribution* with parameter $0 < \gamma < 1$. This distribution can be interpreted as the number of trials until the first success when the trials are independent and have a common success probability γ. If H has a geometric distribution, then its mass function is

$$f_H(a) = \begin{cases} (1 - \gamma)^{a-1}\gamma, & a = 1, 2, 3, \ldots \\ 0, & \text{otherwise} \end{cases}$$

implying that $E[H] = 1/\gamma$ and $Var[H] = (1 - \gamma)/\gamma^2$.

Suppose that a Markov chain enters state i, a state for which $0 < p_{ii} < 1$. The *holding time*, H, is the number of transitions of the Markov chain until the process leaves state i. Use an induction proof to show that the holding time has a geometric distribution with parameter $\gamma = 1 - p_{ii}$.

6.29. Define $f_{jj}^{(n)} \equiv \Pr\{S_n = j, S_{n-1} \in \mathcal{A}, S_{n-2} \in \mathcal{A}, \ldots, S_1 \in \mathcal{A} | S_0 = j\}$, for initial and final state $j \in \mathcal{B}$ and two disjoint subsets of states $\mathcal{A}$ and $\mathcal{B}$. This is the probability that the process returns to state j at time step n after being confined to states in $\mathcal{A}$ for the first $n-1$ time steps. These quantities are called *first-return probabilities*.

Let $\mathbf{R}_{\mathcal{BB}}^{(n)}$ be the $m_{\mathcal{B}} \times m_{\mathcal{B}}$ matrix with elements $f_{jj}^{(n)}$ on the diagonal (we will not interpret the off-diagonal elements). Show that

$$\mathbf{R}_{\mathcal{BB}}^{(n)} = \begin{cases} \mathbf{P}_{\mathcal{BB}}, & n = 1 \\ \mathbf{P}_{\mathcal{BA}}\mathbf{P}_{\mathcal{AB}}, & n = 2 \\ \mathbf{P}_{\mathcal{BA}}\mathbf{P}_{\mathcal{AA}}^{n-2}\mathbf{P}_{\mathcal{AB}}, & n = 3, 4, \dots \end{cases}$$

Now let $\mathcal{A} = \{1, 3, 4\}$ and $\mathcal{B} = \{2\}$ in the customer behavior model (Case 6.1). Compute $f_{22}^{(5)}$, the probability that a customer, whose last transaction was a fetch, requests the next fetch after requesting four other transactions.

6.30. An (r, s) inventory system involves the periodic review of the level of inventory of some discrete unit. If the inventory position at a review is found to be below r units, then enough additional units are ordered to bring the inventory position up to s units. When the inventory position at a review is found to be above r units, no additional units are ordered. One possible goal is to select the values of r and s that minimize inventory cost.

Let $\{S_n; n = 0, 1, 2, \dots\}$ represent the inventory position just after a review at period n, and let $\{X_n; n = 0, 1, 2, \dots\}$ represent the demand for units of inventory in period t. Suppose the inventory position S_n changes in the following way:

$$S_{n+1} = \begin{cases} s, & S_n - X_n < r \\ S_n - X_n, & S_n - X_n \geq r \end{cases}$$

The initial inventory position is s, so that the inventory position is initially at its maximum. The demand $\{X_n; n = 0, 1, 2, \dots\}$ is modeled as a sequence of independent, time-stationary Poisson random variables with mean 2 units.

Determine which of the following policies has the largest long-run expected inventory position (and therefore is most costly in terms of the cost of holding inventory).

policy	r	s
A	4	10
B	3	12

6.31. Formally show that Equation (6.5) is a consequence of the Markov property. Formally show that Equation (6.6) is a consequence of the time-stationarity property.

6.32. Use the Markov and time-stationarity properties to derive the Chapman-Kolmogorov equation, Equation (6.18).

6.33. Read the article by R. A. Russell and R. Hickle. 1986. Simulation of a CD portfolio. *Interfaces* **16**, 49–54. Describe how you could develop a Markov chain model to replace their simulation.

CONTINUOUS-TIME PROCESSES

In Chapter 6 we analyzed the state changes of a stochastic process, ignoring the actual times at which the changes occur. The state-change process alone is sufficient to answer many questions about the performance of a system, but not all of them. For example, in a model of a hospital emergency room that defines the state of the system to be the number of patients waiting for a doctor, it is important to know how many patients are waiting for a doctor and also *how long* they wait. This chapter and Chapter 8 present methods for modeling, analysis and simulation of continuous-time stochastic processes. The results we obtain for continuous-time processes build on those we obtained for discrete-time Markov chains in Chapter 6.

7.1 THE CASE OF THE SOFTWARE SELLOUT

Case 7.1. A mail-order software distributor, Floppy Express, must manage the inventory of software products that it sells. Currently management decisions are based on experience, but a consultant has suggested that modeling the inventory system could aid Floppy Express in making better decisions. The operations manager of Floppy Express has agreed to allow the consultant to develop a model for one popular word-processing package, HyperWord, to demonstrate the value of modeling.

Floppy Express receives requests for software via telephone, e-mail and fax, and although HyperWord has a relatively consistent level of demand, the arrival of requests cannot be predicted with certainty. An additional complication is that the time required to obtain copies of HyperWord from the manufacturer is subject to some variation.

There are two fundamental questions in most inventory problems: "When should more items be ordered?" and "How many items should be ordered when an order is placed?" The answers depend on the demand for items, in this case software. The mail-order software business is very competitive, so if a customer wants to order HyperWord but Floppy Express does not have it, then the customer will simply order it from some other distributor. Therefore, it is important to have enough copies in inventory. On the other hand, new versions of HyperWord are released periodically, making old versions obsolete. There is also a cost to Floppy Express of having its money and storage space committed to copies of HyperWord. Therefore, it is also important not to have too many copies in inventory. One role of a model is to help balance the risk of losing sales against the cost of maintaining inventory.

7.2 NOTATION AND REVIEW

This chapter makes extensive use of the results in Chapter 6, particularly the computation of n-step and steady-state probabilities for a Markov chain with one-step transition matrix $\mathbf{P}$. We review these results in Section 7.2.1 for readers who either skipped Chapter 6 or who want a review of discrete-time Markov chains. We also exploit certain properties of the geometric and the exponential distributions. Those properties are reviewed in Section 7.2.2.

7.2.1 Markov Chain Review

A discrete-time Markov chain is a stochastic process $\{S_n; n = 0, 1, 2, \ldots\}$ with discrete state space $\mathcal{M}$. For convenience, let $\mathcal{M}$ be $\{1, 2, \ldots\}$, or a subset of it. The defining characteristic of a Markov chain is that there exist probabilities p_{ij} such that

$$p_{ij} = \Pr\{S_{n+1} = j \mid S_n = i, S_{n-1} = a, \ldots, S_0 = z\}$$

for all i, j in $\mathcal{M}$ and all values of n. In other words, if $S_0, S_1, \ldots, S_n$ represents the sample path of the process up to time n and this path is known, then the probability distribution of the next state of the process S_{n+1} depends only on the last state S_n, and not on other previous states or the particular value of n. The quantities p_{ij} are called *one-step transition probabilities*, and they can be organized into a one-step transition matrix

$$\mathbf{P} \equiv \begin{pmatrix} p_{11} & p_{12} & \cdots & p_{1m} \\ p_{21} & p_{22} & \cdots & p_{2m} \\ \vdots & \vdots & \vdots & \vdots \\ p_{m1} & p_{m2} & \cdots & p_{mm} \end{pmatrix}$$

The rows of this matrix represent the last state of the process, while the columns represent the next state of the process. Each row of this matrix will sum to 1 since the process must go somewhere on each time step.

Here is a small example from Chapter 6: A sequence of components is produced and each component may be acceptable or defective. Let the state space $\mathcal{M} = \{1, 2\}$ represent acceptable and defective, respectively, so that S_n represents the quality of the $(n + 1)$st component produced (including the 0th component, represented by S_0). If the components are inspected in the order that they are produced, then the quality of successive components may be dependent. Therefore, a Markov chain model such as

$$\mathbf{P} = \begin{pmatrix} 0.995 & 0.005 \\ 0.495 & 0.505 \end{pmatrix}$$

might be appropriate. We interpret $p_{12} = 0.005$ to mean that if the last component produced was acceptable (state 1), then the probability that the next component produced is defective (state 2) is 0.005. Similarly, $p_{22} = 0.505$ is the probability that a defective component will follow a defective component. Being a Markov chain implies that these probabilities do not depend on the quality of other, previous components or on how many components have been produced.

The idea of a one-step transition probability can be extended to define an n-step transition probability $p_{ij}^{(n)} \equiv \Pr\{S_n = j | S_0 = i\}$. Let $\mathbf{P}^{(n)}$ be the $m \times m$ matrix with elements $p_{ij}^{(n)}$. In Chapter 6 we showed that

$$\mathbf{P}^{(n)} = \mathbf{P}^n, \; n = 1, 2, \ldots$$

In words, the n-step transition matrix is obtained by raising the one-step transition matrix to the nth power.

For example, to obtain $p_{12}^{(3)}$, the probability that the fourth component is defective given that the first component is not (remember that we start time at $n = 0$), we compute

$$\mathbf{P}^{(3)} = \begin{pmatrix} p_{11}^{(3)} & p_{12}^{(3)} \\ p_{21}^{(3)} & p_{22}^{(3)} \end{pmatrix}$$

$$= \mathbf{P}^3$$

$$= \begin{pmatrix} 0.995 & 0.005 \\ 0.495 & 0.505 \end{pmatrix}^3$$

$$= \begin{pmatrix} 0.99125 & 0.00875 \\ 0.86625 & 0.13375 \end{pmatrix}$$

Therefore, $p_{12}^{(3)} = 0.00875$.

For some Markov chains $\pi_j \equiv \lim_{n \to \infty} p_{ij}^{(n)}$ exists and is independent of the initial state i. The π_j are called *steady-state probabilities*, and they represent the limiting probability of finding the process in state j as time goes to infinity or, equivalently, the long-run fraction of the time the process spends in state j. The key condition for their existence is that the Markov chain is *irreducible*, meaning that no subset of its state space also forms a Markov chain.

Let $\pi' = (\pi_1, \pi_2, \ldots, \pi_m)$. Then π can be obtained by solving the system of equations

$$\pi' = \pi'\mathbf{P}$$
$$1 = \pi'\mathbf{1}$$

In the production example,

$$(\pi_1, \pi_2) = \pi'\mathbf{P}$$

$$= (\pi_1, \pi_2) \begin{pmatrix} 0.995 & 0.005 \\ 0.495 & 0.505 \end{pmatrix}$$

and

$$1 = \pi'\mathbf{1}$$

$$= (\pi_1, \pi_2) \begin{pmatrix} 1 \\ 1 \end{pmatrix}$$

The complete system of equations is therefore

$$\pi_1 = 0.995\pi_1 + 0.495\pi_2$$
$$\pi_2 = 0.005\pi_1 + 0.505\pi_2$$
$$1 = \pi_1 + \pi_2$$

The solution is $\pi_1 = 0.99$ and $\pi_2 = 0.01$. The steady-state probability π_2 can be interpreted as the long-run fraction of components that are defective or as the limiting probability that a component is defective.

In this chapter we study stochastic processes $\{Y_t; t \geq 0\}$ that may change state at any point in continuous time t. However, we will find that the *sequence* of state changes of Y_t, ignoring the actual times that they occur, form a Markov chain.

7.2.2 Properties of the Exponential and Geometric Distributions

Let X be a geometrically distributed random variable with parameter γ, which implies that $E[X] = 1/\gamma$. Recall that γ is interpreted as the probability that an independent trial is a success, so that $1/\gamma$ is the expected number of independent trials required to obtain the first success, including the successful trial. The cdf of X is

$$F_X(a) = \begin{cases} 0, & a < 1 \\ 1 - (1 - \gamma)^{\lfloor a \rfloor}, & 1 \leq a \end{cases}$$

with associated mass function $f_X(a) = \gamma(1 - \gamma)^{a-1}$ for $a = 1, 2, \ldots$.

Let Z be an exponentially distributed random variable with parameter λ, which implies that $E[X] = 1/\lambda$. The cdf of Z is

$$F_Z(b) = \begin{cases} 0, & b < 0 \\ 1 - e^{-\lambda b}, & 0 \leq b \end{cases}$$

with associated density function $f_Z(b) = \lambda e^{-\lambda b}$ for $b \geq 0$.

Later in the chapter we will need to evaluate probability statements of the form

$$\Pr\{X > n + a | X > n\} \text{ and } \Pr\{Z > t + b | Z > t\}$$

In words, when we know that X (respectively, Z) is greater than n (respectively, t), what is the probability that X (respectively, Z) is at least an additional a (respectively, b) larger?

By directly applying the definition of conditional probability, we can show that the

$$\Pr\{X > n + a | X > n\} = \Pr\{X > a\} = (1 - \gamma)^a = 1 - F_X(a) \tag{7.1}$$

for $a \geq 0$, and the

$$\Pr\{Z > t + b | Z > t\} = \Pr\{Z > b\} = e^{-\lambda b} = 1 - F_Z(b) \tag{7.2}$$

for $b \geq 0$ (you are asked to derive these results in Exercise 6). This is the so-called *memoryless property* of the geometric and exponential distributions.

For the geometric distribution, the memoryless property implies that when a success has not occurred during the first n trials, the probability distribution of the number of remaining trials until the first success has the same geometric distribution as when we first started the trials. This makes sense because the additional trials are independent of the first n trials. If Z represents the time that we wait for something to occur, then the memoryless property of the exponential distribution implies that when the "something" has not occurred by time t, the probability distribution of the remaining time until it does occur has the same exponential distribution as when we first started waiting for it to occur. This property was critical to the analysis of the Poisson arrival-counting process in Chapter 5. The geometric and exponential distributions are the unique discrete and continuous distributions, respectively, that have the memoryless property.

Now let $Z_1, Z_2, \ldots, Z_k$ be independent, exponentially distributed random variables with respective parameters $\lambda_1, \lambda_2, \ldots, \lambda_k$, and let $H = \min\{Z_1, Z_2, \ldots, Z_k\}$. The cdf of H is

$$
\begin{aligned}
F_H(t) &\equiv \Pr\{H \leq t\} \\
&= 1 - \Pr\{H > t\} \\
&= 1 - \Pr\{Z_1 > t, Z_2 > t, \ldots, Z_k > t\} \tag{7.3} \\
&= 1 - \Pr\{Z_1 > t\} \Pr\{Z_2 > t\} \cdots \Pr\{Z_k > t\} \tag{7.4} \\
&= 1 - e^{-\lambda_1 t} e^{-\lambda_2 t} \cdots e^{-\lambda_k t} \\
&= 1 - e^{-\left(\sum_{i=1}^{k} \lambda_i\right)t} = 1 - e^{-\lambda_H t} \tag{7.5}
\end{aligned}
$$

where $\lambda_H \equiv \sum_{i=1}^{k} \lambda_i$. Equation (7.3) follows because each Z_i individually must be greater than t for the minimum to be greater than t; and Equation (7.4) follows from the independence of the Z_i. The final result, Equation (7.5), establishes that the minimum of k independent, exponentially distributed random variables is also exponentially distributed with parameter λ_H, the sum of the individual parameters. Later in the chapter we let $Z_1, Z_2, \ldots, Z_k$ represent the time remaining until k different system events occur, so that H is the remaining time until the *next* system event occurs.

One of the Z_i will be the smallest one. The probability that it is Z_1 is

$$\Pr\{Z_1 \leq Z_2, Z_1 \leq Z_3, \ldots, Z_1 \leq Z_k\} = \Pr\{Z_1 \leq \min(Z_2, Z_3, \ldots, Z_k)\}$$

$$= \Pr\left\{Z_1 \leq \min_{j \neq 1} Z_j\right\}$$

Let $M = \min_{j \neq 1} Z_j$, the minimum of the Z_j excluding Z_1. We know from Equation (7.5) that M is exponentially distributed with parameter $\lambda_M = \sum_{i=2}^{k} \lambda_i$, and it is independent of Z_1 by definition. By conditioning on the value of M and applying the law of total probability, we have

$$\Pr\left\{Z_1 \leq \min_{j \neq 1} Z_j\right\} = \Pr\{Z_1 \leq M\}$$

$$= \int_0^\infty \Pr\{Z_1 \leq M | M = t\} \lambda_M e^{-\lambda_M t}\, dt$$

$$= \int_0^\infty \Pr\{Z_1 \leq t\} \lambda_M e^{-\lambda_M t}\, dt$$

$$= \int_0^\infty \left(1 - e^{-\lambda_1 t}\right) \lambda_M e^{-\lambda_M t}\, dt$$

$$= \int_0^\infty \lambda_M e^{-\lambda_M t}\, dt - \int_0^\infty e^{-\lambda_1 t} \lambda_M e^{-\lambda_M t}\, dt$$

$$= 1 - \frac{\lambda_M}{\lambda_1 + \lambda_M} \int_0^\infty (\lambda_1 + \lambda_M) e^{-(\lambda_1 + \lambda_M) t}\, dt$$

$$= 1 - \frac{\lambda_M}{\lambda_1 + \lambda_M}$$

$$= \frac{\lambda_1}{\sum_{j=1}^{k} \lambda_j}$$

where we use the fact that both $\int_0^\infty \lambda_M e^{-\lambda_M t}\, dt = 1$ and $\int_0^\infty (\lambda_1 + \lambda_M) e^{-(\lambda_1 + \lambda_M) t}\, dt = 1$, since they are integrals of the density functions of exponential distributions over their entire domain. There is nothing special about Z_1, so the result generalizes to any of the Z_i:

$$\Pr\left\{Z_i \leq \min_{j \neq i} Z_j\right\} = \frac{\lambda_i}{\sum_{j=1}^{k} \lambda_j} \tag{7.6}$$

for $i = 1, 2, \ldots, k$. Thus, the probability that Z_i is the smallest of $Z_1, Z_2, \ldots, Z_k$ is proportional to its parameter, λ_i.

In Exercise 20 you are asked to show—by a similar argument—that

$$\Pr\left\{Z_i \leq \min_{j\neq i} Z_j, H > t\right\} = \Pr\left\{Z_i \leq \min_{j\neq i} Z_j\right\} \Pr\{H > t\} = \left(\frac{\lambda_i}{\lambda_H}\right) e^{-\lambda_H t}$$

$$(7.7)$$

In other words, *which* Z_i is the minimum and the actual *value* of the minimum are independent random variables.

7.3   SIMULATING THE SOFTWARE SELLOUT

In this section we develop a model for HyperWord inventory, as described in Case 7.1.

The consultant for Floppy Express decides that there are two critical uncertainties in managing the inventory of HyperWord: the demand for HyperWord and the time from when an order is placed until it is received from the software manufacturer, which is called the *lead time*. Based on historical data the demand is approximated by an arrival process with independent, time-stationary interarrival-time gaps, $G_1, G_2, \ldots$ having common cdf F_G (Chapter 5). The lead times are modeled as independent, time-stationary random variables, $R_1, R_2, \ldots$ having common cdf F_R, where R_i is the amount of time from when the ith order is placed until it is received. The demands and lead times are taken to be independent of each other since the customers and the software manufacturer have no direct knowledge of what the other one is doing. This is a rough-cut model; a more refined model might account for aspects of the system, such as the possibility that demand increases when a new version of HyperWord is released or that customers order more than one copy at a time.

There are many possible inventory policies that the consultant could consider. One that is easily implemented is an (r, s) policy, which works as follows: Whenever the inventory level drops to the *reorder point* of r units, order up to the *stock level* s (that is, order $s - r$ additional units).[1] For example, a $(100, 500)$ policy orders $500 - 100 = 400$ copies of HyperWord whenever the inventory level drops to 100 copies. Such a policy can be completely automated, provided that the inventory level is continuously tracked. The modeling-and-analysis question then becomes determining values for r and s. The consultant will only consider stock levels and reorder points such that $s - r > r$, implying that each time an order arrives, the inventory level rises above the reorder point, r.

To evaluate the performance of a particular (r, s), inventory policy the consultant formulates a stochastic-process model as follows: Let S_n represent the inventory level (copies of HyperWord) just after the nth change in inventory

---

[1]The standard notation for such a policy is "(s, S)," but we use (r, s) to make clear that the stock level is not a random variable.

level, which could be caused by a customer request or the arrival of an order from the manufacturer. The state space of S_n is therefore $\mathcal{M} = \{0, 1, \ldots, s\}$. There are two system events of interest, a customer demand, e_1, and receiving an order from the manufacturer, e_2, with corresponding clocks C_1 and C_2. The consultant starts the simulation with s copies of HyperWord in inventory, so that the system-event functions are as given below:

$e_0()$ **(initial inventory level)**

     $S_0 \leftarrow s$ (initially s copies)

     $C_1 \leftarrow F_G^{-1}(\text{random}())$ (set clock for first demand)
     $C_2 \leftarrow \infty$ (no order arrival pending)

$e_1()$ **(demand from customer)**

     if $\{S_n > 0\}$ then   (if copy in inventory)
         $S_{n+1} \leftarrow S_n - 1$ (reduce inventory one copy)
     endif

     if $\{S_{n+1} = r\}$ then   (if reorder point reached)

         $C_2 \leftarrow T_{n+1} + F_R^{-1}(\text{random}())$ (set clock for lead time)
     endif

     $C_1 \leftarrow T_{n+1} + F_G^{-1}(\text{random}())$ (set clock for next demand)

$e_2()$ **(order arrives from manufacturer)**

     $S_{n+1} \leftarrow S_n + (s - r)$ (add order to inventory)

Floppy Express typically has to manage inventories in the hundreds or thousands of copies, but for the purpose of illustrating the model, suppose that the reorder point is $r = 2$ copies, the stock level is $s = 5$ copies, and time is measured in days. We use this policy to illustrate our results throughout the chapter. A possible sample path generated by simulating 20 days is shown in Figure 7.1. The output process is $\{Y_t; t \geq 0\}$, the inventory level at continuous time t days, where $Y_t \leftarrow S_n$ when $T_n \leq t < T_{n+1}$. Notice that the state decreases in steps of 1 copy, corresponding to a customer demand, and increases in steps of $s - r = 5 - 2 = 3$ copies, corresponding to the arrival of an order from the manufacturer.

There are several performance measures of interest to Floppy Express. For instance, the time-average value of Y_t is the average inventory level, which is a measure of how much money and space is committed to HyperWord inventory. Another important performance measure is lost sales, which occur when there is no inventory. Let

$$K_t = \mathcal{I}(Y_t = 0)$$

be an indicator of when $\{Y_t = 0\}$, meaning that K_t takes the value 1 when there is

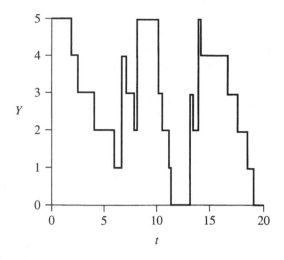

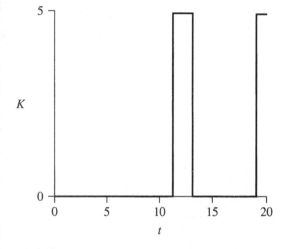

FIGURE 7.1
A sample path of the Floppy Express inventory simulation; inventory level Y_t above, indicator of zero inventory K_t below.

no inventory on hand and is 0 otherwise. Figure 7.1 also shows the sample path of K_t corresponding to the sample path of Y_t. The time-average value of K_t up to time 20 days is

$$\bar{K} = \frac{\int_0^{20} K_t \, dt}{20}$$

The random variable $\bar{K}$ is the fraction of time that there is no inventory. The area under the curve, $\int_0^{20} K_t \, dt$, can be computed as the simulation progresses by updating the area after executing each system event, as described in Chapter 4:

$$\text{area} \leftarrow \text{area} + (T_n - T_{n-1})\mathcal{I}(S_{n-1} = 0)$$

We extended the simulation in Figure 7.1 out to time 1000 days and found that the fraction of time that there was no inventory was 0.296, or roughly 30%

of the time. Whether or not this is too large depends on other factors, discussed later in the chapter, but if it is too large, then different values of (r, s) can be tried until the fraction is acceptably small.

Of course, the numerical result 0.296 depends on the distributions chosen for F_G and F_R, and the particular random-number seeds or streams used. We selected an exponential distribution with parameter $\lambda = 1$ for the interarrival-time gap between demands (that is, it is a Poisson arrival process with rate 1 demand per day) and an exponential distribution with parameter $\mu = 1/2$ for the lead time (implying that the expected lead time is 2 days).

We could have chosen any distributions, and the system logic embodied in e_0, e_1 and e_2 would be the same, although the numerical results could be quite different. The reason we focus on exponential distributions is that many performance measures, including the fraction of time that there is no inventory, can be derived via mathematical analysis when exponential distributions are appropriate. This makes the exponential distribution a *convenient choice*, although it will not be the *best choice* in all applications. Even when it is not the best choice, however, using exponential distributions for a rough-cut mathematical analysis is often a good idea. Other distributions may ultimately be required if the behavior of the system inputs differs substantially from the exponential model, and simulation may therefore be required for the analysis.

7.4 SAMPLE PATHS OF THE SOFTWARE SELLOUT

In this section we examine the probability structure of sample paths of $\{Y_t; t \geq 0\}$, the inventory level for HyperWord at Floppy Express, when the clock settings come from independent exponential distributions. Here is what we will find: *The state changes (inventory-level changes) can be described by a discrete-time Markov chain whose one-step transition matrix is easy to derive. And the actual time the process spends in each state (at each inventory level) is exponentially distributed with a parameter that is also easy to determine.* This particularly simple structure allows us to obtain exact sample-path performance measures via mathematical analysis, rather than relying on estimates obtained from simulation.

We start by examining the discrete-time, state-change process $\{S_n; n = 0, 1, \ldots\}$ underlying the continuous-time output process $\{Y_t; t \geq 0\}$. For Floppy Express, Y_t represents the inventory level at continuous time t, while S_n represents the inventory level after the nth change in level. There are some important features of the state-change process to notice:

- When either system event e_1 or e_2 occurs, the new state of the process, S_{n+1}, depends on the previous state of the process, S_n, and which system event occurred; nothing else matters.
- The set of pending system events depends only on the current state of the process. Specifically, when the state-change process S_n is in any state

$\{0, 1, \ldots, s\}$, then there is a customer demand event e_1 pending; but only in states $\{0, 1, \ldots, r\}$ is there an order-arrival event e_2 pending.

For the moment let us accept on faith that the state-change process $\{S_n; n = 0, 1, \ldots\}$ is in fact a Markov chain and try to formulate its one-step transition matrix for the case $r = 2$ and $s = 5$. If S_n equals $5, 4$ or 3, then we know that S_{n+1} equals $4, 3$ or 2, respectively, because the only pending system event is a demand and the inventory level decreases by 1 whenever a demand occurs. Similarly, if $\{S_n = 0\}$, then we know that $\{S_{n+1} = 3\}$ because the only possible state change occurs when an order arrives, and the inventory level increases by $s - r = 3$ whenever an order arrives. Therefore, a portion of the one-step transition matrix for S_n is

$$\mathbf{P} = \begin{pmatrix} p_{00} & p_{01} & p_{02} & p_{03} & p_{04} & p_{05} \\ p_{10} & p_{11} & p_{12} & p_{13} & p_{14} & p_{15} \\ p_{20} & p_{21} & p_{22} & p_{23} & p_{24} & p_{25} \\ p_{30} & p_{31} & p_{32} & p_{33} & p_{34} & p_{35} \\ p_{40} & p_{41} & p_{42} & p_{43} & p_{44} & p_{45} \\ p_{50} & p_{51} & p_{52} & p_{53} & p_{54} & p_{55} \end{pmatrix} = \begin{pmatrix} 0 & 0 & 0 & 1 & 0 & 0 \\ - & - & - & - & - & - \\ - & - & - & - & - & - \\ 0 & 0 & 1 & 0 & 0 & 0 \\ 0 & 0 & 0 & 1 & 0 & 0 \\ 0 & 0 & 0 & 0 & 1 & 0 \end{pmatrix}$$

(remember that the state space is $\{0, 1, 2, 3, 4, 5\}$). Since these transition probabilities are completely determined by the value of S_n and do not depend on the values of $S_{n-1}, S_{n-2}, \ldots$, the Markov property applies when $S_n \in \{0, 3, 4, 5\}$.

When the process is in state 1, both a demand and the arrival of an order are pending. If the next system event is a customer demand, then $\{S_{n+1} = 0\}$, but if it is the arrival of an order from the manufacturer, then $\{S_{n+1} = 4\}$. The state change that occurs depends on whether the time remaining until the pending demand is less than the time remaining until the pending order arrival. The transition probability p_{14} is the probability that the order arrival occurs first, and it is this probability that we need to determine.

Let Z_1 represent the time remaining until the next demand, and let Z_2 represent the time remaining until the next order arrival, when $\{S_n = 1\}$. When the clock for the next demand was set, the interarrival-time gap had distribution F_G. Similarly, when the clock for the next order arrival was set, the lead time had distribution F_R. Unfortunately, the clocks were almost certainly not set at the same time. Therefore, knowing F_G and F_R is not enough information to determine the distributions of Z_1 and Z_2, since they are the *remaining times*. To determine the distributions of Z_1 and Z_2 in general, we also need to know how much time has passed since these clocks were set.

This difficulty is illustrated in Figure 7.2. Notice that at time T_{n-1} the inventory position changed from 3 to 2 due to a demand. Then since the reorder point is 2, an order was placed with lead time R. Now at time T_n another demand caused the state of the process to change from 2 to 1, and the interarrival-time gap until the next demand is G. Therefore, at time T_n the time remaining until the next demand is $Z_1 = G$. However, the time remaining until the next order arrival is $Z_2 < R$, because some time has passed since this clock was set.

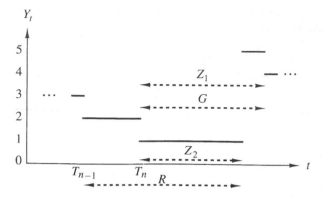

FIGURE 7.2
Time until the next system event after time T_n in the inventory model. Since $\{Z_2 \le Z_1\}$ occurs, the next system event is an order arrival.

Thus, $\Pr\{Z_2 \le Z_1\} \ne \Pr\{R \le G\}$, *unless F_G and F_R are independent exponential distributions so that the memoryless property applies* (Equation (7.2)).

To be more specific, suppose the interarrival-time gap between demands has an exponential distribution with parameter λ and the lead time is exponentially distributed with parameter μ. Then the memoryless property implies that Z_1—the time remaining until the next demand given that a demand has not occurred—is exponentially distributed with parameter λ, and Z_2—the time remaining until the next order arrival given that an order arrival has not occurred—is exponentially distributed with parameter μ. Applying Equation (7.6), the probability that the next state change is due to an order arrival is

$$p_{14} \equiv \Pr\{S_{n+1} = 4 \,|\, S_n = 1\} = \Pr\{Z_2 \le Z_1\} = \frac{\mu}{\lambda + \mu}$$

Similarly, the probability that the next state change is due to a demand is $p_{10} = \lambda/(\lambda + \mu)$. No other transitions are possible from state 1.

The same reasoning can be used to show that $p_{21} = \lambda/(\lambda + \mu)$ and $p_{25} = \mu/(\lambda + \mu)$. These transition probabilities do not depend on $S_{n-1}, S_{n-2}, \ldots$, provided we know S_n, so the Markov property applies. Thus, the complete one-step transition matrix for $\{S_n; n = 0, 1, \ldots\}$, provided F_G and F_R are independent exponential distributions, is

$$\mathbf{P} = \begin{pmatrix} 0 & 0 & 0 & 1 & 0 & 0 \\ \frac{\lambda}{\lambda+\mu} & 0 & 0 & 0 & \frac{\mu}{\lambda+\mu} & 0 \\ 0 & \frac{\lambda}{\lambda+\mu} & 0 & 0 & 0 & \frac{\mu}{\lambda+\mu} \\ 0 & 0 & 1 & 0 & 0 & 0 \\ 0 & 0 & 0 & 1 & 0 & 0 \\ 0 & 0 & 0 & 0 & 1 & 0 \end{pmatrix} \qquad (7.8)$$

and we have established that the state-change process S_n is a Markov chain. Notice that we did not *define* S_n to be a Markov chain as we did in Chapter 6; *it is a Markov chain as a result of the logic of the state changes and of F_G and*

F_R being independent exponential distributions. For this reason we refer to S_n as the *embedded Markov chain* within the continuous-time process $\{Y_t; t \geq 0\}$. If the distributions of the clock settings had not been exponential, then the state-change process would not have been a Markov chain, in general.

Adapting Equation (7.8) to the case $\lambda = 1$ and $\mu = 1/2$, we obtain

$$
\mathbf{P} = \begin{pmatrix}
0 & 0 & 0 & 1 & 0 & 0 \\
\frac{2}{3} & 0 & 0 & 0 & \frac{1}{3} & 0 \\
0 & \frac{2}{3} & 0 & 0 & 0 & \frac{1}{3} \\
0 & 0 & 1 & 0 & 0 & 0 \\
0 & 0 & 0 & 1 & 0 & 0 \\
0 & 0 & 0 & 0 & 1 & 0
\end{pmatrix}
$$

The analysis tools from Chapter 6 can be applied to this Markov chain to derive performance measures. The only limitation is that the embedded Markov chain S_n describes only the state changes of the inventory level, and does not account for the amount of time spent in each state. In fact, S_n spends exactly one time unit (transition) in each state because $p_{ii} = 0$ for all i in (7.8), implying that a transition from a state to itself is impossible. For this reason we next focus on the amount of (continuous) time spent in each state.

Let H_i be a random variable representing the time Y_t spends in state i on a single visit to state i. This random variable is called the *holding time* in state i. The holding times are the lengths of the horizontal line segments in Figure 7.2. Since F_G and F_R are independent exponential distributions, we can determine the distribution of H_i. For example, when S_n is equal to 5, 4 or 3, the holding time $H_i, i = 3, 4, 5$, is simply the time until the next demand, which is exponentially distributed with parameter λ. Similarly, when $\{S_n = 0\}$, the holding time H_0 is the time until the next order arrives, which is exponentially distributed with parameter μ. The memorylessness property implies that this is true any time we find the process in these states.

Now when $\{S_n = 1\}$, the holding time $H_1 = \min\{Z_1, Z_2\}$, where Z_1 and Z_2 are the remaining times until the next demand and order arrival, respectively. The memoryless property tells us that Z_1 and Z_2 are exponentially distributed with parameters λ and μ, respectively. From Equation (7.5), H_1 is therefore exponentially distributed with parameter $\lambda + \mu$. For example, if $\lambda = 1$ and $\mu = 1/2$ then the expected holding time at inventory level 1 is $E[H_1] = 1/(\lambda + \mu) = 2/3$ of a day. The holding time in state 2 has the same distribution. The parameters for all of the holding-time distributions are given in Table 7.1. Equation (7.7) implies that the holding time in each state is independent of which state change actually occurs next.

In summary, we have established an alternative description of the sample paths of the Floppy Express inventory system when the clock settings are independent and exponentially distributed: The inventory level is a Markov chain with transition matrix (7.8), the actual holding time at each inventory level is an

TABLE 7.1
**Parameters of the exponential holding-time
distributions for the inventory example**

state	holding-time parameter	expected holding time
0	μ	$\frac{1}{\mu}$
1	$\lambda + \mu$	$\frac{1}{\lambda + \mu}$
2	$\lambda + \mu$	$\frac{1}{\lambda + \mu}$
3	λ	$\frac{1}{\lambda}$
4	λ	$\frac{1}{\lambda}$
5	λ	$\frac{1}{\lambda}$

exponentially distributed random variable whose parameter is given in Table 7.1, and the state changes are independent of the holding times. A stochastic process with this structure is called a *continuous-time Markov process*. In the next section we generalize the definition of a Markov process and describe how to parameterize it.

7.5   MARKOV PROCESSES

In this section we describe the generic Markov process, a continuous-time, discrete-state stochastic process $\{Y_t; t \geq 0\}$ with state space $\mathcal{M} = \{1, 2, \ldots, m\}$, where m can go to ∞. We generically number the state space $1, 2, \ldots, m$ so that formulas involving summations—such as $\sum_{i=1}^{m}$—can be written simply. In applications the states may be numbered in any convenient manner. For instance, in the inventory model of Case 7.1 it is convenient to renumber the state space as $\mathcal{M} = \{0, 1, \ldots, s\}$, representing the number of copies of HyperWord in inventory.

---

*A continuous-time, discrete-state stochastic process $\{Y_t; t \geq 0\}$ is a **time-stationary Markov process** if it has the following properties:*

- *At any time $t \geq 0$, the set of pending system events depends only on the state of the process.*
- *The next state of the process after time t depends only on the state of the process at time t and which of the pending system events occurs first.*
- *The clock settings for all system events are independent, time-stationary, exponentially distributed random variables.*

A Markov process is represented by an initial-state probability vector **p** and a *generator matrix* **G**. The initial-state vector is $\mathbf{p}' = (p_1, p_2, \ldots, p_m)$ where

$p_i \equiv \Pr\{Y_0 = i\}$. The generator matrix has the generic form

$$
\mathbf{G} \equiv
\begin{pmatrix}
-g_{11} & g_{12} & \cdots & g_{1m} \\
g_{21} & -g_{22} & \cdots & g_{2m} \\
\vdots & \vdots & \vdots & \vdots \\
g_{m1} & g_{m2} & \cdots & -g_{mm}
\end{pmatrix}
\tag{7.9}
$$

Similar to the transition matrix of a Markov chain, each row of $\mathbf{G}$ represents the current state of the process, while each column represents a potential next state of the process. However, the g_{ij} are not probabilities, but rather they are *transition rates* related to the clock settings. Specifically:

- For $i \neq j$, g_{ij} is the parameter of the clock for the system event that changes the state of the process from i to j, and it can be interpreted as the transition rate from i to j. If $g_{ij} = 0$, then a direct transition from state i to state j is not possible; otherwise, $0 < g_{ij} < \infty$.
- $g_{ii} = \sum_{j \neq i} g_{ij}$ is the parameter of the holding time, H_i, in state i, and it can be interpreted as the overall transition rate out of state i. The negative sign in front of g_{ii} is a useful convention that we exploit later. We only consider cases in which $g_{ii} \leq c$ for some $c < \infty$ and all i.

For example, the generator for the HyperWord inventory model is

$$
\mathbf{G} =
\begin{pmatrix}
-\mu & 0 & 0 & \mu & 0 & 0 \\
\lambda & -(\lambda + \mu) & 0 & 0 & \mu & 0 \\
0 & \lambda & -(\lambda + \mu) & 0 & 0 & \mu \\
0 & 0 & \lambda & -\lambda & 0 & 0 \\
0 & 0 & 0 & \lambda & -\lambda & 0 \\
0 & 0 & 0 & 0 & \lambda & -\lambda
\end{pmatrix}
\tag{7.10}
$$

The off-diagonal entries g_{ij} are either the parameter of the demand interarrival-time gaps, λ, if a transition from i to j is caused by a demand; the parameter of the lead time, μ, if a transition from i to j is caused by the arrival of an order from the manufacturer; or 0 if a direct transition from i to j is not possible. The diagonal entries are (the negative of) the parameters of the exponential holding times for each state. The sum of the entries in each row is 0, and this is true in general.

The following sections describe how the generator matrix characterizes the behavior of a Markov process and also how the generator is parameterized. We also justify interpreting g_{ij} as a transition rate.

7.5.1 Probability Structure of a Markov Process

In this section we describe how the generator $\mathbf{G}$ and initial-state vector $\mathbf{p}$ specify the probability structure of a Markov process. Let $\{Y_t; t \geq 0\}$ be a Markov process with generator matrix $\mathbf{G}$, and let $\{S_n; n = 0, 1, \ldots\}$ be the embedded state-change process; that is, S_n is the state of $\{Y_t; t \geq 0\}$ just after the nth state change. Let H_i represent the holding time when the process enters state i.

The process $\{Y_t; t \geq 0\}$ has the following structure:

- *The state-change process $\{S_n; n = 0, 1, \ldots\}$ is a Markov chain with one-step transition matrix $\mathbf{P}$. If $g_{ii} > 0$ then the transition probabilities p_{ij} are*

$$p_{ij} = \frac{g_{ij}}{g_{ii}}, \; j \neq i$$

$$p_{ii} = 0 \tag{7.11}$$

If $g_{ii} = 0$, then the transition probabilities p_{ij} are

$$p_{ij} = 0, \; j \neq i$$

$$p_{ii} = 1 \tag{7.12}$$

and state i is an absorbing state.
- *The holding time on each visit to state i, H_i, is exponentially distributed with parameter g_{ii}, implying that the expected holding time in state i on each visit is $1/g_{ii}$ and the cdf of the holding time is*

$$F_{H_i}(b) = \begin{cases} 0, & b < 0 \\ 1 - e^{-g_{ii}b}, & 0 \leq b \end{cases} \tag{7.13}$$

The next state after state i is independent of the value of H_i.

With the addition of $\mathbf{p}$ to describe how the initial state is chosen, this characterization completely defines the movement of the Markov process from state to state and the length of time it spends in each state.

Since we have referred to the stochastic process $\{Y_t; t \geq 0\}$ as a Markov process, we should be clear about the sense in which it is "Markov." In continuous time, the Markov property is

$$\Pr\{Y_{t+\Delta t} = j | Y_t = i \text{ and } Y_a \text{ for all } a < t\} = \Pr\{Y_{t+\Delta t} = j | Y_t = i\}$$

for all $\Delta t \geq 0$. The condition "$Y_t = i$ and Y_a for all $a < t$" means that we are given the entire sample path of the process up to and including time t. The Markov property implies that only the state of the process at time t, Y_t, matters for probability statements about the future of the process, and not the path that got us to Y_t.

The time-stationarity property in continuous time is

$$\Pr\{Y_{t+\Delta t} = j \mid Y_t = i\} \text{ is the same for all } t \geq 0$$

for fixed $\Delta t \geq 0$. That is, only the time increment Δt matters, not the reference time t, when making probability statements about the future of the process. The continuous-time Markov process $\{Y_t; t \geq 0\}$ that we defined in this section *satisfies the Markov and stationarity properties*, although we did not prove it directly. This is why we refer to Y_t as a "Markov" process.

Perhaps surprisingly, *any* continuous-time, discrete-state process that satisfies the Markov and stationarity properties in continuous time *must be* a Markov process with some generator **G**. Therefore, the Markov chain transitions and the exponentially distributed holding times of a Markov process can be viewed as *consequences* of the Markov and stationarity properties, rather than arising by construction. A derivation of continuous-time Markov processes that starts from the Markov and stationarity properties can be found in, for instance, Clarke and Disney (1985, Chapter 11) or Çinlar (1975, Chapter 8). From a modeling perspective, it is useful to know that a system that approximately satisfies these properties can be treated as a Markov process.

7.5.2 Parameterizing Markov Processes

In this section we describe three alternative ways to parameterize a Markov process generator **G**. Although they are equivalent, the best one to use depends on the information available in the application at hand.

7.5.2.1 SYSTEM EVENTS AND CLOCKS. When we formulate our stochastic-process model in terms of system events and clocks—as we did for the HyperWord inventory example—then the most natural way to parameterize the generator matrix **G** is to extract the necessary data from the logic of the system-event functions and the distributions of the clock settings. The rules are as follows:

1. For $i \neq j$, if a direct transition from state i to state j *is not* possible, then set $g_{ij} = 0$.
2. For $i \neq j$, if a direct transition from state i to j *is* possible, then set g_{ij} equal to the parameter of the clock that triggers this transition.
3. If state i is an absorbing state, then set $g_{ij} = 0$ for all j so that row i is all zeros.
4. Always set $g_{ii} = \sum_{j \neq i} g_{ij}$.

For example, recall Case 4.1, the Case of the Leaky Bit Bucket, where the goal was to estimate the expected time to failure (TTF) for Bit Bucket Computers. A Bit Bucket computer system consists of a primary computer and a backup computer. The expected time until failure for a computer is 720 hours, while the expected time to repair a computer is 26 hours. Although the distributions of the time to failure and time to repair are not exponential, we will approximate them as exponential distributions by matching the expected values to obtain a rough-cut Markov process model. In Exercise 9 you will evaluate this approximation.

The state space of the process is $\mathcal{M} = \{0, 1, 2\}$, representing the number of failed computers. When the process is in state 0 (no failed computers), then the only possible transition is to state 1 when a computer fails. Therefore, $g_{01} = 1/720$, which is the parameter of an exponential distribution with expected value 720. When the process is in state 1, the next state change can be due to either another computer failure or a computer repair. Therefore, $g_{12} = 1/720$ and $g_{10} = 1/26$. Since we are interested in the time until the system fails (both computers down), we let state 2 be an absorbing state. The generator for this Markov process model is thus

$$G = \begin{pmatrix} -\frac{1}{720} & \frac{1}{720} & 0 \\ \frac{1}{26} & -\left(\frac{1}{720} + \frac{1}{26}\right) & \frac{1}{720} \\ 0 & 0 & 0 \end{pmatrix} \tag{7.14}$$

There are no failed computers at installation, so we can set $\mathbf{p}' = (1, 0, 0)$ implying that $\Pr\{Y_0 = 0\} = 1$.

7.5.2.2 TRANSITION RATES. In some contexts it is useful to interpret the off-diagonal entries of G as *rates*. For example, the Poisson arrival process with parameter λ (Chapter 5) is a Markov process with state space $\mathcal{M} = \{0, 1, 2, \ldots\}$ and generator

$$G = \begin{pmatrix} -\lambda & \lambda & 0 & 0 & 0 & \cdots \\ 0 & -\lambda & \lambda & 0 & 0 & \cdots \\ 0 & 0 & -\lambda & \lambda & 0 & \cdots \\ \vdots & \vdots & \vdots & \vdots & \vdots & \ddots \end{pmatrix} \tag{7.15}$$

Notice that the state (number of arrivals) increases by 1 at each transition, and the holding time in each state is exponentially distributed with parameter λ. Since λ is the *arrival rate* for a Poisson process, it is reasonable to interpret g_{ij} as the *transition rate* from state i to state j. And since $g_{ii} = \sum_{j \neq i} g_{ij}$, we can interpret g_{ii} as the overall transition rate out of state i. This perspective will be used extensively in Chapter 8.

Analogous to the transition-probability diagram for a Markov chain (Section 6.4.3), we can construct a *transition-rate diagram* for a Markov process. The transition-rate diagram for the HyperWord inventory model is displayed in Figure 7.3. Each node $\bigcirc$ represents a state of the system and is numbered accordingly. A directed arc $\longrightarrow$ connects state i to state j if a direct transition from i to j is possible. The transition rate g_{ij} is written next to the arc and enclosed in a box $\Box$ to distinguish it as a rate, *not a probability*.

Every Markov process contains an embedded Markov chain, so the transition-rate diagram has a corresponding transition-probability diagram. The transition-probability diagram for the HyperWord inventory model is also displayed in Figure 7.3.

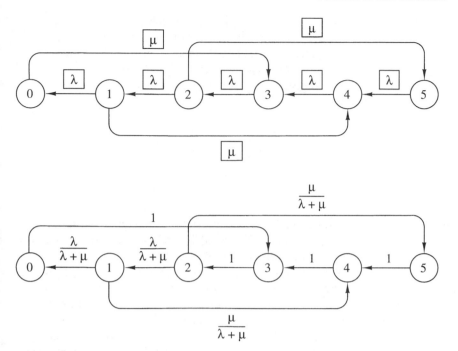

FIGURE 7.3
Transition-rate diagram for the HyperWord inventory model above. Transition-probability diagram for the embedded Markov chain below.

7.5.2.3 MARKOV CHAIN AND HOLDING TIMES. We have emphasized that there is a discrete-time Markov chain embedded within a continuous-time Markov process, and we have shown how to derive its transition matrix **P**. However, if data are instead available for the embedded Markov chain and holding times, then Equations (7.11) and (7.12) can be reversed to find the generator matrix. Specifically, if the holding time in state i is exponentially distributed with parameter ψ_i (expected value $1/\psi_i$) and the probability of a transition from state i to state j is p_{ij}, then

$$
\begin{aligned}
g_{ii} &= \psi_i \\
g_{ij} &= g_{ii} p_{ij} = \psi_i p_{ij}, \ j \neq i
\end{aligned}
\tag{7.16}
$$

We make the assumption that $p_{ii} = 0$ unless state i is an absorbing state, in which case $g_{ii} = g_{ij} = 0$. In other words, in continuous time we cannot recognize the transition from a state to itself unless the process is absorbed in that state.

For example, suppose that we are modeling the lifetime of a component that can be fully functional (state 1), degraded but functional (state 2), failed and repairable (state 3), or failed and unrepairable (state 4). The state space of the process is $\mathcal{M} = \{1, 2, 3, 4\}$. When a fully functional component changes state it may become either degraded; failed and repairable; or failed and unre-pairable. Similarly, a degraded component may become failed and repairable; or failed and unrepairable. Repaired components always become fully functional, but unrepairable components are useless. A discrete-time Markov chain with the following transition matrix (derived from repair and replacement records) might describe these state changes:

$$
P = \begin{pmatrix}
0 & \frac{8}{10} & \frac{1}{10} & \frac{1}{10} \\
0 & 0 & \frac{1}{2} & \frac{1}{2} \\
1 & 0 & 0 & 0 \\
0 & 0 & 0 & 1
\end{pmatrix}
$$

For instance, when a fully functional component (state 1) changes state, it becomes degraded with probability 8/10 (state 2), failed but repairable with probability 1/10 (state 3), or failed and unrepairable with probability 1/10 (state 4).

Suppose that the expected time a component spends fully functional is ex-ponentially distributed with expected value $1/\psi_1 = 5$ days, the expected time a component stays in the degraded state is exponentially distributed with expected value $1/\psi_2 = 8$ days, the expected time to repair a component is exponentially distributed with expected value $1/\psi_3 = 1$ day, and the expected time a compo-nent remains unrepairable is infinite, since the component becomes useless. Then Equation (7.16) implies that

$$
G = \begin{pmatrix}
-\psi_1 & \frac{8}{10}\psi_1 & \frac{1}{10}\psi_1 & \frac{1}{10}\psi_1 \\
0 & -\psi_2 & \frac{1}{2}\psi_2 & \frac{1}{2}\psi_2 \\
\psi_3 & 0 & -\psi_3 & 0 \\
0 & 0 & 0 & 0
\end{pmatrix} = \begin{pmatrix}
-\frac{10}{50} & \frac{8}{50} & \frac{1}{50} & \frac{1}{50} \\
0 & -\frac{2}{16} & \frac{1}{16} & \frac{1}{16} \\
1 & 0 & -1 & 0 \\
0 & 0 & 0 & 0
\end{pmatrix}
$$

7.6   ANALYSIS OF MARKOV PROCESS SAMPLE PATHS

In this section we show how to compute sample-path performance measures for a Markov process with state space $\mathcal{M} = \{1, 2, \ldots, m\}$. There are several ways to do this, and we present two. The first approach, called *uniformization*, exploits the tools presented in Chapter 6 for the analysis of Markov chains. Another ap-proach based on systems of *Kolmogorov differential equations* is also presented. We provide the key computational results first, followed by examples and deriva-tions.

7.6.1   Performance Measures

Analogous to the n-step transition probabilities that were the basis for computing performance measures of Markov chains, the performance measures for a Markov process are based on the *transition functions*

$$p_{ij}(t) \equiv \Pr\{Y_t = j | Y_0 = i\}$$

for $t \geq 0$. The transition function $p_{ij}(t)$ is the probability that the process is in state j at (continuous) time t, given that it started in state i at time 0. Using the initial-state probabilities $p_i = \Pr\{Y_0 = i\}$, for $i = 1, 2, \ldots, m$, and the law of total probability, the unconditional *probability function* is

$$p_j(t) \equiv \Pr\{Y_t = j\} = \sum_{i=1}^{m} p_{ij}(t) p_i$$

We can organize the transition functions $p_{ij}(t)$ for all i, j into a transition-function matrix

$$\mathbf{P}(t) \equiv \begin{pmatrix} p_{11}(t) & p_{12}(t) & \cdots & p_{1m}(t) \\ p_{21}(t) & p_{22}(t) & \cdots & p_{2m}(t) \\ \vdots & \vdots & \vdots & \vdots \\ p_{m1}(t) & p_{m2}(t) & \cdots & p_{mm}(t) \end{pmatrix} \tag{7.17}$$

which is analogous to the n-step transition matrix for Markov chains with each row summing to 1. Similarly, we can organize the $p_j(t)$ for all j into a vector

$$\mathbf{p}(t) \equiv \begin{pmatrix} p_1(t) \\ p_2(t) \\ \vdots \\ p_m(t) \end{pmatrix} \tag{7.18}$$

The law of total probability implies that $\mathbf{p}(t)' = \mathbf{p}' \mathbf{P}(t)$.

The Markov and stationarity properties imply a fundamental decomposition property for Markov process sample paths, namely the *Chapman-Kolmogorov equation*

$$p_{ij}(t + \Delta t) = \sum_{h=1}^{m} p_{ih}(t) p_{hj}(\Delta t) \tag{7.19}$$

for any time $t \geq 0$ and time increment $\Delta t > 0$. The Chapman-Kolmogorov equation shows that the probability of the process being in state j at time $t + \Delta t$ can be decomposed into the probability of being in state h at time t, then moving from state h to state j in the remaining time Δt, provided we account for all possible intermediate states, h.

The Chapman-Kolmogorov equation is derived as follows:

$$p_{ij}(t + \Delta t) = \Pr\{Y_{t+\Delta t} = j | Y_0 = i\}$$

$$= \sum_{h=1}^{m} \Pr\{Y_{t+\Delta t} = j, Y_t = h | Y_0 = i\} \tag{7.20}$$

$$= \sum_{h=1}^{m} \Pr\{Y_{t+\Delta t} = j | Y_t = h, Y_0 = i\} \Pr\{Y_t = h | Y_0 = i\} \tag{7.21}$$

$$= \sum_{h=1}^{m} \Pr\{Y_{t+\Delta t} = j | Y_t = h\} \Pr\{Y_t = h | Y_0 = i\} \tag{7.22}$$

$$= \sum_{h=1}^{m} \Pr\{Y_{\Delta t} = j | Y_0 = h\} \Pr\{Y_t = h | Y_0 = i\} \tag{7.23}$$

$$= \sum_{h=1}^{m} p_{hj}(\Delta t) p_{ih}(t) \tag{7.24}$$

$$= \sum_{h=1}^{m} p_{ih}(t) p_{hj}(\Delta t)$$

Notice that Equation (7.20) follows because the process must be in some state h at the intermediate time t; Equation (7.21) is an application of the law of total probability; Equations (7.22) and (7.23) follow from the Markov and stationarity properties, respectively; and Equation (7.24) is derived by recalling the definition of $p_{ij}(t)$.

7.6.2   Time-Dependent Performance Measures

To derive an expression for $\mathbf{P}(t)$, it is convenient to define an auxiliary matrix

$$\boxed{\mathbf{Q} = \mathbf{I} + \frac{1}{g^*}\mathbf{G}} \tag{7.25}$$

where $g^* = \max_i\{g_{ii}\}$ and $\mathbf{I}$ is the $m \times m$ identity matrix. A physical interpretation of $\mathbf{Q}$ will be given in Section 7.6.6.

The following are our key results:

The transition matrix $\mathbf{P}(t)$ is expressible as the infinite series

$$\mathbf{P}(t) = e^{-g^* t} \sum_{n=0}^{\infty} \mathbf{Q}^n \frac{(g^* t)^n}{n!} \tag{7.26}$$

or as the solution to the system of linear differential equations

$$\frac{d\mathbf{P}(t)}{dt} = \mathbf{P}(t)\mathbf{G} = \mathbf{G}\mathbf{P}(t) \tag{7.27}$$

with the boundary condition $\mathbf{P}(0) = \mathbf{I}$, where $d\mathbf{P}(t)/dt$ is the matrix with i, jth element $dp_{ij}(t)/dt$.

Equation (7.26), which is the uniformization expression, is not directly useful for calculations because of the infinite summation. However, we can truncate the summation at n^* to obtain an approximation

$$\mathbf{P}(t) \approx \tilde{\mathbf{P}}(t) = e^{-g^*t} \sum_{n=0}^{n^*} \mathbf{Q}^n \frac{(g^*t)^n}{n!} \tag{7.28}$$

if n^* is large enough. We use a $\tilde{\ }$ to denote an approximation based on truncating an infinite summation. The approximation (7.28) always underestimates $p_{ij}(t)$. That is,

$$\tilde{\mathbf{P}}(t) \le \mathbf{P}(t)$$

element by element. Fortunately, the error from truncating the summation at n^* is bounded by

$$\text{error} = p_{ij}(t) - \tilde{p}_{ij}(t) \le 1 - e^{-g^*t} \sum_{n=0}^{n^*} \frac{(g^*t)^n}{n!} \tag{7.29}$$

The terms in the error bound (7.29) are computed as a by-product of evaluating (7.28). As a rule of thumb we choose $n^* \approx 2g^*t$, but we can (and should) iteratively increase n^* until the error bound is sufficiently small.

We now present an algorithm for computing $\tilde{\mathbf{P}}(t)$. The algorithm makes use of the fact that

$$\mathbf{Q}^n \frac{(g^*t)^n}{n!} = \left(\mathbf{Q}^{n-1} \frac{(g^*t)^{n-1}}{(n-1)!} \right) \left(\mathbf{Q} \frac{g^*t}{n} \right)$$

so that we can iteratively compute each term in the summation (7.28) by updating the previous term. At iteration n in Step 2 of the algorithm uniformize, the scalar $a = (g^*t)^n/n!$, the matrix $\mathbf{A} = \mathbf{Q}^n$, the matrix $\mathbf{B} = \sum_{j=0}^{n} \mathbf{Q}^j (g^*t)^j/j!$, and the scalar $b = \sum_{j=0}^{n} (g^*t)^j/j!$

algorithm uniformize

1. $\mathbf{A} \leftarrow \mathbf{I}$
   $a \leftarrow 1$
   $\mathbf{B} \leftarrow \mathbf{A}$
   $b \leftarrow a$
2. for $n \leftarrow 1$ to n^*
   do
   $\qquad \mathbf{A} \leftarrow \mathbf{A}\mathbf{Q}$
   $\qquad a \leftarrow ag^*t/n$
   $\qquad \mathbf{B} \leftarrow \mathbf{B} + a\mathbf{A}$
   $\qquad b \leftarrow b + a$
   enddo
3. $\widetilde{\mathbf{P}}(t) \leftarrow e^{-g^*t}\mathbf{B}$
   error $\leftarrow 1 - e^{-g^*t}b$

As a practical matter the Kolmogorov differential equations (7.27) also require a numerical solution, except in the simplest models when we can solve them in closed form. We do not discuss closed-form solution of systems of differential equations in this book, nor do we describe the state of the art in their numerical solution. Instead we describe Euler's method of numerical solution, *which should not be used in practice but does provide intuition about how more accurate methods work.*

The transition function $p_{ij}(t)$ can be expanded as a Taylor series

$$p_{ij}(t + \Delta t) = p_{ij}(t) + \frac{dp_{ij}(t)}{dt}\Delta t + \frac{d^2 p_{ij}(t)}{dt^2}\frac{\Delta t^2}{2!}$$
$$+ \cdots \approx p_{ij}(t) + \frac{dp_{ij}(t)}{dt}\Delta t$$

where the approximation is reasonable if the increment Δt is small enough so that $p_{ij}(t)$ is nearly linear between t and $t + \Delta t$. In matrix form

$$\mathbf{P}(t + \Delta t) \approx \mathbf{P}(t) + \frac{d\mathbf{P}(t)}{dt}\Delta t$$
$$= \mathbf{P}(t) + (\mathbf{P}(t)\mathbf{G})\Delta t$$
$$= \mathbf{P}(t)(\mathbf{I} + \mathbf{G}\Delta t)$$

where we use the substitution $d\mathbf{P}(t)/dt = \mathbf{P}(t)\mathbf{G}$ from the Kolmogorov equations (7.27). Euler's method uses this approximation to iteratively update the value of $\mathbf{P}(t)$ in time steps of Δt. In the algorithm below a ¨ indicates Euler's approximation.

algorithm Euler

1. $\ddot{\mathbf{P}}(0) \leftarrow \mathbf{I}$
$\quad t \leftarrow 0$
$\quad \mathbf{C} \leftarrow \mathbf{I} + \mathbf{G}\Delta t$
2. $\ddot{\mathbf{P}}(t + \Delta t) \leftarrow \ddot{\mathbf{P}}(t)\mathbf{C}$
$\quad t \leftarrow t + \Delta t$
3. repeat 2

The reason that Euler's method is not used in practice is that Δt must typically be quite small for the approximation to be accurate, but a substantial numerical rounding error accumulates when it is too small. More accurate methods are based on the same principles as Euler's method, however.

The choice between uniformization and the Kolmogorov equations is problem dependent, and may be influenced by the available software. Uniformization can be faster in some cases and uses simple matrix calculations, but it may be more cumbersome when we require $\mathbf{P}(t)$ for many values of t.

7.6.3   Time-Dependent Example

To illustrate the calculation of performance measures for a Markov process via uniformization, we adapt the generator matrix (7.10) for HyperWord inventory to the case of $\lambda = 1$ and $\mu = 2$. Therefore

$$\mathbf{Q} = \mathbf{I} + \frac{1}{g^*}\mathbf{G}$$

$$= \begin{pmatrix} 1\ 0\ 0\ 0\ 0\ 0 \\ 0\ 1\ 0\ 0\ 0\ 0 \\ 0\ 0\ 1\ 0\ 0\ 0 \\ 0\ 0\ 0\ 1\ 0\ 0 \\ 0\ 0\ 0\ 0\ 1\ 0 \\ 0\ 0\ 0\ 0\ 0\ 1 \end{pmatrix} + \frac{1}{3/2}\begin{pmatrix} -\frac{1}{2} & 0 & 0 & \frac{1}{2} & 0 & 0 \\ 1 & -\frac{3}{2} & 0 & 0 & \frac{1}{2} & 0 \\ 0 & 1 & -\frac{3}{2} & 0 & 0 & \frac{1}{2} \\ 0 & 0 & 1 & -1 & 0 & 0 \\ 0 & 0 & 0 & 1 & -1 & 0 \\ 0 & 0 & 0 & 0 & 1 & -1 \end{pmatrix}$$

$$= \begin{pmatrix} \frac{2}{3} & 0 & 0 & \frac{1}{3} & 0 & 0 \\ \frac{2}{3} & 0 & 0 & 0 & \frac{1}{3} & 0 \\ 0 & \frac{2}{3} & 0 & 0 & 0 & \frac{1}{3} \\ 0 & 0 & \frac{2}{3} & \frac{1}{3} & 0 & 0 \\ 0 & 0 & 0 & \frac{2}{3} & \frac{1}{3} & 0 \\ 0 & 0 & 0 & 0 & \frac{2}{3} & \frac{1}{3} \end{pmatrix}$$

Notice that $\mathbf{Q}$ is indeed a Markov chain transition matrix, but it is different from the matrix $\mathbf{P}$ for the embedded Markov chain.

If we are interested in the inventory level after $t = 4$ days, then we can compute

$$\mathbf{P}(4) \approx \widetilde{\mathbf{P}}(4) = e^{-(3/2)(4)} \sum_{n=0}^{12} \mathbf{Q}^n \frac{[(3/2)(4)]^n}{n!}$$

$$\approx \begin{pmatrix} 0.280 & 0.108 & 0.166 & 0.249 & 0.111 & 0.078 \\ 0.262 & 0.111 & 0.169 & 0.254 & 0.119 & 0.076 \\ 0.223 & 0.110 & 0.170 & 0.260 & 0.145 & 0.084 \\ 0.216 & 0.115 & 0.167 & 0.253 & 0.152 & 0.088 \\ 0.188 & 0.123 & 0.176 & 0.255 & 0.159 & 0.091 \\ 0.140 & 0.117 & 0.181 & 0.267 & 0.185 & 0.101 \end{pmatrix}$$

where we truncated the summation at $n^* = 2g^*t = 2(3/2)(4) = 12$ terms and displayed the results to three decimal places. If the inventory level at time $t = 0$ is 5, then the last row of $\widetilde{\mathbf{P}}(4)$ is the probability distribution of the inventory level after 4 days. For example, $\Pr\{Y_4 = 0 | Y_0 = 5\} \approx \widetilde{p}_{50}(4) = 0.140$ is (approximately) the probability that there is no HyperWord inventory after 4 days, given we started with 5 copies. And the expected inventory position after 4 days is

$$E[Y_4 | Y_0 = 5] = \sum_{j=0}^{5} j \Pr\{Y_4 = j | Y_0 = 5\} \approx \sum_{j=0}^{5} j \widetilde{p}_{5j}(4)$$

$$= 0(0.140) + 1(0.117) + 2(0.181) + 3(0.267) + 4(0.185) + 5(0.101)$$

$$= 2.525$$

copies of HyperWord.

There are two sources of error in this calculation: truncation of the infinite summation and numerical roundoff. From Equation (7.29) the underestimation from truncating the summation at $n^* = 12$ terms is less or equal to

$$\text{error} = 1 - e^{-(3/2)(4)} \sum_{n=0}^{12} \frac{[(3/2)(4)]^n}{n!} \approx 0.009$$

In other words, the true probability $p_{50}(4)$ could be as great as $0.140 + 0.009 = 0.149$. If this level of error is too large, then n^* can be increased. The remaining error is due to numerical error in evaluating (7.28). Because of these two types of error the probabilities in each row do not sum exactly to 1, as they should in theory.

To perform the same calculation using the Kolmogorov equations and Euler's approximation, we took the step size to be $\Delta t = 1/10$ day. Thus,

$$\mathbf{C} = \mathbf{I} + \mathbf{G}\Delta t$$

$$= \begin{pmatrix} 1 & 0 & 0 & 0 & 0 & 0 \\ 0 & 1 & 0 & 0 & 0 & 0 \\ 0 & 0 & 1 & 0 & 0 & 0 \\ 0 & 0 & 0 & 1 & 0 & 0 \\ 0 & 0 & 0 & 0 & 1 & 0 \\ 0 & 0 & 0 & 0 & 0 & 1 \end{pmatrix} + \begin{pmatrix} -\frac{1}{2} & 0 & 0 & \frac{1}{2} & 0 & 0 \\ 1 & -\frac{3}{2} & 0 & 0 & \frac{1}{2} & 0 \\ 0 & 1 & -\frac{3}{2} & 0 & 0 & \frac{1}{2} \\ 0 & 0 & 1 & -1 & 0 & 0 \\ 0 & 0 & 0 & 1 & -1 & 0 \\ 0 & 0 & 0 & 0 & 1 & -1 \end{pmatrix} \frac{1}{10}$$

$$= \begin{pmatrix} \frac{19}{20} & 0 & 0 & \frac{1}{20} & 0 & 0 \\ \frac{1}{10} & \frac{17}{20} & 0 & 0 & \frac{1}{20} & 0 \\ 0 & \frac{1}{10} & \frac{17}{20} & 0 & 0 & \frac{1}{20} \\ 0 & 0 & \frac{1}{10} & \frac{9}{10} & 0 & 0 \\ 0 & 0 & 0 & \frac{1}{10} & \frac{9}{10} & 0 \\ 0 & 0 & 0 & 0 & \frac{1}{10} & \frac{9}{10} \end{pmatrix}$$

Performing the first iteration of the algorithm Euler yields

$$\ddot{\mathbf{P}}\left(0 + \frac{1}{10}\right) = \ddot{\mathbf{P}}(0)\mathbf{C} = \mathbf{IC}$$

$$= \begin{pmatrix} 1 & 0 & 0 & 0 & 0 & 0 \\ 0 & 1 & 0 & 0 & 0 & 0 \\ 0 & 0 & 1 & 0 & 0 & 0 \\ 0 & 0 & 0 & 1 & 0 & 0 \\ 0 & 0 & 0 & 0 & 1 & 0 \\ 0 & 0 & 0 & 0 & 0 & 1 \end{pmatrix} \begin{pmatrix} \frac{19}{20} & 0 & 0 & \frac{1}{20} & 0 & 0 \\ \frac{1}{10} & \frac{17}{20} & 0 & 0 & \frac{1}{20} & 0 \\ 0 & \frac{1}{10} & \frac{17}{20} & 0 & 0 & \frac{1}{20} \\ 0 & 0 & \frac{1}{10} & \frac{9}{10} & 0 & 0 \\ 0 & 0 & 0 & \frac{1}{10} & \frac{9}{10} & 0 \\ 0 & 0 & 0 & 0 & \frac{1}{10} & \frac{9}{10} \end{pmatrix}$$

$$= \begin{pmatrix} \frac{19}{20} & 0 & 0 & \frac{1}{20} & 0 & 0 \\ \frac{1}{10} & \frac{17}{20} & 0 & 0 & \frac{1}{20} & 0 \\ 0 & \frac{1}{10} & \frac{17}{20} & 0 & 0 & \frac{1}{20} \\ 0 & 0 & \frac{1}{10} & \frac{9}{10} & 0 & 0 \\ 0 & 0 & 0 & \frac{1}{10} & \frac{9}{10} & 0 \\ 0 & 0 & 0 & 0 & \frac{1}{10} & \frac{9}{10} \end{pmatrix}$$

The second iteration gives

$$\ddot{\mathbf{P}}\left(\frac{2}{10}\right) = \begin{pmatrix} \frac{361}{400} & 0 & \frac{1}{200} & \frac{37}{400} & 0 & 0 \\ \frac{9}{50} & \frac{289}{400} & 0 & \frac{1}{100} & \frac{7}{80} & 0 \\ \frac{1}{100} & \frac{17}{100} & \frac{289}{400} & 0 & \frac{1}{100} & \frac{7}{80} \\ 0 & \frac{1}{100} & \frac{7}{40} & \frac{81}{100} & 0 & \frac{1}{200} \\ 0 & 0 & \frac{1}{100} & \frac{9}{50} & \frac{81}{100} & 0 \\ 0 & 0 & 0 & \frac{1}{100} & \frac{9}{50} & \frac{81}{100} \end{pmatrix}$$

Finally, after 40 iterations the solution to three decimal places is

$$\ddot{\mathbf{P}}(4) = \begin{pmatrix} 0.278 & 0.110 & 0.167 & 0.251 & 0.115 & 0.079 \\ 0.262 & 0.113 & 0.171 & 0.254 & 0.121 & 0.079 \\ 0.224 & 0.111 & 0.174 & 0.263 & 0.144 & 0.084 \\ 0.220 & 0.114 & 0.168 & 0.258 & 0.153 & 0.088 \\ 0.194 & 0.123 & 0.175 & 0.256 & 0.161 & 0.092 \\ 0.146 & 0.121 & 0.183 & 0.267 & 0.183 & 0.100 \end{pmatrix}$$

Notice that $\ddot{p}_{50}(4) = 0.146$, which is close to, but different from, $\widetilde{p}_{50}(4)$. A more accurate numerical-integration algorithm gives $p_{50}(4) = 0.142$, to three decimal places, showing that uniformization did indeed underestimate the true probability, but both methods were close.

Figure 7.4 shows a plot of $\ddot{p}_{55}(t)$ and $\ddot{p}_{50}(t)$ for $t = 0, 0.1, 0.2, \ldots, 4.0$. At time $t = 0$ the probability of having 5 copies of HyperWord in inventory is 1, but it drops rapidly to 0.100 by time $t = 4$ days. On the other hand, the probability of having no HyperWord inventory, $\ddot{p}_{50}(t)$, is nearly 0 for the first 2 days and then rises to 0.146 by time $t = 4$ days. In the next section we show that these probabilities become constant (no longer changing with t) as time goes to infinity.

Notice that if we are only interested in the transition functions with initial state 5, then rather than solve (mathematically or numerically) a system of $6^2 = 36$ Kolmogorov differential equations, we need only solve the system of six differential equations for $p_{5j}(t)$. Specifically, we need only the last row of

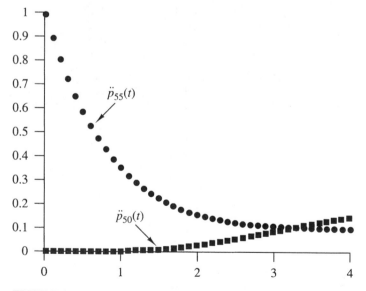

FIGURE 7.4
Plot of the Euler approximation of the transition functions $p_{55}(t)$ and $p_{50}(t)$ for the inventory example.

$d\mathbf{P}(t)/dt = \mathbf{P}(t)\mathbf{G}$, which is

$$\left(\frac{dp_{50}(t)}{dt} \; \frac{dp_{51}(t)}{dt} \; \frac{dp_{52}(t)}{dt} \; \frac{dp_{53}(t)}{dt} \; \frac{dp_{54}(t)}{dt} \; \frac{dp_{55}(t)}{dt}\right) =$$

$$(p_{50}(t) \; p_{51}(t) \; p_{52}(t) \; p_{53}(t) \; p_{54}(t) \; p_{55}(t)) \begin{pmatrix} -\frac{1}{2} & 0 & 0 & \frac{1}{2} & 0 & 0 \\ 1 & -\frac{3}{2} & 0 & 0 & \frac{1}{2} & 0 \\ 0 & 1 & -\frac{3}{2} & 0 & 0 & \frac{1}{2} \\ 0 & 0 & 1 & -1 & 0 & 0 \\ 0 & 0 & 0 & 1 & -1 & 0 \\ 0 & 0 & 0 & 0 & 1 & -1 \end{pmatrix}$$

or

$$\frac{dp_{50}(t)}{dt} = -\frac{1}{2}p_{50}(t) + p_{51}(t)$$
$$\frac{dp_{51}(t)}{dt} = -\frac{3}{2}p_{51}(t) + p_{52}(t)$$
$$\frac{dp_{52}(t)}{dt} = -\frac{3}{2}p_{52}(t) + p_{53}(t)$$
$$\frac{dp_{53}(t)}{dt} = \frac{1}{2}p_{50}(t) - p_{53}(t) + p_{54}(t)$$
$$\frac{dp_{54}(t)}{dt} = \frac{1}{2}p_{51}(t) - p_{54}(t) + p_{55}(t)$$
$$\frac{dp_{55}(t)}{dt} = \frac{1}{2}p_{52}(t) - p_{55}(t)$$

7.6.4 Time-Independent (Long-Run) Performance Measures

The transition function $\mathbf{P}(t)$ provides time-dependent performance measures for specific values of t. We are sometimes interested in time-independent (long-run) performance measures; specifically, the limiting probability

$$\vec{p}_{ij} = \lim_{t \to \infty} p_{ij}(t)$$

The results in this section rely on some subtle conditions on the elements of $\mathbf{G}$ that are described in Section 7.10.

Analogous to Markov chains, the interpretation and calculation of $\vec{p}_{ij}$ depends on the classification of the states i and j. Fortunately, classification depends on only the embedded Markov chain.

The states of a Markov process are classified as recurrent or transient if the corresponding states of the embedded Markov chain are recurrent or transient, respectively. There is no concept of, or problem with, periodicity in continuous time.

See Section 6.7 for the classification of states of a Markov chain into recurrent and transient states.

Suppose that the state space $\mathcal{M}$ is reducible into a set of transient states, $\mathcal{T}$, and irreducible, recurrent sets of states $\mathcal{R}_1, \mathcal{R}_2, \mathcal{R}_3, \ldots$. Our goal is to compute the limiting probabilities $\vec{p}_{ij}$. Two cases are easy:

- $\vec{p}_{ij} = 0$ if $j \in \mathcal{T}$. The limiting probability of being in state j is 0 if j is a transient state, regardless of where the process starts. This makes sense, because even if the process reaches state j, it will eventually leave state j and never return, since $\vec{p}_{jj} = 0$ for a transient state.
- $\vec{p}_{ij} = 0$ if $i \in \mathcal{R}_1$ and $j \in \mathcal{R}_2$. The limiting probability of being in state j given the process starts in state i is 0 if i and j are in different irreducible sets of states. This makes sense because a process that is initially in one irreducible set of states will never leave that set of states.

From here on we focus on a single set of irreducible, recurrent states, denoted $\mathcal{R}$ to keep the notation simple. The more interesting cases are when $i, j \in \mathcal{R}$ (i and j are both in the same irreducible set of states) and when $i \in \mathcal{T}$ and $j \in \mathcal{R}$ (i is transient and j is recurrent). The following results are derived after providing examples. For convenience, suppose that $\mathcal{R} = \{1, 2, \ldots, m_{\mathcal{R}}\}$, so that $\mathbf{G}_{\mathcal{R}\mathcal{R}}$ is the portion of the generator matrix restricted to states in $\mathcal{R}$.

- For $i, j \in \mathcal{R}$, $\vec{p}_{ij} = \pi_j > 0$, independent of i. The vector $\pi'_{\mathcal{R}} = (\pi_1, \pi_2, \ldots, \pi_{m_{\mathcal{R}}})$ is obtained by solving the system of linear equations

$$\boxed{\begin{aligned} 0 &= \pi'_{\mathcal{R}} \mathbf{G}_{\mathcal{R}\mathcal{R}} \\ 1 &= \pi'_{\mathcal{R}} \mathbf{1} \end{aligned}} \qquad (7.30)$$

This system of equations will always have $m_{\mathcal{R}} + 1$ equations and $m_{\mathcal{R}}$ unknowns, and any one of the equations other than $1 = \pi'_{\mathcal{R}} \mathbf{1}$ can always be eliminated.

The π_j are called *steady-state probabilities*. They can be interpreted as the probability of finding the process in state j after a very long time or as the long-run fraction of time that the process spends in state j. The vector $\pi_{\mathcal{R}}$ is called the *steady-state distribution*, and it is a true probability distribution since $\pi'_{\mathcal{R}} \mathbf{1} = \sum_{j=1}^{m_{\mathcal{R}}} \pi_j = 1$.

- For $i \in \mathcal{T}$, and $\mathcal{R} = \{j\}$ an absorbing state, $\vec{p}_{ij} = \alpha_{ij} \geq 0$. The vector $\alpha'_{\mathcal{T}\mathcal{R}} = (\alpha_{ij}; i \in \mathcal{T})$ is obtained from

$$\boxed{\alpha_{\mathcal{T}\mathcal{R}} = (\mathbf{I} - \mathbf{P}_{\mathcal{T}\mathcal{T}})^{-1} \mathbf{P}_{\mathcal{T}\mathcal{R}}} \qquad (7.31)$$

The α_{ij} are called *absorption probabilities*. They can be interpreted as the probability that the process is ultimately "absorbed" into state j. Notice that these probabilities depend only on the embedded Markov chain, **P**, because the state changes alone determine into which state a process is absorbed.

The system of equations (7.30) is a compact expression for the steady-state probabilities, but it is not always the most convenient way to form the equations. Although we do not emphasize it in this book, the so-called balance equation approach is sometimes an easier way to write the equations, and it provides a useful interpretation of them. We describe balance equations next.

Extracting the jth equation from $0 = \pi'_{\mathcal{R}} G_{\mathcal{R}\mathcal{R}}$, we have

$$0 = \sum_{i \neq j} \pi_i g_{ij} - \pi_j g_{jj}$$

Making the substitution $g_{jj} = \sum_{i \neq j} g_{ji}$, we can rewrite this equation as a balance equation

$$\underbrace{\sum_{i \neq j} \pi_i g_{ij}}_{\text{rate into } j} = \underbrace{\pi_j \sum_{i \neq j} g_{ji}}_{\text{rate out of } j}$$

Recalling that g_{ij} is the transition rate from state i *into* state j and that π_i is the long-run probability of finding the process in state i, we can interpret the left-hand side of the equation as the long-run transition rate into state j from all other states. Similarly, since g_{ji} is the transition rate *out of* state j into state i, we can interpret the right-hand side as the long-run transition rate out of state j. For the idea of "steady state" to make sense, these rates must balance over the long run.

Notice that the balance equations are merely a way of rewriting $0 = \pi'_{\mathcal{R}} G_{\mathcal{R}\mathcal{R}}$ to give each equation a physical interpretation. The validity of the equations still depends critically on the process being a continuous-time Markov process, and they are not valid for continuous-time processes in general.

7.6.5 Time-Independent Example

To illustrate the calculation of performance measures for a Markov process, we adapt the generator matrix (7.10) for HyperWord inventory to the case of $\lambda = 1$ and $\mu = 2$. Examining the embedded Markov chain for this problem (Equation (7.8)), we find that the states form one irreducible set of recurrent states. Therefore, it makes sense to compute the steady-state probabilities:

$$0 = \pi'G$$

$$= (\pi_0, \pi_1, \pi_2, \pi_3, \pi_4, \pi_5) \begin{pmatrix} -\frac{1}{2} & 0 & 0 & \frac{1}{2} & 0 & 0 \\ 1 & -\frac{3}{2} & 0 & 0 & \frac{1}{2} & 0 \\ 0 & 1 & -\frac{3}{2} & 0 & 0 & \frac{1}{2} \\ 0 & 0 & 1 & -1 & 0 & 0 \\ 0 & 0 & 0 & 1 & -1 & 0 \\ 0 & 0 & 0 & 0 & 1 & -1 \end{pmatrix}$$

and

$$1 = \pi'\mathbf{1}$$

$$= (\pi_0, \pi_1, \pi_2, \pi_3, \pi_4, \pi_5) \begin{pmatrix} 1 \\ 1 \\ 1 \\ 1 \\ 1 \\ 1 \end{pmatrix}$$

The complete system of equations is therefore

$$
\begin{aligned}
0 &= -\tfrac{1}{2}\pi_0 + \pi_1 \\
0 &= \qquad -\tfrac{3}{2}\pi_1 + \pi_2 \\
0 &= \qquad\qquad -\tfrac{3}{2}\pi_2 + \pi_3 \\
0 &= \tfrac{1}{2}\pi_0 \qquad\qquad - \pi_3 + \pi_4 \\
0 &= \qquad \tfrac{1}{2}\pi_1 \qquad\qquad - \pi_4 + \pi_5 \\
0 &= \qquad\qquad \tfrac{1}{2}\pi_2 \qquad\qquad - \pi_5 \\
1 &= \pi_0 + \pi_1 + \pi_2 + \pi_3 + \pi_4 + \pi_5
\end{aligned}
$$

and any one of the first five equations can be eliminated. Rewritten as balance equations, they are

j	rate into j		rate out of j
0	$1\pi_1$	$=$	$\tfrac{1}{2}\pi_0$
1	$1\pi_2$	$=$	$\left(1+\tfrac{1}{2}\right)\pi_1$
2	$1\pi_3$	$=$	$\left(1+\tfrac{1}{2}\right)\pi_2$
3	$\tfrac{1}{2}\pi_0 + 1\pi_4$	$=$	π_3
4	$\tfrac{1}{2}\pi_1 + 1\pi_5$	$=$	π_4
5	$\tfrac{1}{2}\pi_2$	$=$	π_5

and $\sum_{j=0}^{5}\pi_j = 1$ (recall that demands occur at a *rate* of 1 per day, while orders arrive at a *rate* of $1/2$ per day, when an order is pending).

Solving either system of equations gives

$$\pi = \begin{pmatrix} \tfrac{8}{35} \\ \tfrac{4}{35} \\ \tfrac{6}{35} \\ \tfrac{9}{35} \\ \tfrac{5}{35} \\ \tfrac{3}{35} \end{pmatrix} \approx \begin{pmatrix} 0.229 \\ 0.114 \\ 0.171 \\ 0.257 \\ 0.143 \\ 0.086 \end{pmatrix}$$

Therefore, the long-run fraction of the time that there is no inventory on hand is $\pi_0 = 8/35 \approx 0.229$. Recall that the simulation estimate of the fraction of the time that there is no inventory over 1000 days was 0.296 (Section 7.3), which is very close to π_0 as it should be. In fact, this is what we mean by time-independent, long-run results: the results we would obtain if the simulation could generate an infinitely long sample path. Over the long run the rate of lost sales is $\lambda \pi_0 = (1)(8/35) = 0.229$ sale per day, since demands occur at a rate of 1 per day but there are no copies available 8/35ths of the time.

Contrast these results with the Kolmogorov approximation for the probability that there is no inventory on hand at time t days, $\ddot{p}_{50}(t)$. Another way to think about the steady-state probabilities π is that if we continued the plot in Figure 7.4 for much larger t, then the $*$'s would get closer and closer to 0.229, while the $\bullet$'s would get closer and closer to 0.086. Steady state means that the *probability* of finding the process in a given state becomes constant, even though the behavior of the system is still stochastic.

7.6.6 Derivations

Our goal in this section is to derive Equations (7.26), (7.27) and (7.30). This section can be skipped by readers interested only in applications.

To derive the Kolmogorov differential equations, recall that the Chapman-Kolmogorov equation (7.19) states that

$$p_{ij}(t + \Delta t) = \sum_{h=1}^{m} p_{ih}(t) p_{hj}(\Delta t)$$

for any $t \geq 0$ and $\Delta t > 0$. The differential equations will be derived by substituting an approximation for $p_{hj}(\Delta t)$ into the Chapman-Kolmogorov equation.

If Δt is very small, then it is plausible that at most one transition will take place during this time interval. If the transition occurs, then it will be to state j with probability p_{hj}, the one-step transition probability from h to j for the embedded Markov chain. Whether or not it occurs depends on the time increment Δt, and clearly the larger Δt is, the more likely it is that there will be a transition. Just how likely depends on the transition rate *out of* state h, which is g_{hh}. Thus, it is plausible that $p_{hj}(\Delta t) \approx p_{hj}[g_{hh} \Delta t]$, for small Δt. Roughly, the probability of a transition from h to j in the time increment Δt is proportional to the probability of a transition out of h—given by $g_{hh} \Delta t$—times the probability that the transition is to state j—given by p_{hj}. Then using the fact that $p_{hj} = g_{hj}/g_{hh}$, we have

$$p_{hj}(\Delta t) \approx p_{hj} g_{hh} \Delta t = g_{hj} \Delta t \tag{7.32}$$

for $h \neq j$ and small Δt. However, it is also possible that no transition occurs during Δt. Then since

$$p_{hh}(\Delta t) = 1 - \sum_{j \neq h} p_{hj}(\Delta t)$$

we have

$$p_{hh}(\Delta t) \approx 1 - \sum_{j \neq h} g_{hj} \Delta t = 1 - g_{hh} \Delta t \qquad (7.33)$$

using our approximation (7.32) and the fact that $g_{hh} = \sum_{j \neq h} g_{hj}$.

Approximations (7.32) and (7.33) are our links between $p_{ij}(t)$ and $\mathbf{G}$. Substituting them into the Chapman-Kolmogorov equation, we obtain

$$p_{ij}(t + \Delta t) \approx \left(\sum_{h \neq j} p_{ih}(t)[g_{hj} \Delta t] \right) + p_{ij}(t)[1 - g_{jj} \Delta t] \qquad (7.34)$$

Subtracting $p_{ij}(t)$ from both sides of (7.34) and dividing by Δt yields

$$\frac{p_{ij}(t + \Delta t) - p_{ij}(t)}{\Delta t} \approx \sum_{h \neq j} p_{ih}(t)g_{hj} - p_{ij}(t)g_{jj} \qquad (7.35)$$

Then taking the limit of both sides of (7.35) as $\Delta t \to 0$ and assuming that the approximation becomes exact as Δt gets smaller (it does), we obtain

$$\frac{dp_{ij}(t)}{dt} = \sum_{h \neq j} p_{ih}(t)g_{hj} - p_{ij}(t)g_{jj}$$

Collecting these differential equations over all i and j and arranging them into matrix form gives the matrix differential equation $d\mathbf{P}(t)/dt = \mathbf{P}(t)\mathbf{G}$. The equation $d\mathbf{P}(t)/dt = \mathbf{G}\mathbf{P}(t)$ is derived similarly.

The boundary condition $\mathbf{P}(0) = \mathbf{I}$ follows from observing that

$$p_{ij}(0) = \Pr\{Y_0 = j | Y_0 = i\} = \begin{cases} 1, & i = j \\ 0, & i \neq j \end{cases}$$

is the only definition that makes sense.

To connect the Kolmogorov differential equations to the uniformization result, notice that if $\mathbf{G}$ was a scalar $\mathbf{G} = -g$, then $\mathbf{P}(t) = p(t)$ and the Kolmogorov differential equation is $dp(t)/dt = -g\,p(t)$ with boundary condition $p(0) = 1$. The solution to this differential equation is

$$p(t) = e^{-gt} = \sum_{n=0}^{\infty} \frac{(-gt)^n}{n!}$$

where we have expressed e^{-gt} in terms of its infinite series expansion. It is therefore plausible that when $\mathbf{G}$ is a matrix, the general solution to the Kolmogorov differential equations is

$$\boxed{\mathbf{P}(t) = e^{\mathbf{G}t} = \sum_{n=0}^{\infty} \mathbf{G}^n \frac{t^n}{n!}} \qquad (7.36)$$

where we define the matrix exponential $e^{\mathbf{G}t}$ to be the series expansion on the right. This solution is in fact correct! And while it is not easy to show directly, it can be shown that

$$\sum_{n=0}^{\infty} \mathbf{G}^n \frac{t^n}{n!} = e^{-g^*t} \sum_{n=0}^{\infty} \mathbf{Q}^n \frac{(g^*t)^n}{n!}$$

In other words, the uniformization result is a different series expansion of $e^{\mathbf{G}t}$. One reason we prefer the uniformization expansion is that the series (7.36) is not numerically stable since $\mathbf{G}$ contains both positive and negative elements—which induces rounding error—while $\mathbf{Q}$ contains no negative elements. A second reason we prefer uniformization is that is has a physical interpretation, which we derive next.

Our approach is to develop a *discrete-time* Markov chain $\{\mathcal{S}_n; n = 0, 1, \ldots\}$ with one-step transition matrix $\mathbf{Q}$ that mimics the *continuous-time* Markov process $\{Y_t; t \geq 0\}$ as closely as possible, including matching the expected time spent in each state. The Markov chain $\mathcal{S}_n$ will be different from the embedded Markov chain $\{S_n; n = 0, 1, \ldots\}$ that describes the state changes in $\{Y_t; t \geq 0\}$, and it will allow us to compute performance measures using the tools we already have available for Markov chains. Three steps are required:

1. **Match time scale.** To make $\{\mathcal{S}_n; n = 0, 1, \ldots\}$ and $\{Y_t; t \geq 0\}$ comparable, we start by measuring time on the same standard scale for both processes. The time scale we choose is $1/g^*$, which is the *minimum expected holding time* in any state for Y_t (recall that $g^* = \max_i\{g_{ii}\}$, and the expected holding time in state i is $1/g_{ii}$). Thus, one transition of $\mathcal{S}_n$ will correspond to $1/g^*$ units of continuous time, while the expected holding time in state i for Y_t will be $E[H_i] = (1/g_{ii})/(1/g^*) = g^*/g_{ii}$ in units of $1/g^*$. By construction $g^*/g_{ii} \geq 1$, so the expected holding time in each state is at least 1 time unit on the standardized scale.

   For instance, in the HyperWord inventory example $g^* = 3/2$, so that each transition of the approximating Markov chain $\mathcal{S}_n$ corresponds to 2/3 of a day.

2. **Match expected time spent in each state.** The expected time spent in state i for the Markov chain $\mathcal{S}_n$ is geometrically distributed with parameter $\gamma = 1 - q_{ii}$, where q_{ii} is the iith element of $\mathbf{Q}$ (yet to be determined). Since the geometric distribution is the unique *discrete* distribution with the memoryless property, it is the appropriate distribution to match the *continuous* exponential distribution of holding time for the Markov process Y_t. The expected value of the geometric distribution is $1/\gamma = 1/(1 - q_{ii})$. Since we want the expected time spent in state i of $\mathcal{S}_n$ to match the expected holding time in state i of Y_t, we set

$$\frac{1}{1 - q_{ii}} = \frac{g^*}{g_{ii}}$$

Solving for q_{ii} yields $q_{ii} = 1 - g_{ii}/g^*$, which determines all of the diagonal elements of $\mathbf{Q}$.

For instance, in the HyperWord inventory example $q_{00} = 1 - g_{00}/g^* = 1 - (1/2)/(3/2) = 2/3$, which is the probability of a transition from 0 copies in inventory to 0 copies in inventory for the approximating process S_n. While a transition from a state to itself may not make physical sense, it is necessary for the time spent in state 0 of S_n to approximate the holding time in state 0 of Y_t.

3. **Match state changes.** Now we need to determine appropriate off-diagonal elements q_{ij} of $\mathbf{Q}$. While it might seem that we want to set $q_{ij} = p_{ij}$, where p_{ij} is the transition probability from the embedded Markov chain S_n, we cannot do so because $q_{ii} \neq p_{ii} = 0$ for the approximating process S_n (see Step 2 above). The transition matrix $\mathbf{Q}$ differs from $\mathbf{P}$ in that we allow transitions from a state to itself in $\mathbf{Q}$, but not in $\mathbf{P}$. What we want to ensure is that the probability that S_n moves from state i to state j, *given that the next transition does take the process out of state i*, is equal to p_{ij}. Therefore, we want

$$
\begin{aligned}
p_{ij} &= \Pr\{S_{n+1} = j | S_n = i, S_{n+1} \neq i\} \\
&= \frac{\Pr\{S_{n+1} = j, S_{n+1} \neq i | S_n = i\}}{\Pr\{S_{n+1} \neq i | S_n = i\}} \\
&= \frac{\Pr\{S_{n+1} = j | S_n = i\}}{\Pr\{S_{n+1} \neq i | S_n = i\}} \\
&= \frac{q_{ij}}{1 - q_{ii}}
\end{aligned}
$$

Solving for q_{ij} and substituting $p_{ij} = g_{ij}/g_{ii}$ gives

$$
q_{ij} = (1 - q_{ii})p_{ij} = \frac{g_{ii}}{g^*}\frac{g_{ij}}{g_{ii}} = \frac{g_{ij}}{g^*}
$$

for $i \neq j$.

For instance, in the HyperWord inventory example $q_{03} = g_{03}/g^* = (1/2)/(3/2) = 1/3$, which is the probability of a transition from 0 copies in inventory to 3 copies in inventory for the approximating process S_n. Notice that $q_{03} \neq p_{03} = 1$.

Collecting these results together we find that $\mathbf{Q}$ can be represented as

$$
\mathbf{Q} = \begin{pmatrix} 1 - g_{11}/g^* & g_{12}/g^* & \cdots & g_{1m}/g^* \\ g_{21}/g^* & 1 - g_{22}/g^* & \cdots & g_{2m}/g^* \\ \vdots & \vdots & \vdots & \vdots \\ g_{m1}/g^* & g_{m2}/g^* & \cdots & 1 - g_{mm}/g^* \end{pmatrix} = \mathbf{I} + \frac{1}{g^*}\mathbf{G} \qquad (7.37)
$$

The expected time spent in each state of $\{S_n; n = 0, 1, \ldots\}$ now matches $\{Y_t; t \geq 0\}$, and the processes move in the same manner from state to state, so the long-run fraction of time spent in each state is the same for both processes. Therefore, to find the steady-state probabilities for Y_t when they exist, we can solve the equations for the steady-state probabilities of the Markov chain $\mathcal{S}_n$:

$$\boldsymbol{\pi}' = \boldsymbol{\pi}'\mathbf{Q} = \boldsymbol{\pi}'\left(\mathbf{I} + \frac{1}{g^*}\mathbf{G}\right) = \boldsymbol{\pi}' + \frac{1}{g^*}\boldsymbol{\pi}'\mathbf{G}$$

Subtracting $\boldsymbol{\pi}'$ from both sides and multiplying through by g^* gives Equation (7.30).

To obtain time-dependent results, it might seem that we could write $\mathbf{P}(t) = \mathbf{Q}^n$, the n-step transition probabilities for $\mathcal{S}_n$, since $\mathcal{S}_n$ approximates Y_t. The problem is that we do not know what value of n to use, since the number of transitions of Y_t by time t is random. To account for the stochastic nature of the transition times, let N_t be a random variable representing the number of transitions of $\mathcal{S}_n$ that occur by continuous time t. The times between transitions of Y_t are exponentially distributed, so we let the times between transitions of $\mathcal{S}_n$ also be exponentially distributed. And since the basic time unit for $\mathcal{S}_n$ is $1/g^*$, we take the expected time between transitions be $1/g^*$. Therefore, N_t has a Poisson distribution with parameter g^*t.

Now applying the law of total probability

$$\mathbf{P}(t) = \sum_{n=0}^{\infty}\mathbf{Q}^n\Pr\{N_t = n\} = \sum_{n=0}^{\infty}\mathbf{Q}^n e^{-g^*t}\frac{(g^*t)^n}{n!}$$

The Markov chain $\{\mathcal{S}_n; n = 0, 1, \ldots\}$ and the Poisson random variable N_t are called the *uniformization* of $\{Y_t; t \geq 0\}$. This representation suggests that yet another representation of a Markov process is as a Markov chain whose transition times are determined by the arrivals in a Poisson arrival process. In Exercise 18 you are asked to show that this is not an approximation at all.

7.7 THE CASE OF THE STRESSED-OUT STUDENT

Case 7.2. Orlando S. Underwood is a business student at a large university, and he is paying for his education through a work-study job. Orlando types and answers the telephone for faculty of the Management Science Department. Phone calls receive first priority, so Orlando interrupts his typing whenever the phone rings.

To make better use of Management Science office staff, the Chair of the Department has proposed transferring the responsibility for answering her private phone line from an administrative associate to Orlando. The administrative associate is happy to give up this responsibility, but she is worried that it may significantly reduce the typing that Orlando can accomplish, as well as degrading service for callers. Orlando thinks he spends about three-quarters his time typing and one-quarter of his time answering the phone, but he has never kept detailed records. The Chair has offered Orlando an (unspecified) raise for a careful assessment of the impact of the proposed change.

Orlando first asks how the new system will work. The answer is that Orlando will be given a phone with two lines, one dedicated to the Chair and one for all other calls. If a call comes for the Chair then Orlando must put any other ongoing call on hold, but he will never interrupt a call for the Chair.

Some data are available. The administrative associate keeps a log of calls to the Chair. For instance, over the last 20 school days—each 8 working hours in length—960 calls were logged. The administrative associate thinks she spends about 1 minute with each caller, although the actual time can vary considerably. On the day after he was given this assignment, Orlando noted that he received 80 calls, or a call every 6 minutes on average. The Department Chair reminds Orlando that this average does not include those callers who received a busy signal because the phone is in use. They agree, as an approximation, to say that there is an attempted call every 5 minutes, but some of those calls receive a busy signal. Orlando finds that he spends about 2 minutes with each caller.

We are interested in what will happen to an existing system if we change it, which is a common situation in practice. The outcome of the change is not obvious. We cannot simply subtract the time Orlando will spend answering the Chair's phone from the time he currently spends typing, because the two types of calls will interact due to the priority system (Chair first, then other calls).

Is it necessary to develop a model of the existing system, even though it can be observed directly? There are at least two reasons why the answer is yes: First, if our model of the existing system does a good job representing it, then we can have more confidence that our model of the proposed system will also be *valid*. A second reason is that both models are approximations, and we may obtain a better assessment of the *relative difference* between the new and existing systems by comparing the two models, as opposed to comparing the model of the new system to the existing system itself. This is because both models are based on some of the same approximations, so even if they do not perfectly represent each system individually, they may capture the difference in performance quite well.

In the current system Orlando's status can be represented by two states, typing and answering the phone. Let $\mathcal{M} = \{1, 2\} = \{$typing, phone$\}$ be the state space. To parameterize the generator matrix of a Markov process model, we can use the interpretation that g_{12} is the transition rate from typing to answering the phone, which is 1 call every 5 minutes or $g_{12} = 1/5 = 0.2$ call per minute. The expected holding time in state 2 is 2 minutes per call, so $1/g_{22} = 2$, or $g_{22} = 1/2 = 0.5$ call per minute. Then using the fact that each row of $\mathbf{G}$ must sum to one,

$$\mathbf{G} = \begin{pmatrix} -0.2 & 0.2 \\ 0.5 & -0.5 \end{pmatrix}$$

The steady-state probabilities $\pi' = (\pi_1, \pi_2)$ give the long-run fraction of time spent in each state. In this case $\pi_1 \approx 0.71$, which is a reasonable match to Orlando's feeling that he spends 75% of his time typing. Since callers continue to try to call even when the phone is busy, the rate at which callers receive a busy signal is $(0.2$ call per minute$)\pi_2 \approx (0.2)0.29 \approx 0.06$ call per minute, or about 3.5 calls per hour.

We have formulated a Markov process model without verifying the conditions that make a process a Markov process. Those conditions imply that the arrival of calls is a Poisson process with rate 0.2 call per minute, and the time to handle a call is exponentially distributed with mean 2 minutes. A Poisson process is often a good model of a large number of callers acting independently, and exponentially distributed call durations may be reasonable if most calls are quite brief, but there is occasionally a very long call.

To model the proposed system we might define the state space to be typing, answering the department phone and answering the Chair's phone. This formulation encounters difficulties, however, when we consider what happens after Orlando finishes handling a call to the Chair. Does he return to typing, or does he finish a department call that is on hold? We cannot answer the question without knowing whether or not there is a call on hold, which implies that we need this information in our state space. Let the refined state space be $\mathcal{M} = \{1, 2, 3, 4\} = \{$typing, phone, chair's phone, chair's phone + hold$\}$.

Consider state 1 corresponding to typing. At rate 0.2 Orlando moves to state 2, due to department calls. But at rate 0.1 call per minute he moves to state 3 corresponding to answering the Chair's phone. This rate is derived from the data: 960 calls in (20 days) $\times$ (8 hours/day) $\times$ (60 minutes/hour) = 9600 minutes. Therefore $g_{12} = 0.2$, $g_{13} = 0.1$, and $g_{14} = 0$ since the hold state cannot be entered unless there is an ongoing department call.

In state 2, corresponding to answering a department call, Orlando moves back to typing at rate $g_{21} = 0.5$ call per minute, as established earlier. However, he can also move to state 4 (chair's phone+hold) if a call arrives for the chair. This occurs at rate $g_{24} = 0.1$ call per minute. No other transition is possible in one step.

In state 3, corresponding to answering the Chair's phone with no call on hold, the expected time spent in the state is 1 minute. Therefore, $1/g_{33} = 1$ minute. Since calls to the Chair are never interrupted, the only possible transition out of state 3 is back to typing. Thus, $g_{31} = g_{33} = 1$.

Using similar reasoning, the transition rate out of state 4—which corresponds to answering the Chair's phone while a call is on hold—is $g_{42} = 1$. In other words, after completing the call to the Chair, Orlando returns to the call on hold. The complete generator matrix is therefore

$$
\mathbf{G} = \begin{pmatrix} -0.3 & 0.2 & 0.1 & 0 \\ 0.5 & -0.6 & 0 & 0.1 \\ 1 & 0 & -1 & 0 \\ 0 & 1 & 0 & -1 \end{pmatrix}
$$

The steady-state probabilities are obtained by solving $\pi'\mathbf{G} = \mathbf{0}$ and $\pi'\mathbf{1} = 1$, or

$$
\begin{aligned}
-0.3\pi_1 + 0.5\pi_2 + \pi_3 \quad\quad &= 0 \\
0.2\pi_1 - 0.6\pi_2 \quad\quad + \pi_4 &= 0 \\
0.1\pi_1 \quad\quad - \pi_3 \quad\quad &= 0 \\
0.1\pi_2 \quad\quad - \pi_4 &= 0 \\
\pi_1 + \pi_2 + \pi_3 + \pi_4 &= 1
\end{aligned}
$$

The solution is $\pi' \approx (0.65, 0.26, 0.06, 0.03)$. Interpreting this result, Orlando will spend only 65% of his time typing over the long run in the proposed system, down from 71%. Department calls will be missed whenever Orlando is on any phone, so the rate of missed calls rises slightly from 0.06 call per minute to (0.2

call per minute)$(\pi_2 + \pi_3 + \pi_4) = 0.07$ call per minute. Notice that $\pi_4 = 0.03$ implies that there is a call on hold only 3% of the time. Unless typing time is critical, it does not appear that the burden of answering the Chair's phone will add too much stress to Orlando's job.

To refine the model, Orlando could invest some effort into obtaining a better estimate of the rate of attempts to call the department. Recall that the observed rate of *received* calls was 1 every 6 minutes, which was used as the basis for guessing the rate of attempts. The proper inflation factor depends on how likely a caller is to receive a busy signal. Estimates of this probability could be obtained by asking callers how many attempts they made before getting through, or having someone attempt calls and record how often a busy signal was received.

7.8   THE MARKOV AND STATIONARITY PROPERTIES REVISITED

A Markov process is not an appropriate model for all continuous-time processes. Critical to the construction of a Markov process are the conditions that (a) at any time $t \geq 0$ the set of pending system events depends only on the state of the process, Y_t; (b) the next state of the process after time t depends only on the value of Y_t and which of the pending system events occurs first; and (c) the clock settings for all system events are independent, time-stationary, exponentially distributed random variables.

In Section 7.5.1 we remarked that a process satisfying conditions (a)–(c) has the Markov and stationarity properties in continuous time. In other words, probability statements about future behavior of the process are independent of the past behavior of the process, given we know the current state; and these probability statements do not depend on the current time, but only on how far into the future we wish to look. From a modeling perspective, it is important to recall that the converse is also true: a continuous-time, discrete-state process satisfying the Markov and stationarity properties must be a continuous-time Markov process. Therefore, if the system at hand approximately fulfills these properties, then a Markov process model is appropriate.

We can force conditions (a) and (b) to hold by expanding the state space to contain as much information as required, although this could make the state space enormous and the model practically useless. Condition (c) will not be applicable for many systems, but even then a Markov process model may provide a useful rough-cut approximation. To obtain the approximation, we typically match the expected value of the exponential distribution to the expected value of the distribution we would like to use. The key question is, How does substituting an exponential distribution for another distribution affect the performance measures derived from the model? The following examples provide some answers (see Sections 7.5.2 and 9.2 for additional illustrations).

Case 7.3.   Boodle Bank intends to develop a model of its cash flows. The checks that it receives from depositors and from investments are one source of cash that it will include in the model.

When Boodle redeems a check, it takes an average of 2 days to "clear," meaning that it is 2 days before Boodle receives its money. To develop a Markov process model, Boodle could approximate the time to clear as a random variable C that is exponentially distributed with expected value 2 days. But in reality it takes a minimum of 1 day for a check to clear, so a better model is $C^* = 1 + X$ days, where X is a random variable taking values zero or greater and having expected value 1.

Notice that $\Pr\{C^* \le 1\} = 0 \ne \Pr\{C \le 1\} \approx 0.4$. In words, the actual time to clear is never less than 1 day, while the exponential-distribution approximation has roughly a 0.4 chance of being less. Therefore, the performance measures derived from Boodle's model might indicate that cash flows are available earlier than they will be in reality, and it should treat the results as somewhat optimistic.

Case 7.4. Residents of the state of Ohio can obtain their automobile license plates by submitting a mail-in renewal form to the Bureau of Motor Vehicles during the month of their birth. A model of the overall workload of the registrar's office is desired.

To be concrete suppose that 1 month is 30 days and that the expected time when a renewal form arrives is the fifteenth day of the month. Then we could model the day on which an application arrives as being an exponentially distributed random variable A with expected value 15 days. However, since applications beyond the end of the month are not accepted, a better model is a random variable A^* whose range is only $[0, 30]$ days.

Notice that $\Pr\{A^* > 30\} = 0 \ne \Pr\{A > 30\} \approx 0.14$. In words, an application is never accepted after 30 days, while the exponential-distribution approximation has about a 0.14 chance of being greater. Therefore, a Markov process model might imply that somewhat fewer applications are received during a month than actually will be received.

Case 7.5. Electronic mail (e-mail) arrives to a central computer that distributes the e-mail to the computer workstations of the recipients. The sizes of the messages vary from a few bytes to several thousand bytes.

This is an example of a queueing system. The "customers" are the e-mail messages, and the "server" is the computer that distributes them. Suppose that the expected time between the arrival of messages is approximately a seconds, while the expected time to process and distribute a message is approximately $b < a$ seconds.

If e-mail messages arrived *exactly* a seconds apart and took *exactly* b seconds to process, then there would be no queueing and no message would ever be delayed. Queueing occurs when there is variability around these expected values. A few messages arrive with gaps less than a seconds, or a few large messages take longer than b seconds to process, and suddenly there is a queue. The more the variability, the more likely queueing and delay become.

A Markov process model might treat the interarrival-time gaps as exponentially distributed random variables with expected value a (that is, a Poisson arrival process with arrival rate $1/a$; see Chapter 5). And the time to process a message might be modeled as an exponentially distributed random variable with expected value b. From this model we could derive performance measures such as the expected number of messages in the queue and the expected delay for a message (Chapter 8).

Suppose that a more faithful model of these input processes requires distributions other than the exponential. If the desired distributions are more variable than the exponential, then more queueing and delay should be anticipated. If they are less variable than the exponential, then less queueing and delay should be anticipated. Therefore, we can regard the Markov process results as optimistic or pessimistic, depending on the situation.

A useful way to compare the variability of random variables with the same expected value is via their coefficients of variation (Chapter 3, Equation 3.10). Recall that exponentially distributed random variables have a coefficient of variation of 1. When the means are the same, coefficients of variation larger than 1 typically imply more congestion, while coefficients of variation smaller than 1 typically imply less congestion, than what is predicted by a Markovian queueing model.

The point of these cases is that a Markov process model can differ from a more accurate model in terms of the lower limits, upper limits and variability of the input distributions, but the effect of these differences on the results can often be predicted, particularly in models of service systems.

We do not necessarily have to use exponential distributions to obtain performance measures via mathematical analysis for continuous-time processes. In Section 7.9 we introduce the *semi-Markov process*, a process in which the state transitions follow an embedded Markov chain, but the holding time in each state may have any probability distribution. Although less amenable to mathematical analysis than the Markov process, some useful results are available for the semi-Markov process.

7.9   SEMI-MARKOV PROCESSES

Recall Case 6.1, the Case of the Random Behavior, which required modeling customer sessions for Data Driven, Inc. A customer session consists of some combination of the transactions log on (state 1), fetch a document (state 2), read a page of a document (state 3) and log off (state 4), always beginning with a log on. The model was a Markov chain with one-step transition matrix

$$
\mathbf{P} = \begin{pmatrix} 0 & 0.95 & 0.01 & 0.04 \\ 0 & 0.27 & 0.63 & 0.10 \\ 0 & 0.36 & 0.40 & 0.24 \\ 0 & 0 & 0 & 1 \end{pmatrix}
$$

If the number of customers connected to the computer system is relatively constant or if the system operates near capacity most of the time, then it may be reasonable to alter the model so that the system is closed. In other words, whenever a customer logs off, another customer immediately logs on. The new transition matrix is

$$\mathbf{P} = \begin{pmatrix} 0 & 0.95 & 0.01 & 0.04 \\ 0 & 0.27 & 0.63 & 0.10 \\ 0 & 0.36 & 0.40 & 0.24 \\ 1 & 0 & 0 & 0 \end{pmatrix} \tag{7.38}$$

where $p_{41} = 1$ indicates that a log off is immediately followed by a log on.

The states of the modified Markov chain now constitute to a single irreducible set of recurrent states. Thus, it makes sense to determine the long-run probability of finding a customer thinking about any particular transaction or, interpreted differently, the fraction of all customers who are thinking about each of the four transactions. This probability will depend not only on how transitions occur from state to state—which is what $\mathbf{P}$ describes—but also on how long a customer spends "thinking" about each type of transaction.

Suppose that the think time a customer spends thinking about transaction type i can be modeled as a random variable having some unspecified probability distribution, not necessarily exponential, that depends on the state i but is independent of everything else. Is this enough information to answer the question? In this section we show that it is.

Let $\{Y_t; t \geq 0\}$ be a continuous-time process with state space $\mathcal{M} = \{1, 2, \ldots, m\}$. If the embedded state-change process $\{S_n; n = 0, 1, \ldots\}$ is a Markov chain with transition matrix $\mathbf{P}$ and the holding time in each state is independent of everything except the state of the process, then $\{Y_t; t \geq 0\}$ is called a **semi-Markov process**.

A semi-Markov process is no more difficult to simulate than a Markov process, but it is much more difficult to analyze mathematically. Nevertheless, there are some useful results available. In particular, we can obtain the steady-state probability

$$\pi_i \equiv \lim_{t \to \infty} \Pr\{Y_t = i\}$$

when it exists.

If the embedded Markov chain S_n has a steady-state distribution $\boldsymbol{\xi} = (\xi_1, \xi_2, \ldots, \xi_m)'$ that satisfies

$$\boldsymbol{\xi}' = \boldsymbol{\xi}'\mathbf{P}$$

$$1 = \boldsymbol{\xi}'\mathbf{1}$$

then the steady-state probability π_i for the semi-Markov process Y_t is

$$\pi_i = \frac{\xi_i \tau_i}{\sum_{j=1}^m \xi_j \tau_j} \tag{7.39}$$

where $\tau_i < \infty$ is the expected holding time in state i.

We will not derive Equation (7.39), but it makes intuitive sense: The steady-state probabilities for the embedded Markov chain ξ_i account for the long-run fraction of the time each state is visited. They are weighted by the expected time spent in each state to obtain the steady-state probabilities for the semi-Markov process. Conveniently, the result depends only on the expected holding time in each state, and not the actual distribution of holding time. Notice also that Equation (7.39) provides an alternative method for calculating the steady-state probabilities for a Markov process by setting $\tau_i = 1/g_{ii}$.

For example, in the customer behavior model suppose that the expected times a customer spends thinking about a log on, fetch, read and log off are 12, 15, 30 and 3 seconds, respectively. That is, $(\tau_1, \tau_2, \tau_3, \tau_4) = (12, 15, 30, 3)$. The steady-state probabilities for the Markov chain (7.38) are, to three decimal places,

$$\xi = \begin{pmatrix} 0.132 \\ 0.358 \\ 0.378 \\ 0.132 \end{pmatrix}$$

This gives $\sum_{j=1}^4 \xi_j \tau_j = 18.695$. Therefore, the steady-state probability of finding a customer thinking about a read (state 3), for example, is

$$\pi_3 = \frac{\xi_3 \tau_3}{18.695} = \frac{(0.378)(30)}{18.695} \approx 0.607$$

which is much larger than $\xi_3 = 0.378$, reflecting the long expected holding time in state 3.

7.10   FINE POINTS

In this chapter we put substantial emphasis on the embedded Markov chain S_n within a Markov process Y_t. This is justified for the following reasons:

- The embedded Markov chain and holding times provide intuition about how the sample paths of a Markov process evolve over time.
- We can classify the states of a continuous-time Markov process without introducing any additional concepts beyond those introduced for discrete-time Markov chains.
- Probability statements about the sequence of states a continuous-time Markov process visits, including where it is absorbed, can be derived using the analysis tools for discrete-time Markov chains.

- The embedded Markov chain and holding-time view provide an alternative way to simulate a Markov process: Simulate the state transitions from a Markov chain; then generate the holding times from appropriate exponential distributions.

Stochastic processes in continuous time can exhibit unusual behavior that is not (to our knowledge) found in physical systems that we desire to model. Certain conditions can be imposed to avoid these pathological cases. A complete explanation of the need for these conditions is beyond the purview of this book, but we can list them:

- $\lim_{t \downarrow 0} p_{ij}(t) = \mathcal{I}(i = j)$
- $g_{ii} \leq c$ for all i and some $c < \infty$

To have $p_{ij}(0) = \Pr\{Y_0 = j | Y_0 = i\} = 1$ only if $i = j$ makes sense. The first condition simply ensures that the function $p_{ij}(t)$ is continuous at 0. And since $1/g_{ii}$ is the expected holding in state i, the second condition ensures that no state has an expected holding time of 0 or that its expected holding time approaches 0 arbitrarily closely. For a lucid explanation of why these conditions are required, see Çinlar (1975, Chapter 8).

This chapter focused on Markov process models in which the state space $\mathcal{M}$ is finite; that is, $m < \infty$. There are modeling-and-analysis problems in which there is no natural upper bound on the number of states (see, for instance, Chapter 8 in which the state of the system is the number of customers waiting to be served in a queueing system). As for Markov chains, an infinite state space does not change any of the time-dependent performance measures in Section 7.6.2, but does introduce complications in the determination and interpretation of time-independent performance measures.

As mentioned in Section 7.6.4, the embedded Markov chain in Y_t can be used to classify states as recurrent or transient. If there are an infinite number of recurrent states, then the states can be either positive recurrent or null recurrent, as defined in Section 6.9. Unfortunately, positive- or null-recurrent states of a Markov process *cannot* be distinguished by examining the embedded Markov chain. Fortunately, the additional test required is easy: If the steady-state equations (7.30) have no solution, then the recurrent states (as determined by the embedded Markov chain) are null recurrent; if (7.30) has a solution, then they are positive recurrent.

7.11 EXERCISES

7.1. For a Markov process $\{Y_t; t \geq 0\}$ with state space $\mathcal{M} = \{1, 2\}$ and generator matrix

$$\mathbf{G} = \begin{pmatrix} -2 & 2 \\ 5 & -5 \end{pmatrix}$$

compute π_1 and π_2.

7.2. Consider the Markov processes with state space $\mathcal{M} = \{1, 2, 3, 4, 5\}$ and generator matrices given below. For each one, (i) derive the embedded Markov chain transition matrix; (ii) classify the states; and (iii) compute the steady-state probabilities for each irreducible set of recurrent states.

(a)

$$G = \begin{pmatrix} -1 & 1 & 0 & 0 & 0 \\ 2 & -5 & 2 & 1 & 0 \\ 1 & 0 & -4 & 0 & 3 \\ 3 & 1 & 0 & -6 & 2 \\ 0 & 0 & 0 & 5 & -5 \end{pmatrix}$$

(b)

$$G = \begin{pmatrix} -1 & 1 & 0 & 0 & 0 \\ 2 & -4 & 2 & 0 & 0 \\ 1 & 0 & -4 & 0 & 3 \\ 0 & 0 & 0 & -2 & 2 \\ 0 & 0 & 0 & 5 & -5 \end{pmatrix}$$

(c)

$$G = \begin{pmatrix} 0 & 0 & 0 & 0 & 0 \\ 2 & -4 & 2 & 0 & 0 \\ 1 & 0 & -4 & 0 & 3 \\ 0 & 0 & 0 & -2 & 2 \\ 0 & 0 & 0 & 0 & 0 \end{pmatrix}$$

7.3. For each of the Markov processes in Exercise 2, (i) write the Kolmogorov differential equations and (ii) fill in numerical values for the uniformization expression (7.26).

7.4. Develop a simulation of Case 7.2 that could be used to analyze the situation for *any* distributions of time between calls and time to answer calls.

7.5. The Football State University motor pool maintains a fleet of vans to be used by faculty and students for travel to conferences, field trips, etc. Requests to use a van occur at about 8 per week on average, and a van is used for an average of 2 days (but there is quite a bit of variability around both numbers). If someone requests a van and one is not available, then other transportation, not provided by the motor pool, must be found. The motor pool currently has 4 vans, but due to university "restructuring," it has been asked to reduce its fleet. In order to argue against the proposal, the director of the Motor Pool would like to predict how many requests for vans will be denied if the fleet is reduced from 4 to 3.
   (a) Formulate a model of the 3-van system. Carefully state all of your approximations.
   (b) Provide an estimate for the rate at which requests are denied.
   (c) Provide an estimate for the average number of vans in use.

7.6. Prove the memoryless properties of the geometric and exponential distributions (Equations (7.1) and (7.2)) by directly applying the definition of conditional probability.

7.7. Simulate the HyperWord inventory model for 1000 days with $\lambda = 1$, $\mu = 1/2$ and $(r, s) = (3, 7)$. Add code to compute $\bar{Y}$, the time-average inventory level, and $\bar{K}$, the fraction of time that there is no inventory. Notice that $\lambda \bar{K}$ is an estimate of the *lost-sales rate*, since λ is the demand rate (units/day) and $\bar{K}$ is the fraction of time that there is no inventory to satisfy the demand. Compare your estimate of the lost-sales rate to the long-run lost-sales rate $\lambda \pi_0$, and your estimate of the time-average inventory level to the long-run expected inventory level, $\sum_{j=0}^{7} j \, \pi_j$.

7.8. For the HyperWord inventory model with $\lambda = 1$ and $\mu = 1/2$, evaluate the policies $(r, s) = (3, 7)$ and $(2, 7)$ and argue which one is better.

7.9. For the Markov process approximation of the Bit Bucket Computers model developed in Section 7.5.2, compute the probability that failure occurs in less than 2 years. Compare this to the simulation estimate of 0.02 obtained in Chapter 4. Is the assumption of exponential distributions an optimistic or pessimistic approximation? Explain why.

7.10. Two automated testing machines work together testing circuit boards. Each one is independently subject to failure. The failure rate of an automated testing machine, when it is in use, is 0.01 per hour, with the actual time to failure being exponentially distributed. The time required to repair an automated testing machine is also exponentially distributed with mean 24 hours, and only one machine can be repaired at a time. When one of the automated testing machines has failed, the other handles all of the work, which increases its failure rate to 0.02 per hour.

(*a*) What is the long-run fraction of the time that both testing machines are not working?

(*b*) What is the long-run fraction of time that at least one testing machine is not working?

7.11. For a Markov process with generator matrix **G**, let $\mathcal{T}$ be the set of transient states (assume that there are $m_{\mathcal{T}} > 0$ transient states). Let μ_{ij} denote the expected total time the process spends in transient state j given $\{Y_0 = i\}$. Show that

$$g_{ii} \mu_{ij} = \mathcal{I}(i = j) + \sum_{k \in \mathcal{T}, k \neq i} g_{ik} \mu_{kj}$$

where $\mathcal{I}(i = j)$ is the indicator function taking the value 1 if $i = j$ and 0 otherwise. Taken over all $i, j \in \mathcal{T}$, this is a system of $m_{\mathcal{T}}^2$ equations in the $m_{\mathcal{T}}^2$ unknowns μ_{ij}. Hint: Argue instead that

$$\mu_{ij} = \frac{\mathcal{I}(i = j)}{g_{ii}} + \sum_{k \in \mathcal{T}, k \neq i} p_{ik} \mu_{kj}$$

(where p_{ik} is a one-step transition probability from the embedded Markov chain) by conditioning on the first transition out of state i.

7.12. For a Markov process that consists of a single, irreducible set of m recurrent states and generator matrix $\mathbf{G}$, let V_{ij} be the total time for the process to move from state i to state j for the first time. Let $v_{ij} = E[V_{ij}]$ be the expected time to move from state i to j. Show that

$$g_{ii} v_{ij} = 1 + \sum_{k \neq i,j} g_{ik} v_{kj}$$

Taken over all i, j, this is a system of m^2 equations in the m^2 unknowns v_{ij}. Hint: Argue instead that

$$v_{ij} = \frac{1}{g_{ii}} + \sum_{k \neq i,j} p_{ik} v_{kj}$$

(where p_{ik} is a one-step transition probability from the embedded Markov chain) by conditioning on the first transition out of state i.

7.13. (This exercise is repeated from Exercise 27 in Chapter 6, with an embellishment.) In this problem the goal is to model the use of automated teller machines (ATMs) at a bank.

After inserting their card into an ATM, there are three types of transactions that customers may perform: deposit, withdrawal, and obtain account information. The bank believes that 50% of all customers start with a withdrawal, 40% start with a deposit, and the remainder start by requesting account information.

After completing a transaction, 90% of the customers complete their business (obtain their card and leave); those who do not complete their business are equally likely to select one of the other two types of transactions (for example, if they just made a withdrawal and they do not complete their business, then they are equally likely next to select a deposit or request account information). This pattern continues until their business finally is completed.

Suppose that we have the following additional information about *how long* a customer spends on each transaction. All times are modeled as exponentially distributed random variables.

transaction	expected time
decide on first transaction	30 seconds
withdrawal	1 minute
deposit	2 minutes
account information	1 minute

(a) Derive a Markov process model capable of answering the questions below. Be sure to define your state space, time index and generator matrix.

(b) Do you think the assumption of exponentially distributed transaction times is appropriate? Why or why not?

(c) What is the probability that a customer takes longer than 4 minutes at the ATM? Calculate a numerical result.

(d) Use the result in Exercise 11 to calculate the expected time a customer spends at the ATM.

7.14. Phone calls arrive at a voice-mail system at a rate of 20 per hour according to a Poisson process (the voice-mail system routes calls to appropriate operators based on pushing buttons on the phone; understanding this is not important for working the problem).

The voice-mail system can handle only one call at a time (in other words, it is a one-server system with no queueing). If a second call arrives while the voice-mail system is busy, then it is routed to a human operator; there is only one human operator. If the human operator is also busy, then the call is lost. The mean time for the voice-mail system to route a call is 2 minutes (once it has routed the call, it is free to take another call). The human operator processes calls at a rate of 10 per hour. All times are exponentially distributed.

(*a*) Derive a Markov process model capable of answering the questions below. Be sure to define your state space, time index and generator matrix.

(*b*) Over the long run, what fraction of the time is the human operator busy?

(*c*) What is the probability that an arriving call will be routed by the voice-mail system?

7.15. The problem is to compare four different computer systems that handle airline reservations.[2] The single measure of performance is the expected time to system failure, E[TTF]. The system works if either of two computers works, and the computers are repaired one at a time when they fail. Computer failures are rare, repair times are fast, and the resulting E[TTF] is large. The four systems arise from variations in parameters affecting the time-to-failure and time-to-repair distributions as given in the following table (all rates are in failures or repairs per minute):

system	failure rate λ	repair rate μ
A	9.0×10^{-3}	9.0
B	9.5×10^{-3}	9.5
C	10.0×10^{-3}	10.0
D	10.5×10^{-3}	9.5

Suppose that the time to failure and time to repair are exponentially distributed random variables. Find the alternative with the largest E[TTF]. (Hints: First find the embedded Markov chain and determine the expected number of *transitions* until system failure. Then find the expected holding time between transitions. Or use the result in Exercise 11 to calculate the expected time until system failure directly.)

---

[2]This exercise is based on D. Goldsman, B. L. Nelson and B. Schmeiser. 1991. Methods for selecting the best system. *1991 Winter Simulation Conference Proceedings* (B. L. Nelson, W. D. Kelton and G. M. Clark, eds.), 177–186.

7.16. Since a continuous-time Markov process can be represented as a Markov chain governing the state transitions, combined with exponentially distributed holding times in each state, it can also be *simulated* from this perspective. In other words, the state transitions can be generated from the embedded Markov chain (Chapter 6), and the holding time (time until the next transition) can be generated from the appropriate exponential distribution depending on the current state. This is known as a *discrete-time conversion*.

(a) Develop a simulation of HyperWord inventory (Case 7.1) that takes this perspective.

(b) Using the model developed in Part 16a, estimate the expected time until the inventory level reaches 0 for the first time in the following way: At each transition, use the *expected holding time* in each state instead of randomly generating a holding time.

(c) List all advantages you can think of for using the discrete-time conversion. Consider the possibility that s and r might be large.

7.17. Modify the simulation of the inventory system for HyperWord (Case 7.1) in the following ways:

(a) To allow the number of copies of HyperWord demanded to be a random variable having a Poisson distribution with expected value 1 copy. Notice that this model allows for the possibility that a customer will choose to order no copies, perhaps after finding out the price of the product.

(b) To keep track of the number of lost sales.

(c) To allow for the possibility that $s - r \leq r$.

7.18. Let X be a random variable having a geometric distribution with parameter γ. Let $J_1, J_2, \ldots$ be independent, time-stationary, exponentially distributed random variables with parameter λ, which are independent of X. Use the law of total probability to derive the cdf of

$$H = \sum_{i=1}^{X} J_i$$

Explain why this result supports our use of uniformization to represent Y_t.

7.19. Prove that $\tilde{\mathbf{P}}(t) \leq \mathbf{P}(t)$ element by element, and derive the error bound (7.29). Hint: First establish the signs of the terms that are truncated.

7.20. Derive Equation (7.7). (Hint: $\Pr\left\{Z_1 \leq \min_{j \neq 1} Z_j, H > t\right\} = \Pr\{t < Z_1 \leq M\}$.)

7.21. Consider the semi-Markov process model described in Section 7.9, based on the transition matrix in Equation (7.38). Suppose that the time spent thinking in each state is modeled as being exponentially distributed, which makes the process a Markov process.

(a) Formulate the generator matrix for this Markov process. Compute the steady-state probabilities using the generator matrix and confirm that they are the same as determined using Equation (7.39).

(b) Plot $p_{13}(t)$ as a function of t (recall that $p_{13}(t)$ is the probability that a customer is reading a page at time t, given the customer started the session with a log on). Verify that it converges to π_3.

(*c*) Formulate the generator matrix for this Markov process if state 4 (log off) is once again an absorbing state. Plot $p_{14}(t)$ as a function of t (recall that $p_{14}(t)$ is the probability that a customer has logged off by or before time t, given the customer started the session with a log on).

7.22. Use the renewal-reward theorem (Section 5.9.1) to provide an alternative derivation of the steady-state probabilities for a semi-Markov process. (Hint: Fix a state, say state 1. The nth visit to state 1 receives a reward R_n, which is the holding time in state 1 and has expected value τ_1. The random variable G is the time between visits to state 1, including the time spent in state 1 on each cycle. The expected number of transitions between visits to state 1 is $1/\xi_1$, of which ξ_j/ξ_1 are to state j, each with expected holding time τ_j.)

7.23. Show that Equation (7.39) is an equivalent method for calculating the steady-state probabilities for a Markov process by setting $\tau_i = 1/g_{ii}$.

7.24. Read the article by Swersey, A. J., L. Goldring and E. D. Geyer, Sr. 1993. Improving fire department productivity: Merging fire and emergency medical units in New Haven. *Interfaces* **23**, 109–129. Describe their "spatial queueing model" in terms of a continuous-time Markov process.

CHAPTER
8

QUEUEING PROCESSES

A system that is characterized by customers contending for a resource, called a server, is a *queueing system*. Many diverse types of systems can be viewed as queueing systems by broadly defining "customer" and "server." For example, The Darker Image (Case 2.1) is a queueing system in which the customers are patrons with copying jobs and the servers are the copying machines. The maintenance of Bit Bucket's computers (Case 4.1) is a queueing system in which the customers are failed computers and the servers are the repair technicians. Manufacturing, health-care delivery and communication systems may also be viewed as queueing systems. This chapter presents methods for modeling and analysis of queueing systems, emphasizing long-run (steady-state) performance measures and Markovian queueing processes. Markovian queueing processes are a special case of the continuous-time Markov processes introduced in Chapter 7.

8.1 THE CASE OF THE LAST PARKING SPACE ON EARTH

Case 8.1. Planning for construction of the proposed "Massive Mall"—the largest shopping mall and indoor golf course in the world—includes determining the amount of customer parking to provide. The developers of Massive Mall have told the planners to "give us enough parking for everybody!" The planners must decide what this is supposed to mean in practice, since the number of parking spaces must be some finite number.

221

The planners for Massive Mall have a reliable estimate of the number of parking spaces needed for mall employees, but the number of customers desiring parking at any one time is variable. For the purpose of modeling and analysis, the planners have decided to pretend that parking is unlimited and then investigate how many spaces are sufficient to satisfy demand a large fraction of the time. They have forecasts of customer traffic to the mall, and they also have data from other large malls about the amount of time customers spend shopping. With this input they would like to determine the distribution of parking-space use.

8.2   NOTATION AND REVIEW

This chapter makes use of the summation of certain series and recursive relationships between the terms of certain probability distributions. We review these in Section 8.2.1. Most of our models of queueing systems are special cases of the continuous-time Markov processes described in Chapter 7. Section 8.2.2 is designed for readers who either skipped Chapter 7 or who want a brief review of continuous-time Markov processes. Readers familiar with this material may proceed directly to Section 8.3.

8.2.1   Series and Recursions

Several of the performance measures that we will derive for queueing models involve a geometric series. For a constant r with $|r| < 1$, the summation of the geometric series $1 + r + r^2 + r^3 + \cdots$ is

$$\sum_{i=0}^{\infty} r^i = \frac{1}{1-r} \tag{8.1}$$

The condition $|r| < 1$ is required for the summation to be finite. By differentiating (8.1) with respect to r, we obtain the series $1 + 2r + 3r^2 + 4r^3 + \cdots$ whose summation is

$$\frac{d}{dr} \sum_{i=0}^{\infty} r^i = \sum_{i=0}^{\infty} i\, r^{i-1} = \frac{d}{dr}\left(\frac{1}{1-r}\right) = \frac{1}{(1-r)^2} \tag{8.2}$$

Additional derivatives can be taken to further extend this idea. For any value of $r \neq 1$ and $0 < m < \infty$, the summation of the finite geometric series $1 + r + r^2 + r^3 + \cdots + r^m$ is

$$\sum_{i=0}^{m} r^i = \frac{1 - r^{m+1}}{1-r} \tag{8.3}$$

One additional series we need is the power series expansion of e^b. For any real b,

$$e^b = 1 + b + \frac{b^2}{2!} + \frac{b^3}{3!} + \cdots = \sum_{i=0}^{\infty} \frac{b^i}{i!} \qquad (8.4)$$

The following relationships are useful for evaluating performance measures of queueing models. Recall that a random X variable having a Poisson distribution with parameter $\eta > 0$ has mass function

$$p_X(c) \equiv \Pr\{X = c\} = \frac{e^{-\eta}\eta^c}{c!}$$

for $c = 0, 1, \ldots$. Therefore

$$\sum_{c=0}^{\infty} \frac{e^{-\eta}\eta^c}{c!} = 1$$

since p_X is a mass function. A recursive relationship that is useful for computing Poisson probabilities is

$$p_X(c+1) = \frac{e^{-\eta}\eta^{c+1}}{(c+1)!} = \frac{e^{-\eta}\eta^c}{c!}\frac{\eta}{c+1} = p_X(c)\frac{\eta}{c+1}$$

for $c = 1, 2, \ldots$.

For integers $0 \leq i \leq k$, the combinatorial term

$$\binom{k}{i} \equiv \frac{k!}{(k-i)!i!}$$

is the number of combinations of k items taken i at a time, when the order of the items does not matter; these terms appear in the binomial distribution (Section 5.2), for instance. Two useful relationships between combinatorial terms are

$$\binom{k}{i} = \binom{k}{k-i} \qquad (8.5)$$

and

$$\binom{k}{i+1} = \binom{k}{i}\frac{(k-i)}{(i+1)} \qquad (8.6)$$

Equation (8.6) is valid for $i = 0, 1, \ldots, k-1$.

8.2.2   Markov Process Review

The analysis of Markov process sample paths depends on certain properties of the exponential distribution, so we review these properties before defining a Markov process.

Recall that an exponentially distributed random variable Z with parameter g has expectation $E[Z] = 1/g$ and cdf

$$F_Z(b) = \begin{cases} 0, & b < 0 \\ 1 - e^{-gb}, & 0 \le b \end{cases}$$

By directly applying the definition of conditional probability, we can show that the

$$\Pr\{Z > t + b | Z > t\} = \Pr\{Z > b\} = e^{-gb} = 1 - F_Z(b) \qquad (8.7)$$

for $b \ge 0$. This is the so-called *memoryless property* of the exponential distribution. If Z represents the time that we wait for something to occur, then the memoryless property implies that when the "something" has not occurred by time t, the probability distribution of the remaining time until it does occur has the same exponential distribution as when we first started waiting. This property was critical to the analysis of Poisson arrival-counting processes in Chapter 5. The exponential distribution is the unique continuous distribution that has the memoryless property.

Now let $Z_1, Z_2, \ldots, Z_k$ be independent, exponentially distributed random variables with respective parameters $g_1, g_2, \ldots, g_k$, and let $H = \min\{Z_1, Z_2, \ldots, Z_k\}$. In Section 7.2 we showed that the cdf of H is

$$F_H(t) \equiv \Pr\{H \le t\} = 1 - e^{-gt}$$

where $g \equiv \sum_{i=1}^k g_i$. In words, the minimum of k independent, exponentially distributed random variables is also exponentially distributed. In a Markov process $Z_1, Z_2, \ldots, Z_k$ represent the time remaining until one of k pending system event occurs, so that H is the remaining time until the *next* system event occurs. In Section 7.2 we also showed that the probability that Z_i is the smallest of $Z_1, Z_2, \ldots, Z_k$ is g_i/g.

We are now ready to define a continuous-time Markov process: It is a stochastic processes $\{Y_t; t \ge 0\}$ with discrete state space $\mathcal{M}$. For convenience we let $\mathcal{M} = \{0, 1, 2, \ldots\}$, since later in this chapter the state space will represent the number of customers in a queueing system. A defining characteristic of a Markov process is that all of the clock settings that determine when the system events occur are independent, exponentially distributed random variables. Therefore, at any point in time the remaining time until each pending system events occurs is exponentially distributed, due to the memoryless property. Further, the remaining time until the *next* system event occurs is the minimum of these independent, exponentially distributed random variables and is therefore also exponentially distributed.

Suppose that at time t the Markov process is in state i, and there is some pending system event that will change the state of the process to state $j \neq i$. Let

g_{ij} be the parameter of the exponential distribution that sets the clock for this event. Then $1/g_{ij}$ is the expected time for the process to move from i to j, and we refer to g_{ij} at the *transition rate* from i to j. The remaining time until the first system event occurs is exponentially distributed with parameter $g_{ii} \equiv \sum_{j \neq i} g_{ij}$, so that we can refer to g_{ii} as the *overall transition rate* out of state i.

Here is a small example: Suppose a system has two components that work in series, meaning that if one component fails, the system fails. Let the time until component i fails be modeled as an exponentially distributed random variable with parameter λ_i, implying that the expected time to failure is $1/\lambda_i$, for $i = 1, 2$. Let the time to repair component i be exponentially distributed with parameter μ_i, for $i = 1, 2$. Define the stochastic process $\{Y_t; t \geq 0\}$ with state space $\mathcal{M} = \{0, 1, 2\}$ to represent the status of the system at time t, where state 0 represents the system being fully functional, and state i represents the system having failed due to a failure of component i, for $i = 1, 2$. We assume that when one component has failed, the other component is not in use and therefore cannot fail until the system is functioning again.

Whenever this process is in state 0, there are two pending systems events, the failure of component 1 which moves the process to state 1, and the failure of component 2 which moves the process to state 2. The time until each system event occurs is exponentially distributed, so that the time until the first one fails is also exponentially distributed with parameter $\lambda_1 + \lambda_2$. Thus, $g_{01} = \lambda_1$, $g_{02} = \lambda_2$, and $g_{00} = \lambda_1 + \lambda_2$. When the process is in state 1, there is one pending system event, completing repairs on component 1 which moves the process to state 0. Thus, $g_{10} = \mu_1$, $g_{12} = 0$, and $g_{11} = \mu_1$. Notice that $g_{12} = 0$ because a transition from state 1 to 2 is not possible (the rate of this transition is 0). Similarly, $g_{20} = \mu_2$, $g_{21} = 0$, and $g_{22} = \mu_2$.

This information can be displayed in a *transition-rate diagram* like Figure 8.1. In the diagram each node $\bigcirc$ represents a system state, each directed arc $\longrightarrow$ a possible transition, and the transition rate g_{ij} is written next to the arc and enclosed in a box $\square$.

Many performance measures of continuous-time Markov processes can be derived from the steady-state probabilities $\pi_j \equiv \lim_{t \to \infty} \Pr\{Y_t = j\}$, where π_j can be interpreted either as the probability of finding the process in state j after a

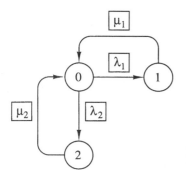

FIGURE 8.1
Transition-rate diagram for the reliability example.

long period of time or as the long-run fraction of time that the process is in state j. These probabilities are similar to the steady-state probabilities for discrete-time Markov chains (Chapter 6), except that we now account for how long the process spends in each state, not just which states it enters. We next develop a way to compute the π_j.

When the process is in state i, it is moving into state j at rate g_{ij}. Over the long run the rate at which the process is moving into state j from state i is $g_{ij}\pi_i$, the transition rate from i to j times the probability of finding the process in state i. Therefore, the overall transition rate into state j must be

$$\text{rate into } j \; = \sum_{i \neq j} g_{ij}\pi_i$$

Similarly, the overall transition rate out of state j must be

$$\text{rate out of } j = g_{jj}\pi_j = \pi_j \sum_{i \neq j} g_{ji}$$

If a steady state is going to exist, then over the long run the rate into state j must equal the rate out of state j. This leads to the so-called balance equation

$$\overset{\text{rate into } j}{\overbrace{\sum_{i \neq j} g_{ij}\pi_i}} = \overset{\text{rate out of } j}{\overbrace{\pi_j \sum_{i \neq j} g_{ji}}}$$

Combined with the fact that $\sum_j \pi_j = 1$ to form a probability distribution, we have a system of equations that can be solved for π_j.

For instance, the balance equations for the reliability example are

j	rate into j		rate out of j
0	$\mu_1\pi_1 + \mu_2\pi_2$	$=$	$(\lambda_1 + \lambda_2)\pi_0$
1	$\lambda_1\pi_0$	$=$	$\mu_1\pi_1$
2	$\lambda_2\pi_0$	$=$	$\mu_2\pi_2$

and $\pi_0 + \pi_1 + \pi_2 = 1$. We can rewrite these equations as

$$
\begin{aligned}
-(\lambda_1 + \lambda_2)\pi_0 + \mu_1\pi_1 + \mu_2\pi_2 &= 0 \\
\lambda_1\pi_0 - \mu_1\pi_1 \qquad\qquad &= 0 \\
\lambda_2\pi_0 \qquad\quad - \mu_2\pi_2 &= 0 \\
\pi_0 + \quad \pi_1 + \quad \pi_2 &= 1
\end{aligned}
$$

or in matrix form

$$\pi'\mathbf{G} = \mathbf{0}$$

$$\pi'\mathbf{1} = 1 \tag{8.8}$$

where

$$\mathbf{G} = \begin{pmatrix} -g_{00} & g_{01} & g_{02} \\ g_{10} & -g_{11} & g_{12} \\ g_{20} & g_{21} & -g_{22} \end{pmatrix} = \begin{pmatrix} -(\lambda_1 + \lambda_2) & \lambda_1 & \lambda_2 \\ \mu_1 & -\mu_1 & 0 \\ \mu_2 & 0 & -\mu_2 \end{pmatrix}$$

and $\pi' = (\pi_0, \pi_1, \pi_2)$. The solution to this system of equations is

$$\pi = \begin{pmatrix} \pi_0 \\ \pi_1 \\ \pi_2 \end{pmatrix} = \frac{1}{\mu_1\mu_2 + \lambda_1\mu_2 + \lambda_2\mu_1} \begin{pmatrix} \mu_1\mu_2 \\ \lambda_1\mu_2 \\ \lambda_2\mu_1 \end{pmatrix}$$

The matrix $\mathbf{G}$ is called the *generator matrix* of the Markov process, and it has the general form

$$\mathbf{G} = \begin{pmatrix} -g_{00} & g_{01} & g_{02} & \cdots \\ g_{10} & -g_{11} & g_{12} & \cdots \\ g_{20} & g_{21} & -g_{22} & \cdots \\ \vdots & \vdots & \vdots & \vdots \end{pmatrix}$$

In row i, the off-diagonal element g_{ij}, $i \neq j$, is the transition rate out of state i in to state j, while the diagonal element $-g_{ii} = -\sum_{j\neq i} g_{ij}$ is the *negative of* the overall transition rate out of state i. Thus, each row of $\mathbf{G}$ sums to 0. The system of equations (8.8) is the general formula for the steady-state probabilities, when they exist.

Another example of a continuous-time Markov process is the Poisson arrival process with arrival rate λ (Chapter 5); it has state space $\mathcal{M} = \{0, 1, 2, \ldots\}$ representing the total number of arrivals, and generator

$$\mathbf{G} = \begin{pmatrix} -\lambda & \lambda & 0 & 0 & 0 & \cdots \\ 0 & -\lambda & \lambda & 0 & 0 & \cdots \\ 0 & 0 & -\lambda & \lambda & 0 & \cdots \\ \vdots & \vdots & \vdots & \vdots & \vdots & \ddots \end{pmatrix}$$

The steady-state probabilities do not exist for this process, however, because the state (number of arrivals) increases without bound as time goes to infinity.

8.3 A QUEUEING MODEL FOR THE LAST PARKING SPACE ON EARTH

In their analysis, the planners for Massive Mall, Case 8.1, envision an infinite-size parking facility so that they can study how much of it is regularly used. Since we are focusing on Markovian queueing models in this chapter, we will develop a Markov process model of parking. However, the appropriateness of a model for the situation at hand must always be justified. If a Markovian queueing model is not appropriate, it may still provide a rough-cut analysis that can be refined if necessary (see Chapter 9, Section 9.2 for further discussion of rough-cut modeling and Section 8.10 of this chapter for non-Markovian queues). This is often the role of Markovian queueing models.

Let $\{Y_t; t \geq 0\}$ be a continuous-time Markov process representing the number of customer cars parking at Massive Mall at time t hours, measured from some starting time 0. The state space of Y_t is $\mathcal{M} = \{0, 1, 2, \ldots\}$, since the planners are treating the facility size as (conceptually) infinite. Let $G_1, G_2, \ldots$ be an independent, time-stationary sequence of car interarrival-time gaps, and let $X_1, X_2, \ldots$ be an independent and time-stationary sequence of parking times. In other words, X_i is the time that the ith car to arrive occupies its parking space. Modeling these sequences as independent is plausible, since most customers choose their arrival times and time spent shopping independently of other customers. Time stationarity, though, is a rough approximation.

If the G_i are exponentially distributed with parameter λ, then the car-arrival process is a Poisson arrival process with arrival rate λ cars per hour, and a portion of the generator matrix for Y_t is

$$\mathbf{G} = \begin{pmatrix} - & \lambda & - & - & \cdots \\ - & - & \lambda & - & \cdots \\ - & - & - & \lambda & \cdots \\ \vdots & \vdots & \vdots & \vdots & \ddots \end{pmatrix}$$

Marketing forecasts predict that the arrival rate will be $\lambda = 1000$ cars per hour.

When the process Y_t is in state $i > 0$, it indicates that there are i cars in the parking facility at time t. Let $Z_1, Z_2, \ldots, Z_i$ be the remaining time until each of these cars depart, so that $H = \min\{Z_1, Z_2, \ldots, Z_i\}$ is the time that the first car departs. If the occupancy times, X, are exponentially distributed with parameter μ, then the memoryless property of the exponential distribution and Equation (7.5) imply that H is exponentially distributed with parameter $i\mu$. Therefore, $g_{i,i-1} = i\mu$, and the complete generator matrix is

$$\mathbf{G} = \begin{pmatrix} -\lambda & \lambda & 0 & 0 & 0 \cdots \\ \mu & -(\lambda+\mu) & \lambda & 0 & 0 \cdots \\ 0 & 2\mu & -(\lambda+2\mu) & \lambda & 0 \cdots \\ 0 & 0 & 3\mu & -(\lambda+3\mu) & \lambda \cdots \\ \vdots & \vdots & \vdots & \vdots & \vdots \ddots \end{pmatrix} \tag{8.9}$$

Studies at other large malls have found that the expected time that a space is occupied is $1/\mu = 3$ hours, so that $\mu = 1/3$ car per hour. The 0s in $\mathbf{G}$ imply that, in this model, multiple cars do not arrive or depart precisely simultaneously.

The planners are interested in how many parking spaces will be used regularly. The steady-state probability π_j for the Markov process Y_t is the long-run probability of finding j cars in the facility. Therefore, the quantity

$$\sum_{j=0}^{\infty} j \, \pi_j$$

is the steady-state expected number of cars in the facility, a measure of how many cars are there on average. To look at the extreme use of parking spaces, the planners might also want to find the minimum number of parking spaces, c^*, such that

$$\Pr\{\text{number in lot} > c^*\} = \sum_{j=c^*+1}^{\infty} \pi_j \le \alpha$$

where $0 < \alpha < 1$. The quantity c^* is the smallest facility size that will accommodate all of the cars desiring a space with probability at least $1 - \alpha$, over the long run. Stated differently, a parking facility of size c^* spaces will be inadequate no more than $\alpha 100\%$ of the time. To make the facility effectively infinite, α should be small, say $\alpha = 0.001$, so that there is enough parking with probability 0.999.

The steady-state probabilities, π_j, can be calculated using the results in Chapter 7. In the following sections we adapt those general results to the case of queueing processes and define performance measures that are often of interest in queueing systems.

A Poisson arrival process and exponential parking times may be reasonable for modeling Massive Mall parking, except that the arrival process is certainly not time stationary. For instance, the arrival rate is 0 when the mall is closed. Mathematical analysis of queueing models with nonstationary arrival processes is often difficult. The time-stationary model proposed here may be useful for planning purposes, however, if long-run conditions are actually reached during the course of a day, or if we set λ equal to the peak daily arrival rate so that the planning results are conservative because the actual rate is always less than or equal to λ.

8.4 MARKOVIAN QUEUEING PROCESSES

In this section we define the generic queueing process and calculate performance measures.

8.4.1 The Birth-Death Process

Our generic queueing model is a continuous-time Markov process $\{Y_t; t \ge 0\}$ with state space $\mathcal{M} = \{0, 1, \ldots\}$ and generator matrix

$$\mathbf{G} = \begin{pmatrix} -\lambda_0 & \lambda_0 & 0 & 0 & \cdots \\ \mu_1 & -(\lambda_1 + \mu_1) & \lambda_1 & 0 & \cdots \\ 0 & \mu_2 & -(\lambda_2 + \mu_2) & \lambda_2 & \cdots \\ \vdots & \vdots & \vdots & \vdots & \ddots \end{pmatrix} \tag{8.10}$$

The state space represents the number of customers in the *system* at time t, where the "system" includes all customers receiving service or waiting for service. The term *queue* refers to only those customers waiting for, but not yet receiving, service. In many cases $\mathcal{M}$ is finite, but we allow the possibility of

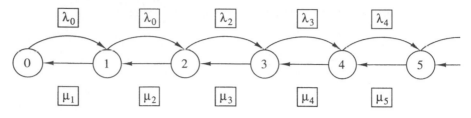

FIGURE 8.2
Transition-rate diagram for the generic queueing model, also called the birth-death process.

a (conceptually) infinite-size system, such as the Massive Mall parking facility. A Markov process with generator (8.10) is sometimes called a *birth-death process* because the state of the system can only increase or decrease by 1 at each transition, similar to a population changing due to a "birth" or a "death," respectively. Figure 8.2 shows the transition-rate diagram for our generic queueing model.

Suppose that the process is in state $i > 0$, meaning that there are i customers in the system. Then the state of the system can next increase by 1 due to an arrival or decrease by 1 due to a customer departing the system. The structure of the generator $\mathbf{G}$ and the properties of Markov process sample paths (Chapter 7) imply that the remaining time until the next arrival is exponentially distributed with parameter λ_i (expected value $1/\lambda_i$), the remaining time until the next departure is exponentially distributed with parameter μ_i (expected value $1/\mu_i$), and the remaining time until *something* happens is exponentially distributed with parameter $\lambda_i + \mu_i$ (stated differently, the expected holding time in state i is $1/(\lambda_i + \mu_i)$). In the context of Markovian queues, λ_i is called the *arrival rate* in state i, while μ_i is called the *service rate* in state i. Both are measured in units of customers per time. We will only consider queueing processes in which the embedded Markov chain implied by $\mathbf{G}$ forms a single, irreducible set of recurrent states.

8.4.2   Performance Measures

Since the queueing process $\{Y_t; t \geq 0\}$ is a Markov process, all of the performance-measure calculations presented in Section 7.6 apply. Computing time-dependent results is difficult unless $\mathcal{M}$ is finite, for obvious reasons. However, we can exploit the special structure of the generator (8.10) to develop a general formula for $\pi_j = \lim_{t\to\infty} \Pr\{Y_t = j\}$, the steady-state probability of finding j customers in the system, even for the case of an infinite state space.

Recall that the steady-state probabilities $\pi' = (\pi_0, \pi_1, \dots)$ satisfy the system of equations

$$0 = \pi'\mathbf{G}$$
$$1 = \pi'\mathbf{1}$$

when they exist. The individual equations are

$$0 = -\lambda_0 \pi_0 + \mu_1 \pi_1$$

$$0 = \lambda_0 \pi_0 - (\lambda_1 + \mu_1) \pi_1 + \mu_2 \pi_2$$

$$0 = \lambda_1 \pi_1 - (\lambda_2 + \mu_2) \pi_2 + \mu_3 \pi_3$$

$$\vdots$$

$$\text{(8.11)}$$

$$1 = \pi_0 + \pi_1 + \pi_2 + \cdots \tag{8.12}$$

Or written as balance equations

j	rate into j		rate out of j
0	$\mu_1 \pi_1$	$=$	$\lambda_0 \pi_0$
$1, 2, \ldots$	$\lambda_{j-1} \pi_{j-1} + \mu_{j+1} \pi_{j+1}$	$=$	$(\lambda_j + \mu_j) \pi_j$

and $\pi_0 + \pi_1 + \cdots = 1$.

This system of equations can be solved recursively in terms of π_0. The first equation implies that $\pi_1 = (\lambda_0/\mu_1)\pi_0$. The second equation implies that

$$\pi_2 = -\frac{\lambda_0}{\mu_2}\pi_0 + \frac{(\lambda_1 + \mu_1)}{\mu_2}\pi_1 = -\frac{\lambda_0}{\mu_2}\pi_0 + \frac{(\lambda_1 + \mu_1)}{\mu_2}\frac{\lambda_0}{\mu_1}\pi_0 = \frac{\lambda_0}{\mu_1}\frac{\lambda_1}{\mu_2}\pi_0$$

In general,

$$\pi_j = \frac{\lambda_0 \lambda_1 \cdots \lambda_{j-1}}{\mu_1 \mu_2 \cdots \mu_j}\pi_0 = \prod_{i=1}^{j}\frac{\lambda_{i-1}}{\mu_i}\pi_0$$

for $j = 1, 2, \ldots$.

To avoid writing $\prod_{i=1}^{j} \lambda_{i-1}/\mu_i$ repeatedly, let $d_0 \equiv 1$ and let

$$\boxed{d_j \equiv \prod_{i=1}^{j}\frac{\lambda_{i-1}}{\mu_i}}$$

for $j = 1, 2, \ldots$. Then $\pi_j = d_j \pi_0$, and the equation $1 = \boldsymbol{\pi}'\mathbf{1}$ becomes

$$1 = \pi_0 + \pi_1 + \pi_2 + \cdots = d_0\pi_0 + d_1\pi_0 + d_2\pi_0 + \cdots = \pi_0 \sum_{j=0}^{\infty} d_j$$

Solving for π_0 gives

$$\boxed{\pi_0 = \frac{1}{\sum_{j=0}^{\infty} d_j}}$$

and therefore

$$\pi_j = d_j \pi_0 = \frac{d_j}{\sum_{j=0}^{\infty} d_j} \qquad j = 0, 1, 2, \ldots$$

Apparently the steady-state probabilities π_j exist provided that $0 < \sum_{j=0}^{\infty} d_j < \infty$.

In the context of queueing, $\boldsymbol{\pi}' = (\pi_0, \pi_1, \ldots)$ is the steady-state probability distribution of the number of customers in the system; in other words, π_j is the probability that we find j customers in the system after the queueing system has been in operation for a long time.[1] Each π_j can also be interpreted as the long-run fraction of time that there are j customers in the system. A summary measure of the steady-state distribution $\boldsymbol{\pi}$ is the steady-state *expected number of customers in the system*[2]

$$\ell \equiv \sum_{j=0}^{\infty} j \, \pi_j \tag{8.13}$$

If only $s < \infty$ customers can be served simultaneously, so that customers must wait for service when there are more than s customers in the system, then a related performance measure is the steady-state *expected number of customers in the queue*

$$\ell_q \equiv \sum_{j=s+1}^{\infty} (j - s)\pi_j \tag{8.14}$$

This follows because when there are j customers in the system, there are $j - s$ of them in the queue, provided $j \geq s$.

In queueing systems with arrival rate λ in all states and a service process having s identical servers each working at rate μ, the ratio of the arrival rate to the maximum service rate is called the *traffic intensity*, and is denoted

$$\rho \equiv \frac{\lambda}{s\mu}$$

---

[1] Although we focus on steady-state performance measures from here on, time-dependent measures— such as $p_{ij}(t) = \Pr\{Y_t = j | Y_0 = i\}$, the probability of finding j customers in the system at time t—can be obtained using the methods described in Chapter 7, at least when $\mathcal{M}$ is finite.

[2] The quantities ℓ, ℓ_q, w and w_q defined here are frequently denoted L, L_q, W and W_q, respectively, in other books. We use lowercase letters to emphasize that these quantities are *expected values* of random variables, not random variables.

Typically $\rho < 1$ is required for the system to be "stable" (for $\sum_{j=0}^{\infty} d_j < \infty$). If $\rho \geq 1$, then customers are arriving as fast or faster than they can be served. Queueing formulas are often presented in terms of the traffic intensity. When $\rho < 1$, the traffic intensity is also the long-run *utilization* of each server. A related measure is the *offered load*

$$o \equiv \lambda/\mu$$

which is the expected number of busy servers.

One reason the study of queueing is so important is that the relationship between the traffic intensity ρ and a quantity like the expected number of customers in the queue ℓ_q is not what most people imagine. Certainly the expected number in the queue should increase as the ratio of arrival rate to the maximum service rate approaches 1, and it does. What is surprising is that for many queueing systems ℓ_q increases *explosively* as ρ nears 1. We discuss this phenomenon later in the chapter, but you can see the effect by looking ahead to Figure 8.3, which is a graph of ℓ_q as a function of ρ for one particular queueing model.

The quantities ℓ, ℓ_q, ρ and o are system-level performance measures, by which we mean that they measure the workload or congestion in the entire system. Customer-level performance measures account for the time a customer spends in the system. To compute these measures, we first define the *effective arrival rate* to the system, λ_{eff}, as

$$\lambda_{\text{eff}} \equiv \sum_{j=0}^{\infty} \lambda_j \pi_j \tag{8.15}$$

The effective arrival rate is an average arrival rate in the sense that the arrival rate in state j is weighted by the probability of finding the process in state j, π_j.

Let w be the steady-state expected *waiting time*, which is the expected time a customer spends in the system from arrival to departure. The quantities w, ℓ and λ_{eff} are related by

$$\ell = \lambda_{\text{eff}} w \tag{8.16}$$

This relationship, known as *Little's formula*, holds under remarkably broad conditions that amount to the system being stable—meaning that the number of customers in the system does not grow without bound—and that every customer that enters the system, and only the customers that enter the system, also comes out—meaning that no customer is created or destroyed by the system. Proving this result is not easy, but we can give some rough intuition:

For a queueing system to be stable, the rate at which customers depart the system must be equal to the rate at which they enter the system. The departure rate (customers per time) certainly cannot be greater than the arrival rate, since no customers are created. But the departure rate also cannot be less that the arrival rate if the system is to keep pace with the arriving customers. Therefore, λ_{eff} is also the *departure rate* from the queueing system.

In expectation, each customer spends w time units in the system, so the departure rate for an individual customer is $1/w$. But in expectation there are ℓ customers in the system. Therefore, the departure rate for the system as a whole is (number of customers)(departure rate per customer) $= (\ell)(1/w)$. Since λ_{eff} is also the departure rate for the system, we must have $\lambda_{\text{eff}} = \ell/w$, which is Little's formula.

An analogous relationship holds for the expected number of customers in the queue, denoted ℓ_q, and the expected *delay* in the queue, denoted w_q:

$$\ell_q = \lambda_{\text{eff}} w_q \tag{8.17}$$

In the queueing systems we study in this chapter, the service time has expected value $1/\mu$. Since the time in the system is the delay in the queue plus the service time, w and w_q are related by

$$w = w_q + \frac{1}{\mu} \tag{8.18}$$

To illustrate these performance measures, recall the model of the Massive Mall parking facility with generator (8.9). Since the arrival rates are $\lambda_i = \lambda$ for all i and the service rates are $\mu_i = i\mu$ for $i = 1, 2, \ldots$, we have

$$d_j = \frac{\lambda_0 \lambda_1 \cdots \lambda_{j-1}}{\mu_1 \mu_2 \cdots \mu_j} = \frac{\lambda \lambda \cdots \lambda}{\mu \, 2\mu \cdots j\mu} = \frac{(\lambda/\mu)^j}{j!}$$

Thus,

$$\sum_{j=0}^{\infty} d_j = 1 + \frac{(\lambda/\mu)}{1!} + \frac{(\lambda/\mu)^2}{2!} + \cdots = e^{\lambda/\mu}$$

using Equation (8.4) to evaluate the infinite summation. Therefore, $\pi_0 = 1/\sum_{j=0}^{\infty} d_j = e^{-\lambda/\mu}$ and $\pi_j = d_j \pi_0 = e^{-\lambda/\mu}(\lambda/\mu)^j/j!$.

The expected number of cars in the parking facility is

$$\ell = \sum_{j=0}^{\infty} j \, \frac{e^{-\lambda/\mu}(\lambda/\mu)^j}{j!}$$

$$= \sum_{j=1}^{\infty} \frac{e^{-\lambda/\mu}(\lambda/\mu)^j}{(j-1)!}$$

$$= \frac{\lambda}{\mu} \sum_{j=1}^{\infty} \frac{e^{-\lambda/\mu}(\lambda/\mu)^{j-1}}{(j-1)!}$$

$$= \frac{\lambda}{\mu} \sum_{i=0}^{\infty} \frac{e^{-\lambda/\mu}(\lambda/\mu)^{i}}{i!}$$

$$= \frac{\lambda}{\mu} \cdot 1 = \frac{\lambda}{\mu}$$

We could also obtain this result directly by noticing that π_j, $j = 0, 1, \ldots$ is the mass function of a Poisson random variable with parameter λ/μ, so that its expected value is λ/μ. With $\lambda = 1000$ and $\mu = 1/3$ the expected number of parked cars is $\lambda/\mu = 3000$, which is also the offered load o in this model.

Clearly $w_q = \ell_q = 0$, since no one waits in a queue (remember that there are infinite number of parking spaces in the model). From Little's formula the expected time in the system is $w = \ell/\lambda = 1/\mu = 3$ hours, which is just the expected parking time for a car.

To find the parking capacity that is adequate 99.9% of the time, let L be a random variable representing the steady-state number of cars in the system. The probability distribution of L is π, since $\Pr\{L = j\} = \pi_j$. We have just shown that L has a Poisson distribution with parameter 3000, and we need to find c^* such that $\Pr\{L > c^*\} = \sum_{j=c^*+1}^{\infty} \pi_j \leq 0.001$. From the normal approximation to the Poisson distribution (Exercise 2, Chapter 5),

$$\Pr\{L > c^*\} = 1 - \Pr\{L \leq c^*\} \approx 1 - \Phi\left(\frac{c^* + 1/2 - 3000}{\sqrt{3000}}\right)$$

where Φ is the cdf of the standard normal distribution. From tabled values of the standard normal distribution, we find that $\Phi(3.0902) = 0.999$. Therefore, setting $(c^* + 1/2 - 3000)/\sqrt{3000} = 3.0902$, and solving for c^* yields a capacity of $c^* = 3169$ spaces. Although 3169 spaces is not infinite, our analysis indicates that it will be adequate with very high probability. And notice that it is 169 spaces larger than the expected number of spaces in use, $\ell = 3000$ spaces.

8.5 STANDARD FORMULATIONS

This section is a guide to formulating the arrival and service rates of a queueing model for situations that are frequently encountered in practice. Once the form of the rates are determined, they must still be parameterized to obtain numerical results.

8.5.1 Arrival Rates

In this section we describe common forms for the arrival rates, λ_i. A critical distinction is that λ_i is the arrival rate *into* the queueing system when there are i customers already present, which can be different from the arrival rate of *potential* customers to the system, some of whom may not actually enter the system.

8.5.1.1 LARGE CUSTOMER POPULATION

Case 8.2. Kent Sporting Goods plans to open a "superstore" in a major city. A queueing analysis will be used to help determine staffing levels for the store.

A Poisson arrival process is a viable model for customer arrivals to a system such as Kent's. As discussed in Chapter 5, Section 5.5.1, whenever we can argue that the number of arrivals in nonoverlapping intervals of time are independent of each other and that the expected number of arrivals in an interval of time is a constant rate times the length of the interval, then a Poisson process is a plausible model. In a queueing system, choosing a Poisson model also implies that the system is large enough to accommodate all of the customers that want to be there simultaneously. Under these conditions the arrival rate into the store does not depend on the number of customers in the store, so

$$\lambda_i = \lambda, \ i = 0, 1, 2, \ldots$$

where λ is the arrival rate of the Poisson process. Of course, this is a stationary Poisson arrival process model in which the arrival rate also does not depend on time, which may or may not be appropriate for Kents.

8.5.1.2 BALKING

Case 8.3. The management of Sharon and LeRoy's Ice Cream has noticed that when potential customers find that the queue of waiting customers is too long, they sometimes go around the corner and buy ice cream at a grocery store. The management would like to incorporate this phenomenon into its staffing model.

Balking occurs when potential customers arriving at a queueing system choose not to enter it. The effect of balking on a queueing system is equivalent to reducing the arrival rate *into* the system to something less than the rate at which potential customers arrive. One way to model balking is to define a collection of probabilities

$b_i \equiv \Pr\{$potential customer balks when i customers already in the system$\}$

for $i = 0, 1, \ldots$. If potential customers arrive according to a Poisson process with rate λ, then the effect of balking is to decompose the arrival process into two subprocesses, one of customers that enter the system and one of customers that do not enter the system. This is an extension of the idea of decomposing a Poisson process described in Section 5.6.1. If we apply the same reasoning, the arrival rate into the system in state i becomes

$$\lambda_i = (1 - b_i)\lambda, \ i = 0, 1, 2, \ldots$$

Balking frequently occurs in queueing systems, but it may be difficult to estimate b_i directly when customers that balk do not actually enter the system and therefore are not observable.

8.5.1.3 LIMITED CAPACITY

Case 8.4. Academic Computing Services (ACS) at The Ohio State University provides a dial-up facility so that students, faculty and staff can connect to university computers via a modem. Due to limits on hardware capacity, only a finite number of users can be connected at any one time; if a user tries to connect when all slots are in use, then the person receives a busy signal and is disconnected. Some students have complained that this capacity should be increased, so ACS has decided to study the matter.

The effect on a queueing system of finite capacity is to reduce the arrival rate into the system to 0 whenever the system capacity is reached. Notice that this does not mean that customers no longer attempt to arrive, only that there are no arrivals *from the perspective of the queueing system*. If the capacity of the system is n customers, then $\lambda_i = 0$ for $i = n, n + 1, \ldots$. More specifically, if the arrival process of potential customers is Poisson with rate λ, then

$$\lambda_i = \begin{cases} \lambda, & i = 0, 1, \ldots, n - 1 \\ 0, & i = n, n + 1, \ldots \end{cases}$$

Notice that $\sum_{j=0}^{\infty} d_j = \sum_{j=0}^{n} d_j < \infty$ in this case because $\lambda_i = 0$ for $i > n - 1$. Therefore, the summation is finite, and the steady-state probabilities π_j always exist. A performance measure of interest in finite-capacity queueing systems is $\lambda \pi_n$, which is the *rate of losing customers* since π_n is the probability of finding the system full.

8.5.1.4 SMALL CUSTOMER POPULATION

Case 8.5. A fleet of automated guided vehicles (AGVs) moves parts between 10 work centers in a factory. A queueing model will help determine the number of AGVs needed.

If the work centers are the customers and the AGVs are the servers, then the population of potential customers is small relative to the number of customers that might be in the queueing system (a work center is "in the queueing system" if it has a part being moved or is waiting to have a part moved). In the extreme, if all of the work centers have a part to be moved, then all of the potential customers are in the system and the arrival rate is 0. Therefore, the arrival rate depends on the state of the system, unlike the Poisson arrival process. To account for a small population of customers, we model individual customers in terms of the time that each one spends *outside* the system.

Suppose that there are a total of k potential customers ($k = 10$ in Case 8.5), and the time a customer spends outside the system is exponentially distributed with parameter τ. Then when there are i customers in the system, there are $k - i$ potential customers outside the system. Let $Z_1, Z_2, \ldots, Z_{k-i}$ be the remaining time until each of these potential customers chooses to enter the system, so that $H = \min\{Z_1, Z_2, \ldots, Z_{k-i}\}$ is the remaining time until the first of these customers enters the system. Exploiting the memoryless property of the exponential distribution and Equation (7.5), we find that H is exponentially distributed with parameter $(k-i)\tau$, implying that the arrival rate in state i is $(k-i)\tau$. In general,

$$\lambda_i = \begin{cases} (k-i)\tau, & i = 0, 1, \ldots, k-1 \\ 0, & i = k, k+1, \ldots \end{cases}$$

Notice that $\sum_{j=0}^{\infty} d_j = \sum_{j=0}^{k} d_j < \infty$ in this case because $\lambda_i = 0$ for $i > k-1$. Therefore, the summation is finite and the steady-state probabilities π_j always exist.

8.5.2   Service Rates

In this section we describe common forms for the service rates, μ_i. A critical distinction is that μ_i is the rate at which customers *depart* the queueing system when there are i customers already present, which can be different from the rate at which customers actually receive useful service.

8.5.2.1   MULTIPLE IDENTICAL SERVERS

Case 8.6.   Parking is very limited at the University of Minnesota, so cars line up at the entrance to parking lots to wait for an available opening. The university would like to evaluate the effect of adding additional spaces to a particular lot.

Let the parking spaces be servers, the cars be customers, and suppose that the time a car occupies a space is exponentially distributed with parameter μ. If there are s spaces and $i \leq s$ cars in the lot, then there are i "servers" delivering service. Let $Z_1, Z_2, \ldots, Z_i$ be the remaining time until each of these servers completes service (the car departs the lot). Then $H = \min\{Z_1, Z_2, \ldots, Z_i\}$ is the remaining time until the first space becomes available. Exploiting the memoryless property of the exponential distribution and Equation (7.5), H is exponentially distributed with parameter $i\mu$, implying that the service rate in state i is $i\mu$. In general,

$$\mu_i = \begin{cases} i\mu, & i = 0, 1, \ldots, s \\ s\mu, & i = s+1, s+2 \ldots \end{cases}$$

Notice that the service rate is $\mu_i = s\mu$ for $i > s$ because the service rate is limited by the number of servers. Stated differently, if there are $i > s$ cars in the system, then s of them have spaces and the other $i - s$ are waiting in the queue.

8.5.2.2   RENEGING

Case 8.7.   When customers call Fluttering Duck Airline's toll-free number to make reservations, they may be placed in a "hold" queue until an agent is available. Some customers will hang up if they are on hold too long. This phenomenon should be a part of Fluttering Duck Airline's capacity-planning models.

Reneging occurs when customers in a queueing system choose to leave the system prior to receiving service. The effect of reneging on a queueing system is equivalent to increasing the service rate of the system to something greater than the rate at which the servers deliver service (of course, the reneging customers have not actually been "served").

One way to account for reneging is to model the time a customer is willing to spend waiting in the queue prior to starting service. Suppose that this time is modeled as an exponentially distributed random variable with parameter β, and the service time is exponentially distributed with parameter μ. If there are s identical servers and $i > s$ customers in the system, then there are s customers receiving service (who presumably will not renege) and $i - s$ customers who might renege. Let $Z_1, Z_2, \ldots, Z_s$ be the remaining time until the s customers in service complete service, and let $Z_{s+1}, Z_{s+2}, \ldots, Z_i$ be the remaining time until the $i - s$ customers in the queue will renege. Then $H = \min\{Z_1, Z_2, \ldots, Z_i\}$ is the remaining time until some customer departs from the system, either due to a service completion or reneging. Exploiting the memoryless property of the exponential distribution and Equation (7.5), H is exponentially distributed with parameter $s\mu + (i - s)\beta$, implying that the service rate in state i is $s\mu + (i - s)\beta$. In general,

$$\mu_i = \begin{cases} i\mu, & i = 0, 1, \ldots, s \\ s\mu + (i - s)\beta, & i = s + 1, s + 2 \ldots \end{cases}$$

8.6 PARAMETERIZING QUEUEING PROCESSES

The arrival rates and service rates are the fundamental parameters of a birth-death Markovian queueing model. The "rate" interpretation is useful when data are available in the form of a count over a period of time. For example, if 172 customers arrive over a 3-hour period, then $\widehat{\lambda} = 172/3 \approx 31$ customers per hour is an estimate of the arrival rate for a Poisson arrival process. See Section 5.5.2 for more on parameterizing a Poisson arrival process. Also, people experienced with a system can sometimes provide rough estimates of arrival and service rates even when data are not available.

Some care must be taken when estimating service rates, because the service rate is the rate at which service is delivered *when the server is busy*. For example, knowing that 46 customers were served by a single server during an 8-hour day does not necessarily mean that the service rate was $46/8 = 5.75$ customers per hour, because the server may not have been busy during the entire 8 hours. To adjust the estimate, we need to know how much time was actually spent serving customers. If it took only 6.8 hours to serve the 46 customers, then $\widehat{\mu} = 46/6.8 \approx 6.76$ customers per hour is the appropriate estimate. An estimate of the standard error for this rate is the same as for a Poisson arrival process.

Time-study data are the easiest to use and the most reliable when they can be obtained. For example, if the times to serve five customers were observed to be 12, 17, 4, 5 and 2 minutes, then the sample-average service time is $(12 + 17 + 4 + 5 + 2)/5 = 40/5 = 8$ minutes, with estimated standard error $\widehat{se} = \widehat{\sigma}/\sqrt{5} \approx 2.8$ minutes, as described in Section 3.2. Using the interpretation that $1/\mu$ is the expected service time, we have $\widehat{\mu} = 1/8$ customer per minute as an estimate of the service rate. Of course, five observations is a small number from which to estimate a service rate, which is reflected in the large standard error of the estimate of the expected service time.

A similar caution applies to small customer populations in which the arrival rate depends on the state of the system (how many customers are in the system compared to how many are outside the system). A raw count of customer arrivals divided by the total observation period does not account for this dependence.

When modeling a small customer population, one ideally has data on the interval from when a customer departs the system until that customer next enters it. For instance, suppose that there are k total customers and the arrival rates are of the form $(k - i)\tau$ when there are i customers in the system. Then $1/\tau$ is the expected time from when customers depart the system until they next enter it. The sample average of the observed times that *individual customers* spend outside the system is an estimator of $1/\tau$, with the usual standard error of a sample average. Of course, this requires tracking the behavior of individual customers, which may not be feasible. Fortunately, the customers themselves may be able to provide information about how long it has been since they last visited the system if we survey them upon arrival.

8.7   SHORTHAND NOTATION AND EXAMPLES

We present two examples to illustrate how performance measures are calculated for birth-death Markovian queues, after describing a shorthand notation used to specify certain standard queueing models. Many books contain extensive lists of formulas for π, w, ℓ, etc., for these standard models (see, for instance, Winston 1991, Chapter 22). We have not emphasized this notation, or the standard models, because we believe that in practice it is best to formulate the model that is appropriate for the application at hand, rather than to try to fit the application to a standard model. Nevertheless, the notation is so ubiquitous that it is important to recognize it. The general form is

> arrival process/service process/*s*/*n*/*k*/queue discipline

There is nearly universal agreement about the meaning of the first three modifiers (arrival process/service process/*s*), while the last three may have different meanings in different books.

When we consider the special case of Markovian queues, the arrival and service processes are denoted "M/M." More General queues can be denoted "GI/G," which implies that the gaps between arrivals are independent, time-stationary random variables (a renewal arrival process, see Section 5.9), and the service times are independent and time stationary. Other common options for the arrival and service processes are "D" (deterministic), "E_k" (Erlang with k phases), and "PH" (phase-type distribution).

The s denotes the number of servers, n the capacity of the system, k the size of the customer population, and "queue discipline" the order in which customers receive service. If not specified, the default values are $n = k = \infty$ and the queue discipline is first-come-first-served. For example, the queueing model for the Massive Mall parking facility is an M/M/∞ queue, in the standard notation. Two other examples are given below.

8.7.1 The M/M/s Queue

Case 8.8. In Case 2.1 we compared adding a second photocopier to The Darker Image for either general use or as a dedicated self-service copier. The primary performance measure of interest was expected customer delay.

For either of the proposed systems a Poisson arrival process is a tenable model for customer arrivals, provided shop capacity is not a problem. Therefore, $\lambda_i = \lambda$ for $i = 0, 1, \ldots$.

If the new copier is for general use, then the system has $s = 2$ identical servers. If the new copier is dedicated to self-service customers, then the shop effectively becomes two independent queueing systems—one for full-service and one for self-service customers—each with $s = 1$ server. In either case,

$$\mu_i = \begin{cases} i\mu, & i = 0, 1, \ldots, s \\ s\mu & i = s+1, s+2, \ldots \end{cases}$$

Therefore, for $j = 1, 2, \ldots, s$

$$d_j = \frac{\lambda\lambda \cdots \lambda}{\mu\, 2\mu \cdots j\mu} = \frac{(\lambda/\mu)^j}{j!}$$

while for $j = s+1, s+2, \ldots$

$$d_j = \frac{\lambda\lambda \cdots \lambda}{\mu\, 2\mu \cdots s\mu s\mu \cdots s\mu} = \frac{(\lambda/\mu)^j}{s!\, s^{j-s}}$$

To compute the steady-state probabilities, we first need

$$\sum_{j=0}^{\infty} d_j = \sum_{j=0}^{s} \frac{(\lambda/\mu)^j}{j!} + \sum_{j=s+1}^{\infty} \frac{(\lambda/\mu)^j}{s!\, s^{j-s}}$$

$$= \sum_{j=0}^{s} \frac{(\lambda/\mu)^j}{j!} + \frac{(\lambda/\mu)^s}{s!} \sum_{i=1}^{\infty} \left(\frac{\lambda}{s\mu}\right)^i$$

$$= \sum_{j=0}^{s} \frac{(\lambda/\mu)^j}{j!} + \frac{(\lambda/\mu)^s}{s!} \left(\frac{\lambda}{s\mu}\right) \sum_{k=0}^{\infty} \left(\frac{\lambda}{s\mu}\right)^k$$

$$= \sum_{j=0}^{s} \frac{(\lambda/\mu)^j}{j!} + \frac{(\lambda/\mu)^s}{s!} \frac{(\lambda/(s\mu))}{1 - \lambda/(s\mu)} \tag{8.19}$$

$$= \left(\sum_{j=0}^{s} \frac{(s\rho)^j}{j!}\right) + \frac{s^s \rho^{s+1}}{s!(1-\rho)} \tag{8.20}$$

where (8.19) comes from applying Equation (8.1), and (8.20) comes from substituting $\rho \equiv \lambda/(s\mu)$. Then

$$
\pi_0 = \left[\left(\sum_{j=0}^{s} \frac{(s\rho)^j}{j!} \right) + \frac{s^s \rho^{s+1}}{s!(1-\rho)} \right]^{-1}
$$

and $\pi_j = d_j \pi_0$, $j = 1, 2, \ldots$.

The steady-state expected number of customers in the queue is

$$
\begin{aligned}
\ell_q &= \sum_{j=s+1}^{\infty} (j-s)\pi_j \\
&= \pi_0 \sum_{j=s+1}^{\infty} (j-s)\frac{(\lambda/\mu)^i}{s!s^{i-s}} \\
&= \pi_0 \frac{(\lambda/\mu)^s}{s!} \sum_{i=1}^{\infty} i \left(\frac{\lambda}{s\mu} \right)^i \\
&= \pi_0 \frac{(\lambda/\mu)^s}{s!} \left(\frac{\lambda}{s\mu} \right) \sum_{i=1}^{\infty} i \left(\frac{\lambda}{s\mu} \right)^{i-1} \\
&= \pi_0 \frac{(\lambda/\mu)^s}{s!} \frac{(\lambda/(s\mu))}{(1-\lambda/(s\mu))^2}
\end{aligned}
\tag{8.21}
$$

where (8.21) comes from applying Equation (8.2). Making the substitution $\rho \equiv \lambda/(s\mu)$ and noticing that $\pi_0(\lambda/\mu)^s/s! = \pi_s$, we get

$$
\ell_q = \frac{\pi_s \rho}{(1-\rho)^2}
$$

From ℓ_q we can obtain the expected delay in the queue from Little's formula as

$$
w_q = \frac{\ell_q}{\lambda} = \frac{\pi_s \rho}{\lambda(1-\rho)^2}
\tag{8.22}
$$

The expected waiting time in the system is

$$
w = w_q + \frac{1}{\mu}
$$

Finally, the expected number of customers in the system is

$$
\ell = \lambda w = \frac{\pi_s \rho}{(1-\rho)^2} + \frac{\lambda}{\mu} = \ell_q + \frac{\lambda}{\mu}
$$

To gain some intuition, we adapt these results to $s = 1$ server, which is the M/M/1 queue with $\rho = \lambda/\mu$. After some algebraic simplification

$$\pi_j = (1 - \rho)\rho^j$$

for $j = 0, 1, \ldots$, the geometric distribution with parameter ρ. Thus, the steady-state probability of no customers in the system is $\pi_0 = 1 - \rho$, which decreases linearly in the traffic intensity. The expected number in the queue is

$$\ell_q = \frac{\rho^2}{1 - \rho}$$

Figure 8.3 is a plot of ℓ_q as a function of ρ. The important feature to notice is that as the traffic intensity approaches 1, the expected number in the queue increases dramatically. Thus, long delays occur in apparently "balanced" systems where the arrival rate and service rate are nearly the same. This is counterintuitive, since a system with identical arrival and service rates seems perfectly paced. The queueing is caused by the variability in the arrival and service processes, which can be explained as follows: If there were no variability, then a perfectly balanced system would always have the same number of customers in the system as there are servers. When there is variability, the servers are idle some of the

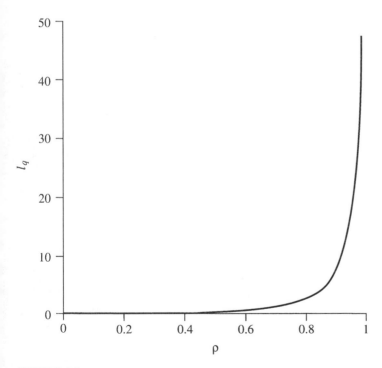

FIGURE 8.3
The expected queue length ℓ_q as a function of traffic intensity $\rho = \lambda/\mu$ for the M/M/1 queue.

time—since service can only take place when there are customers present—and busy at other times. The servers cannot "store up" capacity when they are idle to expend when there is the occasional crush of arrivals, so queueing happens. One benefit of modeling is to determine just how heavily a system can be loaded before congestion is excessive.

Notice also that the formula for ℓ_q is *scale-independent*, meaning that the value of ℓ_q does not change if we measure the arrival rate and service rate in terms of customers per second, per minute, per hour, etc. This is true for both ℓ and ℓ_q in general.

8.7.2 The M/M/s/n/k Queue with s = n

Case 8.9. A computer serves k video display terminals, but only $n < k$ users can connect to the computer at any one time. Users who try to connect when the network is at capacity are disconnected and must try again later.

We view the users as the customers and the n allowed connections as the servers. If the time a user spends between attempts to connect to the computer is exponentially distributed with parameter τ, then the arrival rates are

$$\lambda_i = \begin{cases} (k - i)\tau, & i = 0, 1, \ldots, n - 1 \\ 0, & i = n, n + 1, \ldots \end{cases}$$

reflecting the small customer population and the finite capacity of the system. If the time a user spends connected to the computer is exponentially distributed with expected value $1/\mu$, then the service rates are

$$\mu_i = \begin{cases} i\mu, & i = 0, 1, \ldots, n - 1 \\ n\mu & i = n, n + 1, \ldots \end{cases}$$

Since $\lambda_i = 0$ for $i \geq n$, the M/M/n/n/k queue can be viewed as a finite state-space Markov process with state space $\mathcal{M} = \{0, 1, \ldots, n\}$ in order to obtain time-dependent results. To obtain steady-state results, notice that

$$
\begin{aligned}
d_j &= \frac{\lambda_0 \lambda_1 \cdots \lambda_{j-1}}{\mu_1 \mu_2 \cdots \mu_j} \\
&= \frac{k\tau \, (k - 1)\tau \cdots (k - j + 1)\tau}{\mu \, 2\mu \cdots j\mu} \\
&= \left(\frac{\tau}{\mu}\right)^j \frac{1}{j!} k(k - 1) \cdots (k - j + 1) \\
&= \left(\frac{\tau}{\mu}\right)^j \frac{1}{j!} \frac{k!}{(k - j)!} \\
&= \left(\frac{\tau}{\mu}\right)^j \binom{k}{j}
\end{aligned}
$$

for $i = 1, 2, \ldots, n$, and $d_j = 0$ for $j = n+1, n+2, \ldots$. Thus,

$$
\pi_0 = \frac{1}{\sum_{j=0}^{n} \left(\frac{\tau}{\mu}\right)^j \binom{k}{j}}
$$

$$
\pi_j = \begin{cases} \pi_0 \left(\frac{\tau}{\mu}\right)^j \binom{k}{j}, & j = 1, 2, \ldots, n \\ 0, & j = n+1, n+2, \ldots \end{cases}
$$

Using this result, we can obtain the effective arrival rate into the computer system

$$
\lambda_{\text{eff}} = \sum_{j=0}^{\infty} \lambda_j \pi_j
$$

$$
= \sum_{j=0}^{n} (k-j)\tau \, \pi_0 \left(\frac{\tau}{\mu}\right)^j \binom{k}{j}
$$

$$
= \pi_0 \tau \sum_{j=0}^{n} (k-j) \left(\frac{\tau}{\mu}\right)^j \binom{k}{j}
$$

and the long-run expected number of users connected to it

$$
\ell = \sum_{j=0}^{\infty} j \, \pi_j
$$

$$
= \sum_{j=0}^{n} j \, \pi_0 \left(\frac{\tau}{\mu}\right)^j \binom{k}{j}
$$

$$
= \pi_0 \sum_{j=0}^{n} j \left(\frac{\tau}{\mu}\right)^j \binom{k}{j}
$$

You should verify that $w = \ell/\lambda_{\text{eff}} = 1/\mu$, as it must be since the waiting time in the system is just the time a customer spends connected to the computer.

The quantity π_n is the long-run fraction of the time that the computer is at capacity. In the special case when $1/\tau = 1/\mu$—implying that the expected time between attempts to connect is the same as the time spent connected to the computer—this formula simplifies to

$$
\pi_n = \frac{\binom{k}{n}}{\sum_{j=0}^{n} \binom{k}{j}} \tag{8.23}
$$

To gain some appreciation for the effect of limited capacity, the table below shows the probability (8.23) and the expected number of users connected ℓ for $n = 8$ connections as the number of terminals k increases (Equations (8.5) and (8.6) are useful for evaluating these expressions). Notice that increasing the number of terminals by a factor of 3 (from 8 to 24) increases the probability of being at capacity by a factor of nearly 145 (from 0.004 to 0.578). This illustrates that system performance can be very sensitive to the relative sizes of the customer population and the system capacity.

k	π_8	ℓ
8	0.004	4.0
12	0.130	5.7
16	0.328	6.7
20	0.477	7.1
24	0.578	7.4

8.8   THE CASE OF THE TARDY TICKET

Case 8.10.   A 20-theater movie complex is planned for a new shopping mall. Each individual theater will seat 200 people. During peak business hours a new movie will start (and therefore another one will end) roughly every 6 minutes. Market research predicts that each theater will be 83% full on average.

The problem of interest is how many ticket windows to include in the theater design. Because some patrons arrive just prior to the start of the movie they want to see, the theater developers do not want the time required to purchase a ticket to be too long (an average of 3 minutes or less is desirable). And if they build too few windows initially, the cost of remodeling to add more is substantial. On the other hand, there is the cost of staffing the windows with ticket agents and of purchasing the ticketing machines and cash registers. Together these facts suggest that the developers should be careful to include enough windows, but not more than actually needed.

The time required to sell tickets might seem to be nearly the same from person to person, but observation at other theaters indicates that there can be substantial variability due to the number of tickets purchased, the method of payment (cash or credit card) and whether or not the desired movie is sold out (in which case patrons may select another movie). The typical transaction takes about one minute per person, and the most likely purchase is two tickets.

Given this information, your task is to recommend a number of ticket windows.

The ticket-selling process is naturally viewed as a queueing system in which the patrons are the customers and the ticket windows (each with a ticket agent) are the servers. The analysis will be rough, given the sketchy information available, as is often the case in practice.

Based on the available data, there will be $200(0.83) = 164$ customers per movie on average, but the tickets will be purchased by about $164/2 = 83$ of them since each ticket purchaser typically purchases two tickets. Even though ticket-purchasing customers may arrive at any time before (or even right after the start of) the movie, they must be coming at a rate of 83 every 6 minutes, or approximately 13.83 per minute, to fill the theaters that empty every 6 minutes.

What more can be said about this arrival process? Certainly there will be variability in the times between arrivals since customers do not come on a schedule and each movie will not have exactly 164 customers. But should we think of the base arrival *rate* as being constant over time?

In a single-theater complex a constant-rate model is not appropriate. Customers tend to arrive in surges just prior to the start of a movie, with only a few arriving in the hour or so after the movie's start. Therefore, our standard birth-death queueing model is not useful because it represents the behavior of a time-stationary process. But in the proposed multiple-theater complex, in which a new movie starts frequently, a constant rate may be reasonable.

In this chapter we have emphasized long-run performance of queueing systems. Long-run performance is relevant for the theater complex provided the peak business period lasts for a significant span of time. As a rough rule of thumb, we use long-run performance measures if the time period of interest is at least two orders of magnitude longer than the expected time between arrivals or service completions. In this example service takes on the order of 1 minute, so a peak period of 3 or more hours (180 or more minutes) suggests that long-run performance is a useful measure.

Based on the available data, a service rate of 1 purchase per minute per ticket window can be conjectured. Unless there is some reason to believe that there will be substantial differences among the ticket agents (and there is really no way to know this since the theater does not yet exist), we are forced to treat the ticket agents as identical servers. Over the long run, the offered load is $o =$ arrival rate/service rate $= 13.83/1$. Since the offered load can be interpreted as the expected number of busy servers, this implies that at least 14 windows are required just to keep up with arrivals. But whether or not 14 is sufficient depends on the waiting times that customers experience.

To make statements about the waiting time or the number of customers in a queueing system, the distributions of the interarrival-time gaps and service times must be specified. If we treat them as exponentially distributed, then long-run performance can be computed using the tools in this chapter. If we treat them as having some other distributions, then we must either simulate the system or make use of results for non-Markovian queues (see Section 8.10).

In the present case, treating the arrival process as Poisson seems plausible because it is composed of a large number of ticker buyers acting largely independently. However, the time to purchase tickets may be less variable than the exponential distribution is. Since variability in queueing systems is what leads to congestion, results we obtain from a Markovian queueing model will indicate somewhat more congestion and longer waits than we will actually see. This is acceptable for planning purposes as long as we understand in which direction the error is.

Next we need to characterize the system logic. Is it a single queue for all windows or individual queues for each? Is there any capacity limit on the system? Do customers balk or renege? The answers to each of these questions have an impact on our model.

If there will actually be a single queue or if customers can jockey to another queue when it becomes empty, then the system effectively has a single queue.

On the other hand, if customers are segregated into distinct queues from which it is not easy to jockey, then we might treat each window as an isolated queueing system receiving $1/k$th of the arrivals, when there are k windows. This is an approximation since arriving customers probably choose the shortest queue when they arrive, rather than distributing themselves randomly.

Balking and reneging are not likely if there are adequate windows to keep the wait short. There is limit on the capacity of each movie theater, however, that indirectly affects the ticket-selling process. The effect on the ticket-selling process will be small if theaters do not often fill up or if customers frequently choose another movie when their first choice is not available. Although we will ignore the capacity restriction here, it should be noted as a phenomenon worth further study.

We have implicitly assumed that there will be a fixed number of ticket windows open at all times. However, it is also possible to change the number of open windows depending on the number of customers waiting. Although we will not consider this option here, the effect of state-dependent staffing is to increase the service rate when the number in the system reaches selected critical numbers.

We have described a queueing system with a Poisson arrival process having rate $\lambda = 13.83$ per minute, exponentially distributed service times with rate $\mu = 1$ per minute, $s \geq 14$ identical servers, and unlimited capacity. This is an M/M/s queue, and the results in Section 8.7.1 are relevant.

When $s = 14$, we have $\rho = \lambda/(s\mu) = 13.83/(14 \cdot 1) \approx 0.99$, revealing a very high utilization per ticket window. The probability of an empty system is

$$\pi_0 = \left[\left(\sum_{j=0}^{14} \frac{(14(0.99))^j}{j!} \right) + \frac{14^{14}(0.99)^{14+1}}{14!(1-0.99)} \right]^{-1} \approx 0.107 \times 10^{-6}$$

and the probability of s in the system is

$$\pi_s = \pi_{14} = d_{14}\pi_0 = \frac{(13.83/1)^{14}}{14!}\pi_0 \approx 0.0115$$

Therefore,

$$w_q = \frac{\pi_s \rho}{\lambda(1-\rho)^2} \approx 5.7$$

and

$$w = w_q + \frac{1}{\mu} = 5.7 + 1 = 6.7 \text{ minutes}$$

In other words, the expected time to complete the ticket purchase is 6.7 minutes, which exceeds the desired 3-minute average. A similar calculation with $s = 15$ ticket agents gives $w \approx 1.6$ minutes, revealing that 15 ticket agents will be adequate.

8.9 NETWORKS OF MARKOVIAN QUEUES

Many complex systems can be represented as *networks of queues* in which customers receive service from one or more groups of servers, and each group of servers has its own queue. Stated differently, a network of queues is a collection of individual queueing systems that have common customers. Networks of queues are particularly prevalent in manufacturing, communication and computer-performance applications. In this section we present a handy result for the analysis of networks of Markovian queues.

8.9.1 The Case of the Incredible Shrinking Leviathan

Case 8.11. Leviathan Limited, a very large enterprise indeed, maintains a repair facility that reconditions certain expensive products when they fail. The repair facility consists of a repair station, an inspection station and a combined repair-and-inspection station. The repair station makes the first attempt to repair a product. The product then moves to the inspection station to verify that the repairs have been successful. Products that pass inspection are shipped back to their owner. The small number of products that fail inspection move to the combined repair-and-inspection station where they are repeatedly repaired and inspected until they work properly. The original vision for this configuration was to exploit the efficiency of specialized repair and inspection stations for most products and have a general-purpose station for the particularly difficult cases.

Regrettably, profits have not been good for Leviathan Ltd., so management is searching for ways to make more efficient use of resources. An industrial engineer (IE) has noticed that the combined repair-and-inspection station is not heavily utilized; he speculates that this station can be eliminated by sending products that fail inspection back to the primary repair station. For this to be a viable option, it must not lead to excessive delays, so the IE must predict the impact of the change to support his proposal.

Figure 8.4 shows the repair facility as it is currently configured. The arrows indicate the flow of products through the system. The proposal is to retain only that portion enclosed by the dashed line and then change the product flow to be as shown in Figure 8.5. Clearly the proposed configuration will increase the workload at both the repair and inspection stations, and therefore it will increase delays. The question is, How much of an increase in delay?

Since Leviathan Ltd. produces so many products, each of the stations in the repair facility employs many technicians (servers). But for the purpose of illustration, we will pretend that there is only one repair technician at the repair station and only one inspector at the inspection station. The analysis for multiple servers at each station is an easy extension of the analysis for a single server, and we will indicate later how it is done. The combined repair-and-inspection station is unimportant to our analysis and will not be discussed further.

To develop tools for the analysis of networks of queues, we start by presenting a simple model of the repair station and inspection station as they currently operate (see Figure 8.4). In the sequel we analyze the proposed system configuration.

There are three system inputs we will consider:

- The interarrival-time gaps between the arrival of products to the repair station, $G_1, G_2, \ldots$, which are independent and time stationary with common cdf F_G

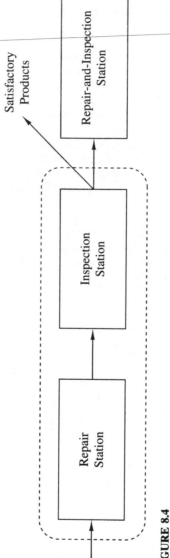

FIGURE 8.4

The repair facility at Leviathan Ltd. as it is currently configured. The portion enclosed by the dashed line is to be retained.

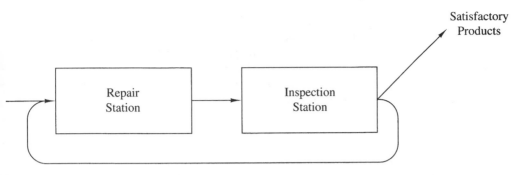

FIGURE 8.5
The proposed repair facility.

- The times to repair the products, $X_1, X_1, \ldots$, which are independent and time stationary with common cdf F_X
- The times to inspect the products, $Z_1, Z_1, \ldots$, which are independent and time stationary with common cdf F_Z

We also suppose that the G_i's, X_i's and Z_i's are mutually independent, which is plausible, although it is possible that the time required to inspect a product is dependent on the time required to repair it.

The system logic for our simple model is as follows:

- Products arrive at the repair station and wait, first-come-first-served, for the repair technician.
- When repairs are complete, products move to the inspection station and wait, first-come-first-served, for the inspector.
- When inspection is complete, products leave the system (since our "system" does not include the combined station).

We know that useful performance measures for individual queueing systems can be derived from information about the number of customers in the system. To extend this idea to Leviathan Ltd.'s network of two queues, let $S_{1,n}$ represent the number of products at the repair station, and let $S_{2,n}$ represent the number of products at the inspection station, just after the nth change of system state. The state space is therefore $\mathcal{M} = \{(i, j) : i = 0, 1, \ldots \ j = 0, 1, \ldots\}$, where the first element i represents the number of products at the repair station, and second element j represents the number of products at the inspection station.

The system events are the arrival of a product to the repair station, the completion of repairs at the repair station, and the completion of inspection at the inspection station. Notice that the completion of repairs at the repair station implies the arrival of a product to the inspection station. This is a characteristic of networks of queues: *a departure from one queueing system is often an arrival to another queueing system in the network.*

Let the clocks corresponding to the three system events be C_1 for the time when the next product arrives at the repair station, C_2 for the time when the next product completes repairs (and arrives at inspection), and C_3 for the time when the next product completes inspection. The system-event functions are given below. Notice that e_2, which models a product completing repairs, also contains the logic for the part arriving at the inspection station.

$e_0()$ **(initialization)**

$\quad\quad S_{1,0} \leftarrow 0$ (initially no products in repair)

$\quad\quad S_{2,0} \leftarrow 0$ (initially no products in inspection)

$\quad\quad C_1 \leftarrow F_G^{-1}(\text{random}())$ (set clock for first product arrival)

$\quad\quad C_2 \leftarrow \infty$ (indicate no pending repair)

$\quad\quad C_3 \leftarrow \infty$ (indicate no pending inspection)

$e_1()$ **(arrival to repair)**

$\quad\quad S_{1,n+1} \leftarrow S_{1,n} + 1$ (one more product at repair)

$\quad\quad \text{if } \{S_{1,n+1} = 1\} \text{ then}$ (if only one product then start)

$\quad\quad\quad\quad C_2 \leftarrow T_{n+1} + F_X^{-1}(\text{random}())$ (set clock for completion)
$\quad\quad \text{endif}$

$\quad\quad C_1 \leftarrow T_{n+1} + F_G^{-1}(\text{random}())$ (set clock for next arrival)

$e_2()$ **(complete repair)**

$\quad\quad S_{1,n+1} \leftarrow S_{1,n} - 1$ (one fewer product at repair)

$\quad\quad \text{if } \{S_{1,n+1} > 0\} \text{ then}$ (if another product then start)

$\quad\quad\quad\quad C_2 \leftarrow T_{n+1} + F_X^{-1}(\text{random}())$ (set clock for completion)
$\quad\quad \text{endif}$

$\quad\quad S_{2,n+1} \leftarrow S_{2,n} + 1$ (one more product at inspection)

$\quad\quad \text{if } \{S_{2,n+1} = 1\} \text{ then}$ (if only one product then start)

$\quad\quad\quad\quad C_3 \leftarrow T_{n+1} + F_Z^{-1}(\text{random}())$ (set clock for completion)
$\quad\quad \text{endif}$

$e_3()$ **(complete inspection)**

$\quad\quad S_{2,n+1} \leftarrow S_{2,n} - 1$ (one fewer product at inspection)

$\quad\quad \text{if } \{S_{2,n+1} > 0\} \text{ then}$ (if another product then start)

$\quad\quad\quad\quad C_3 \leftarrow T_{n+1} + F_Z^{-1}(\text{random}())$ (set clock for completion)
$\quad\quad \text{endif}$

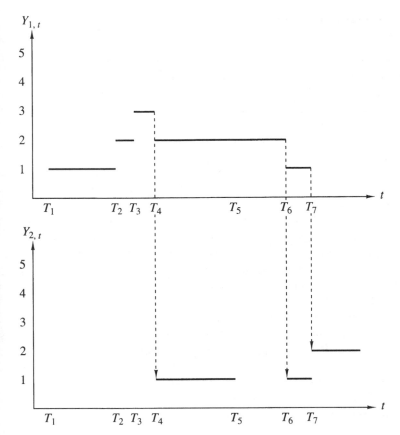

FIGURE 8.6
Sample path of the queueing-network simulation.

If, as usual, we let $Y_{i,t} \leftarrow S_{i,n}$ when $T_n \leq t < T_{n+1}$, then $Y_{1,t}$ is the total number of products at the repair station at continuous-time t, and $Y_{2,t}$ is the corresponding quantity for the inspection station. Given specific distributions F_G, F_X and F_Z, we can simulate this system and extract performance measures from the sample paths. Figure 8.6 shows a portion of a sample path. Notice that each time $Y_{1,t}$ decreases by 1—indicating a departure from the repair station—then $Y_{2,t}$ simultaneously increases by 1—indicating an arrival to the inspection station.

In this chapter we have emphasized making statements about queueing systems as time goes to infinity, which requires long simulations. However, if F_G, F_X and F_Z are all exponential distributions, then our model of the repair facility satisfies the conditions of a Markov process. In the next section we develop a Markov process model of the current and proposed repair facility and derive steady-state performance measures.

8.9.2   Markov Process Model of the Incredible Shrinking Leviathan

Suppose that exponential distributions provide good approximations for F_G, F_X and F_Z in the model of Leviathan Ltd.'s repair facility, and let the respective parameters of these exponential distributions be λ, $\mu^{(1)}$ and $\mu^{(2)}$. Clearly this is not a birth-death process, so we are forced to derive the generator, **G**. For this purpose it is convenient to identify the states as "(i, j)" meaning that $\{Y_{1,t} = i, Y_{2,t} = j\}$ for $i, j = 0, 1, 2, \ldots$. For example, state $(2, 1)$ corresponds to 2 products at the repair station and 1 product at the inspection station.

Consider the state $(0, 0)$, corresponding to an empty system. Since there is no product in the system, the only pending system event is an arrival, which moves the process to state $(1, 0)$.

Next consider a state $(i, 0)$, with $i > 0$. In this case the pending system events are an arrival—which moves the process to state $(i + 1, 0)$—and the completion of a repair—which moves the process to state $(i - 1, 1)$. Similarly, if the process is in state $(0, j)$ with $j > 0$, the pending system events are an arrival—which moves the process to state $(1, j)$—and the completion of an inspection—which moves the process to state $(0, j - 1)$.

Finally, suppose that the process is in state (i, j) with $i > 0$ and $j > 0$. All three system events are pending, so the next state can be $(i + 1, j)$ if the arrival occurs first, $(i - 1, j + 1)$ if the completion of repairs occurs first, or $(i, j - 1)$ if the completion of inspection occurs first. Making use of the fact that arrivals occur at rate λ, repairs at rate $\mu^{(1)}$, and inspections at rate $\mu^{(2)}$, our description above leads to the transition-rate diagram in Figure 8.7. From this diagram we can extract the elements of the generator matrix:

$$
\begin{aligned}
g_{(i,j)(i+1,j)} &= \lambda, & i, j &= 0, 1, \ldots \\
g_{(i,j)(i-1,j+1)} &= \mu^{(1)}, & i &= 1, 2, \ldots; \ j = 0, 1, \ldots \\
g_{(i,j)(i,j-1)} &= \mu^{(2)}, & i &= 0, 1, \ldots; \ j = 1, 2, \ldots
\end{aligned}
$$

and

$$
g_{(i,j)(i,j)} = \begin{cases}
\lambda, & i = j = 0 \\
\lambda + \mu^{(1)}, & i > 0, j = 0 \\
\lambda + \mu^{(2)}, & i = 0, j > 0 \\
\lambda + \mu^{(1)} + \mu^{(2)}, & i, j > 0
\end{cases}
$$

All other elements $g_{(i,j)(k,l)} = 0$.

Let $\pi_{(i,j)}$ be the steady-state probability of finding the process in state (i, j). That is,

$$
\pi_{(i,j)} = \lim_{t \to \infty} \Pr\{Y_{1,t} = i, Y_{2,t} = j\}
$$

For this model it is easier to use the balance-equation approach to derive the steady-state probabilities than to write out $\boldsymbol{\pi}'\mathbf{G} = \mathbf{0}$ directly. For example, if we pick state $(0, 0)$, then we can see from Figure 8.7 that the expected rate into the state is $\mu^{(2)}\pi_{(0,1)}$, while the expected rate out is $\lambda\pi_{(0,0)}$, and these must be equal.

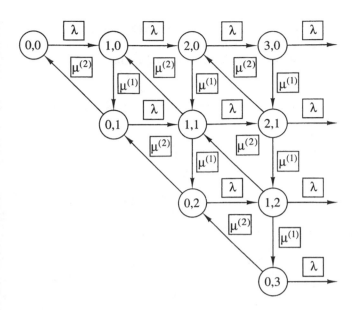

FIGURE 8.7
Transition-rate diagram for the current repair facility.

Altogether,

$$\text{rate in} = \text{rate out}$$

$$\mu^{(2)}\pi_{(0,1)} = \lambda\pi_{(0,0)} \qquad (8.24)$$

$$\lambda\pi_{(i-1,0)} + \mu^{(2)}\pi_{(i,1)} = (\lambda + \mu^{(1)})\pi_{(i,0)}, \; i > 0 \quad (8.25)$$

$$\mu^{(1)}\pi_{(1,j-1)} + \mu^{(2)}\pi_{(0,j+1)} = (\lambda + \mu^{(2)})\pi_{(0,j)}, \; j > 0 \quad (8.26)$$

$$\lambda\pi_{(i-1,j)} + \mu^{(1)}\pi_{(i+1,j-1)} + \mu^{(2)}\pi_{(i,j+1)} = \delta\pi_{(i,j)}, \; i, j > 0 \quad (8.27)$$

where $\delta = \lambda + \mu^{(1)} + \mu^{(2)}$. We can, with some effort, solve these equations for $\pi_{(i,j)}$. But what would we like the answer to be?

Notice that in isolation the repair station is an M/M/1 queue with arrival rate λ, service rate $\mu^{(1)}$ and traffic intensity $\rho_1 = \lambda/\mu^{(1)}$. Therefore, the steady-state probability of finding i products at the repair station is $(1 - \rho_1)\rho_1^i$, and part of the problem is solved.

Also notice that the arrival rate into the inspection station must be λ, because the arrival rate into the inspection station is the departure rate out of the repair station, and the departure rate out of the repair station must be equal to its arrival rate for the system to be stable. So it would be nice if the inspection station is also an M/M/1 queue with traffic intensity $\rho_2 = \lambda/\mu^{(2)}$.

Finally, to make things really convenient, it would be helpful if the number of products at the repair station was independent of the number of products at the inspection station. This is certainly *not* true in general, because departures from the repair station become the arrivals to the inspection. So $\Pr\{Y_{1,t} = i, Y_{2,t} = j\} \neq \Pr\{Y_{1,t} = i\}\Pr\{Y_{2,t} = j\}$ for all t. But perhaps in the limit as time $t \to \infty$, the joint probability of the number of products at each station could be the product of the marginal probabilities. In other words,

$$\pi_{(i,j)} = \lim_{t \to \infty} \Pr\{Y_{1,t} = i, Y_{2,t} = j\} = (1 - \rho_1)\rho_1^i(1 - \rho_2)\rho_2^j \tag{8.28}$$

for $i, j = 0, 1, 2, \ldots$. Fortunately, Equation (8.28) *is correct*! To see this, we will show that (8.28) satisfies balance equation (8.27). In Exercise 22 you are asked to show that (8.28) also satisfies Equations (8.24)–(8.26) and that $\sum_{i=0}^{\infty} \sum_{j=0}^{\infty} \pi_{(i,j)} = 1$.

Substituting (8.28) for $\pi_{(i,j)}$ in Equation (8.27), we obtain

$$\lambda(1 - \rho_1)\rho_1^{i-1}(1 - \rho_2)\rho_2^j + \mu^{(1)}(1 - \rho_1)\rho_1^{i+1}(1 - \rho_2)\rho_2^{j-1} +$$

$$\mu^{(2)}(1 - \rho_1)\rho_1^i(1 - \rho_2)\rho_2^{j+1} = (\lambda + \mu^{(1)} + \mu^{(2)})(1 - \rho_1)\rho_1^i(1 - \rho_2)\rho_2^j$$

Dividing out the common terms yields

$$\lambda\rho_2 + \mu^{(1)}\rho_1^2 + \mu^{(2)}\rho_1\rho_2^2 = (\lambda + \mu^{(1)} + \mu^{(2)})\rho_1\rho_2$$

Then using the fact that $\rho_i = \lambda/\mu^{(i)}$, we obtain

$$\lambda\rho_2 + \lambda\rho_1 + \lambda\rho_1\rho_2 = \lambda\rho_1\rho_2 + \lambda\rho_2 + \lambda\rho_1$$

which verifies that (8.28) satisfies the last balance equation.

The primary consequence of this result is that we can derive steady-state performance measures for each station individually as if it were an isolated M/M/1 queue. For instance, the expected number of customers in the queue at each station is

$$\ell_q^{(i)} = \frac{\rho_i^2}{1 - \rho_i}$$

and the expected delay in queue, from Little's formula, is

$$w_q^{(i)} = \frac{\rho_i^2}{\lambda(1 - \rho_i)}$$

for $i = 1, 2$.

To be more specific, if $\lambda = 5$ products arriving per day, $\mu^{(1)} = 6$ products repaired per day, and $\mu^{(2)} = 8$ products inspected per day, then we obtain the performance measures in Table 8.1. Notice that products can expect to spend about $0.83 + 0.21 \approx 1$ day waiting for repair or inspection to begin. Of course, these values should match what is currently happening in the repair facility, which is one way to validate the model.

To extend the analysis to the proposed system, we must account for products that fail inspection and are returned to the repair station. Suppose that a product fails inspection with probability r, independent of how many times it has been repaired (certainly an approximation). Let $\lambda^{(1)}$ be the *overall* arrival rate to the

TABLE 8.1
Performance measures for the
repair facility as it is currently
configured

performance measure	repair station	inspection station
ℓ_q (customers)	4.17	1.04
w_q (days)	0.83	0.21

repair station, including those products that failed inspection, and let $\lambda^{(2)}$ be the corresponding overall arrival rate to the inspection station. Then

$$\lambda^{(1)} = \lambda + r\lambda^{(2)}$$

In words, the overall rate is the sum of the rate of new arrivals, λ, and the fraction of those products arriving at (and therefore departing from) the inspection station that must be repaired again, $r\lambda^{(2)}$. And since all arrivals to the repair station become arrivals to the inspection station, we have

$$\lambda^{(2)} = \lambda^{(1)}$$

Solving these two equations simultaneously gives

$$\lambda^{(1)} = \lambda^{(2)} = \frac{\lambda}{1-r} \tag{8.29}$$

For example if 10% of products fail inspection, then $r = 0.1$ and the overall arrival rates are

$$\lambda^{(1)} = \lambda^{(2)} = \frac{\lambda}{1-0.1} = \frac{10}{9}\lambda \text{ products per day}$$

which is clearly greater than λ.

Fortunately, the steady-state number of products at the two stations in this model of the proposed system also behave (in the limit) like independent M/M/1 queues with traffic intensities $\varrho_i = \lambda^{(i)}/\mu^{(i)}$ for $i = 1, 2$. Substituting the values $\lambda = 5$, $\mu^{(1)} = 6$ and $\mu^{(2)} = 8$, and using $r = 0.1$, gives $\lambda^{(1)} = \lambda^{(2)} = (10/9)(5) \approx 5.56$, and the results displayed in Table 8.2. Notice that the expected delay at the repair station has increased substantially, from less than 1 day to over 2 days. And this is the expected delay *on each visit* to the repair station, so that products that fail inspection will experience a long delay multiple times.

Perhaps more important is what happens to products that do not fail inspection: In the current configuration their expected time in the system is 0.83 (delay at repair) + 1/6 (repair time) + 0.21 (delay at inspection) + 1/8 (inspection time) ≈ 1.33 days. In other words, products that are successfully repaired on the first try are returned to their owners in about 1.33 days. But in the proposed configuration the expected time in the system is $2.11 + 1/6 + 0.28 + 1/8 \approx 2.68$ days; in other words it has doubled. The dramatic change illustrates the effect of traffic intensity

TABLE 8.2
Performance measures for the proposed repair facility

performance measure	repair station	inspection station
ℓ_q (customers)	11.71	1.58
w_q (days)	2.11	0.28

as it approaches 1. In the current system $\rho_1 \approx 0.83$, but in the proposed system $\varrho_1 \approx 0.93$. As Figure 8.3 illustrates, beyond a traffic intensity of 0.8 the expected queue length, and thus the expected delay, explodes.

The Markov process models we have constructed are special cases of the general class of *open Jackson networks*, and all networks in this class have the property that the steady-state number of customers at each queueing system is independent of the number of customers at each of the other queueing systems. We present the general result for open Jackson networks in the next section.

8.9.3   Open Jackson Networks

The class of open Jackson networks generalizes the example in the previous section in two ways: There can be any number of stations and each station can have multiple servers. We state the general result after defining notation:

- m is the number of stations (queueing systems) in the network.
- s_j is the number of identical servers at station $j = 1, 2, \ldots, m$.
- $\mu^{(j)}$ is the service rate for each server at station $j = 1, 2, \ldots, m$.
- a_{0j} is the *external* arrival rate to station $j = 1, 2, \ldots, m$; that is, the arrival rate from outside the queueing network. Let $\mathbf{A} \equiv (a_{01}, a_{02}, \ldots, a_{0m})'$ be the $m \times 1$ vector of external arrival rates.
- $\lambda^{(j)}$ is the *overall* arrival rate to station $j = 1, 2, \ldots, m$. Let $\mathbf{\Lambda} = (\lambda^{(1)}, \lambda^{(2)}, \ldots, \lambda^{(m)})'$ be the $m \times 1$ vector of overall arrival rates.
- r_{jk} is the probability that a customer departing from station j next goes to station k, for $j, k = 1, 2, \ldots, m$. The probability that the customer departs the network from station j is denoted $r_{j,m+1} = 1 - \sum_{k=1}^{m} r_{jk}$. Let $\mathbf{R}$ be the $m \times m$ matrix with elements r_{jk} for $j, k = 1, 2, \ldots, m$. This matrix is sometimes called the customer *routing matrix*.
- $\varpi(\rho, s, i)$ is the steady-state probability of i customers in the system for an M/M/s queue with traffic intensity ρ (see Section 8.7.1).
- $\pi_{(i_1, i_2, \ldots, i_m)}$ is the steady-state probability of simultaneously finding i_j customers at stations $j = 1, 2, \ldots, m$. More precisely,

$$\pi_{(i_1, i_2, \ldots, i_m)} = \lim_{t \to \infty} \Pr\{Y_{1,t} = i_1, Y_{2,t} = i_2, \ldots, Y_{m,t} = i_m\}$$

where $Y_{j,t}$ is the number of customers at station j at time t.

Given the external arrival rates a_{0j} and routing probabilities r_{kj}, the overall arrival rates $\lambda^{(j)}$ solve the system of equations

$$\lambda^{(j)} = a_{0j} + \sum_{k=1}^{m} r_{kj} \lambda^{(k)}$$

for $j = 1, 2, \ldots, m$. This system of equations can be written in matrix form as

$$(\mathbf{I} - \mathbf{R})' \Lambda = \mathbf{A}$$

for which the solution is

$$\Lambda = \left[(\mathbf{I} - \mathbf{R})' \right]^{-1} \mathbf{A}$$

To make sure the notation is understood, notice that for the proposed repair facility in Section 8.9.2 we have $\mathbf{A} = (a_{01}, a_{02})' = (\lambda, 0)' = (5, 0)'$ and

$$\mathbf{R} = \begin{pmatrix} r_{11} & r_{12} \\ r_{21} & r_{22} \end{pmatrix} = \begin{pmatrix} 0 & 1 \\ r & 0 \end{pmatrix} = \begin{pmatrix} 0 & 1 \\ 0.1 & 0 \end{pmatrix}$$

The fundamental result for Jackson networks is that we can decompose the problem of determining the joint steady-state probabilities for all of the queueing systems in the network into the problem of determining the marginal steady-state probabilities of the m queueing systems individually. This is a tremendous simplification, analogous to the simplification we obtain in analyzing a stochastic process when all of the random variables are independent.

In a network of m stations, where each station is a (potentially) multiple server, infinite-capacity, first-come-first-served queueing system, if the external arrival processes are independent Poisson arrival processes, and the service times are independent and exponentially distributed with the same parameter for all servers at a station, then

$$\pi_{(i_1, i_2, \ldots, i_m)} = \prod_{j=1}^{m} \varpi (\rho_j, s_j, i_j) \tag{8.30}$$

for $i_1, i_2, \ldots, i_m = 0, 1, 2, \ldots$, where

$$\rho_j = \frac{\lambda^{(j)}}{s_j \mu^{(j)}}$$

is the traffic intensity at station j. In other words, the number of customers at each station has the distribution of an independent M/M/s queue, and standard performance measures for an M/M/s queue can be computed for each station in isolation.

From this result for open Jackson networks we see that analyzing the proposed repair facility for Leviathan Ltd. does not become any more difficult if there are multiple servers at each station: We simply use the results for an M/M/s queue with $s > 1$, but otherwise the analysis is unchanged.

8.10 NON-MARKOVIAN QUEUES AND NETWORKS

In this chapter we have developed tools for the analysis of Markovian queueing systems and networks of such queues; in other words, we develop tools for the analysis of queueing systems that can be represented as Markov processes. As you are well aware by now, the Markov property in continuous time implies that the random variables which determine the clock settings (the service times, interarrival-time gaps, time to balk, etc.) are exponentially distributed. Because so many systems of practical importance can be represented as queueing systems and because exponential distributions are not always appropriate, there has been significant interest in non-Markovian queueing systems.

For instance, it is known that the steady-state probability of finding j customers in an M/G/∞ queueing system is the same as the corresponding probability for an M/M/∞ queue; specifically,

$$\pi_j = \frac{e^{-\lambda/\mu} \left(\frac{\lambda}{\mu}\right)^j}{j!} \text{ for } j = 0, 1, \ldots$$

where $1/\mu$ is interpreted as the expected service time. Thus, our conclusions about Massive Mall parking do not depend on exponentially distributed parking times. Unfortunately, results for queueing systems as general as, say, a GI/G/s queue are not available.

If we take the perspective that all stochastic-process models are approximations, then we can view Markovian queueing models as "one-moment approximations" because they assume exponential distributions and the exponential distribution is completely characterized by its "first moment" (expected value).[3] Recall that if X is exponentially distributed with parameter λ, then $\lambda = 1/E[X]$. Therefore, the expected values of the service times, interarrival-time gaps, time to balk, etc., completely determine the approximation.

A one-moment approximation can be quite useful, even when exponential distributions are not entirely appropriate, if we focus on *relative performance* rather than *absolute performance*. An example of relative performance is the percentage decrease in expected delay from employing an additional server, while

---

[3]The *nth moment* of a random variable X is $E[X^n]$, so the first moment is simply $E[X]$. The first and second moments of a random variable determine its variance, because $\text{Var}[X] = E[X^2] - (E[X])^2$.

absolute performance is the actual value of the expected delay. Predictions of the relative performance from changing the queueing system can be good even when the absolute performance is in error.

However, to more accurately approximate general queueing systems, we can incorporate information beyond the first moment. Although the variance of a random variable is a natural choice, it is more common in queueing theory to specify the *squared coefficient of variation*. The coefficient of variation was defined in Section 3.1.3, Equation 3.1., and is simply the standard deviation of a random variable divided by its expected value. Thus, the coefficient of variation is a measure of the relative variability, while the standard deviation and the variance are measures of the absolute variability.

Recall that the coefficient of variation of an exponentially distributed random variable is 1, no matter what its expected value is. The value 1 is considered a benchmark. A coefficient of variation much greater than 1 indicates a highly variable quantity, while a coefficient of variation much less than 1 indicates a quantity with low variability, relative to its expected value.

Let ε be generic for the *squared* coefficient of variation. Specifically, for a random variable X the squared coefficient of variation is

$$\varepsilon \equiv (CV[X])^2 = \left(\frac{\sqrt{Var[X]}}{E[X]} \right)^2 = \frac{Var[X]}{(E[X])^2} \tag{8.31}$$

Notice that $Var[X] = (E[X])^2 \varepsilon$, so knowing the expected value and the squared coefficient of variation of a random variable is equivalent to knowing the expected value and the variance of a random variable. The primary advantage of the coefficient of variation is that it expresses variability in relative terms that are sometimes easier to understand.

The squared coefficient of variation can be estimated from a sample of data. For example, if 7.5, 12.3, 15.1, 2.6, 5.5 are observed service times and service times are believed to be independent and time stationary, then an estimate of the squared coefficient of variation is

$$\widehat{\varepsilon} = \frac{25.69}{(8.6)^2} \approx 0.35$$

which is just the sample variance (Equation 3.17) divided by the square of the sample average of the observed service times.

In the next section we present an approximation for the GI/G/s queue that depends on the squared coefficient of variation of the arrival and service processes. In Section 8.10.2 we incorporate this approximation into an approximation for networks of queues. These approximations are based on the work of Whitt (1983).

8.10.1 A GI/G/s Approximation

Our goal in this section is to approximate the expected customer delay in a GI/G/s queueing system, a queueing system in which the times between arrivals are independent, time-stationary random variables (in other words, a renewal arrival process, see Section 5.9), the service times are also independent, time-stationary random variables, and there are s identical servers. If we can approximate the expected delay, then we can use Equations (8.16)–(8.18) to determine other performance measures, such as the expected number in the queue.

Let G be generic for an interarrival-time gap, and let X be generic for a service time. The inputs to our approximation are:

- $\lambda = 1/E[G]$, the arrival rate in customers per time
- ε_a, the squared coefficient of variation of G
- $\mu = 1/E[X]$, the service rate per server in customers per time
- ε_s, the squared coefficient of variation of X
- s, the number of identical servers

Notice that we do not specify the probability distributions of G and X beyond their first two moments.

Our approximation will be in the form of a correction factor applied to the corresponding M/M/s result. Conceptually,

$$\text{GI/G/}s = (\text{second-moment correction}) \ (\text{M/M/}s)$$

Specifically, let $w_q(\lambda, \varepsilon_a, \mu, \varepsilon_s, s)$ denote our approximation for the expected delay in a GI/G/s queue, and let $w_q(\lambda, \mu, s)$ be the expected delay in an M/M/s queue with arrival rate λ, service rate μ and s servers (see Equation 8.22). Our approximation modifies $w_q(\lambda, \mu, s)$ to account for coefficients of variation other than 1:

$$w_q(\lambda, \varepsilon_a, \mu, \varepsilon_s, s) = \left(\frac{\varepsilon_a + \varepsilon_s}{2} \right) w_q(\lambda, \mu, s) \qquad (8.32)$$

This approximation is known to work well when the traffic intensity $\rho = \lambda/(s\mu)$ is near 1, and when ε_a and ε_s do not differ too dramatically from 1. Notice that (8.32) provides insight into how a queueing system is affected by variability in the arrival and service processes: The expected delay increases or decreases linearly in $\varepsilon_a + \varepsilon_s$, the sum of the variability parameters.

To illustrate the approximation, consider a repair station at which products arrive at a rate of $\lambda = 5$ per day (implying an expected time between arrivals of 0.2 day) and which has a single repair technician who works at a rate of $\mu = 6$ repairs per day (implying an expected repair time of approximately 0.17 day). The traffic intensity is therefore $\rho = \lambda/\mu \approx 0.83$. This queueing system is a portion of the model of Leviathan Ltd. presented in Section 8.9. A one-moment approximation of this system is to model it as an M/M/1 queue for which

$$w_q(5, 6, 1) = \frac{\rho^2}{\lambda(1 - \rho)} = \frac{\left(\frac{5}{6}\right)^2}{5\left(1 - \frac{5}{6}\right)} \approx 0.83 \text{ day}$$

Suppose, however, that the repair station reconditions a diverse collection of products that vary widely in their repair times, so that the standard deviation of the repair time is 0.24 day. This standard deviation is quite large relative to the expected repair time of 0.17 day, which is reflected in the coefficient of variation of the repair time, $0.24/0.17 \approx 1.41$, and also its square $\varepsilon_s = (0.24/0.17)^2 \approx 2.0$. If $\varepsilon_a = 1.4$ is appropriate for the arrival process, then an approximation for the expected delay in queue is

$$w_q(5, 1.4, 6, 2.0, 1) = \left(\frac{1.4 + 2.0}{2}\right)(0.83) = (1.7)(0.83) \approx 1.41 \text{ day}$$

Notice that accounting for the variability in the service process by incorporating a second moment implied a 70% increase in our approximation of the expected delay, relative to the one-moment M/M/1 model. A difference this large could significantly change our assessment of system performance.

A few closing comments about our approximation:

- A nice feature of (8.32) is that it is correct for the M/M/s queue because $\varepsilon_a = \varepsilon_s = 1$ in that case. However, $\varepsilon_a = 1$ does not necessarily imply that the arrival process is Poisson, nor does $\varepsilon_s = 1$ imply that the service times are exponentially distributed. Therefore, $\varepsilon_a = \varepsilon_s = 1$ is not sufficient for the approximation to be correct.

- The formula (8.32) is also correct for the M/G/1 queue, for which $\varepsilon_a = 1$ (see Exercise 19). But again, $\varepsilon_a = 1$ does not necessarily imply that the arrival process is Poisson, so that $\varepsilon_a = 1$ is not sufficient for the approximation to be correct.

- A number of similar approximations for the GI/G/s queueing system have been proposed, and they may be better in certain situations. Approximation (8.32) itself can be improved by a refined correction factor. But *any* approximation based on only two moments has an inherent limitation: There are an unlimited number of probability distributions that have a given first two moments, and not all of them imply the same value of w_q. In other words, no single number could possibly be correct because there are many correct answers for any first two moments. At best, the approximation may be close for the sorts of distributions typically encountered in practice.

- Combining the approximation (8.32) with Equations (8.16)–(8.18) provides approximations for the expected waiting time, expected number in the queue and the expected number in the system. Specifically, $\ell_q = \lambda w_q(\lambda, \varepsilon_a, \mu, \varepsilon_s, s)$, $w = w_q(\lambda, \varepsilon_a, \mu, \varepsilon_s, s) + 1/\mu$ and $\ell = \lambda w$. Unfortunately, we do *not* obtain approximations for the steady-state probability of j customers in the system, π_j, by applying the correction factor.

8.10.2 A Queueing-Network Approximation

Our goal in this section is to obtain approximate performance measures for queueing networks that generalize the open Jackson networks of Section 8.9 in the following ways:

- The external arrival processes need not be Poisson, but they are renewal arrival processes. In other words, the interarrival-time gaps are independent, time-stationary random variables but need not be exponentially distributed.
- The service times at each station need not be exponentially distributed, but are still independent and time stationary.

Our approach for obtaining approximations will mimic the exact result for Jackson networks by (approximately) decomposing the network into a collection of individual GI/G/s queues, as follows:

1. For each station in the network, derive an arrival rate and squared coefficient of variation to approximate the overall arrival process to the station as a renewal arrival process.
2. Derive approximate performance measures for each station in isolation by treating it as an independent GI/G/s queue.
3. Assemble the performance measures for the individual stations to obtain overall performance measures for the network.

Figure 8.8 illustrates how we view a single station j in a network of m stations. The following sections describe Figure 8.8 in detail.

8.10.2.1 ARRIVALS TO STATIONS. The superposition of all arrival processes into station j is approximated by an overall renewal arrival process with arrival rate $\lambda^{(j)}$ and squared coefficient of variation of the interarrival-time gaps $\varepsilon_a^{(j)}$.

Customer arrivals to station j can be from as many as $m + 1$ sources, including a source external to the network (source 0) and customers departing from stations $i = 1, 2, \ldots, m$ (including station j itself). Arrival source i is approxi-

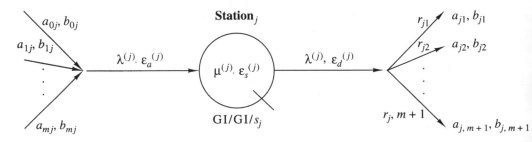

FIGURE 8.8
A view of a single queueing system within a network in isolation.

mated by a renewal arrival process with arrival rate a_{ij} and squared coefficient of variation of the interarrival-time gaps b_{ij}.

8.10.2.2 SERVICE AT STATIONS. Service at station j is delivered by one of s_j identical servers. The service times are characterized by the service rate per server $\mu^{(j)}$ and the squared coefficient of variation of the service-time distribution $\varepsilon_s^{(j)}$.

8.10.2.3 DEPARTURES FROM STATIONS. The customer departure process from station j is approximated by a renewal process with a departure rate that is equal to the arrival rate $\lambda^{(j)}$ and a squared coefficient of variation of the time between departures of $\varepsilon_d^{(j)}$.

The departure process may be decomposed into as many as $m + 1$ subprocesses, with subprocesses $k = 1, 2, \ldots, m$ becoming arrivals to station k, and subprocess $m + 1$ consisting of those customers that depart the network. A departure becomes part of subprocess k with probability r_{jk}, and a customer departs the network with probability $r_{j,m+1} = 1 - \sum_{k=1}^{m} r_{jk}$. Subprocess k is approximated by a renewal process with arrival rate a_{jk} and squared coefficient of variation b_{jk}.

8.10.2.4 THE INPUTS. To approximate the queueing network, values for the following parameters must be known or estimated:

- The arrival rate and squared coefficient of variation of the external arrival process to station j: (a_{0j}, b_{0j}), for $j = 1, 2, \ldots, m$
- The service rate, the squared coefficient of variation of the service time and the number of identical servers at station j: $(\mu^{(j)}, \varepsilon_s^{(j)}, s_j)$, for $j = 1, 2, \ldots, m$
- The probability that a customer departing station j next goes to station k: r_{jk}, for $j, k = 1, 2, \ldots, m$

8.10.2.5 THE APPROXIMATION. A queueing-network approximation is obtained by specifying approximations for the parameters a_{ij}, b_{ij}, $\lambda^{(j)}$, $\varepsilon_a^{(j)}$ and $\varepsilon_d^{(j)}$, for $i, j = 1, 2, \ldots, m$.

Table 8.3 lists one collection of simple approximations that can be shown to be reasonable when the traffic intensity at each station is near 1. Significantly more refined versions can be found in Whitt (1983).

8.10.2.6 AN EXAMPLE. To illustrate the approximation, consider again the proposed repair facility for Leviathan Ltd., Case 8.11, shown in Figure 8.5. The one-moment approximations for this network—derived by modeling it as an open Jackson network—are given in Table 8.2.

Now suppose that we have additional second-moment information about the external arrival process to the repair station and the service times at both the repair station (station 1) and the inspection station (station 2). Specifically,

a_{01}	b_{01}	$\mu^{(1)}$	$\varepsilon_s^{(1)}$	$\mu^{(2)}$	$\varepsilon_s^{(2)}$
5.0	1.4	6.0	2.0	8.0	0.75

TABLE 8.3
A collection of formulas for approximating station j in a network of m queues

	parameter	approximation	remark
(A1)	arrival rate into j	$\lambda^{(j)} = \sum_{i=0}^{m} a_{ij}$	exact
(A2)	arrival variability into j	$\varepsilon_a^{(j)} = \sum_{i=0}^{m} \left(\frac{a_{ij}}{\lambda^{(j)}}\right) b_{ij}$	approximate
(A3)	utilization at j	$\rho_j = \frac{\lambda^{(j)}}{s_j \mu^{(j)}}$	exact
(A4)	offered load at j	$o_j = \frac{\lambda^{(j)}}{\mu^{(j)}}$	exact
(A5)	departure variability from j	$\varepsilon_d^{(j)} = \varepsilon_a^{(j)}$	approximate
(A6)	departure rate from j to k	$a_{jk} = r_{jk}\lambda^{(j)}$	exact
(A7)	departure variability from j to k	$b_{jk} = r_{jk}\varepsilon_d^{(j)} + (1 - r_{jk})$	approximate

Notice that the external arrival process and the service process at the repair station are more variable (have larger squared coefficient of variation) than that of an exponential distribution, while the service process at the inspection station is less variable. Clearly $a_{02} = b_{02} = 0$ since there are no external arrivals to station 2.

The overall arrival rates to each station do not depend on the second-moment information. Therefore, they are determined in the same manner as for an open Jackson network (see Section 8.9.3). For Leviathan Ltd. we previously determined that $\lambda^{(1)} = \lambda^{(2)} \approx 5.56$. The only approximation here is rounding the result to two decimal places. The utilization (traffic intensity) and offered load then follow immediately from (A3) and (A4). In other words, we can obtain overall arrival rates, utilizations and offered loads from only first-moment information, and they are not approximations. Approximations come into play when we derive measures of delay and congestion.

To use the GI/G/s approximation for customer delay, we need the squared coefficient of variation of the arrival process into each station, $\varepsilon_a^{(j)}$, $j = 1, 2$. For station 1 we can use (A2) to write

$$\varepsilon_a^{(1)} = \left(\frac{a_{01}}{\lambda^{(1)}}\right) b_{01} + \left(\frac{a_{11}}{\lambda^{(1)}}\right) b_{11} + \left(\frac{a_{21}}{\lambda^{(1)}}\right) b_{21}$$

$$= \left(\frac{5}{5.56}\right)(1.4) + 0 + \left(\frac{a_{21}}{5.6}\right) b_{21}$$

$$\approx 1.26 + \left(\frac{a_{21}}{5.56}\right) b_{21}$$

But from (A6), $a_{21} = r_{21}\lambda^{(2)} = (0.1)(5.56) = 0.56$, so $a_{21}/5.56 \approx 0.1$. Therefore we can further simplify the expression for $\varepsilon_a^{(1)}$ to

$$\varepsilon_a^{(1)} = 1.26 + (0.1)b_{21} \tag{8.33}$$

The squared coefficient of variation of the departures from station 2 that return to station 1, b_{12}, remains to be determined.

Moving to station 2, it is clear that $\varepsilon_a^{(2)} = \varepsilon_d^{(1)}$, since station 2 receives all of the departures from station 1 (we could also obtain this relationship using (A7)). But (A5) tells us that $\varepsilon_d^{(1)} = \varepsilon_a^{(1)}$, so linking these together

$$\varepsilon_a^{(2)} = \varepsilon_a^{(1)} \tag{8.34}$$

Therefore, if we can determine b_{12}, then we can obtain both $\varepsilon_a^{(1)}$ and $\varepsilon_a^{(2)}$.

Using (A7) we have

$$b_{21} = r_{21}\varepsilon_d^{(2)} + (1 - r_{21}) = (0.1)\varepsilon_d^{(2)} + 0.9$$

But since $\varepsilon_d^{(2)} = \varepsilon_a^{(2)}$ according to (A5), and $\varepsilon_a^{(2)} = \varepsilon_a^{(1)}$ from (8.34), we can write

$$b_{21} = (0.1)\varepsilon_a^{(1)} + 0.9$$

Substituting (8.33) into this expression gives

$$b_{21} = (0.1)(1.26 + (0.1)b_{21}) + 0.9$$

Solving for b_{12} gives $b_{21} \approx 1.04$. Therefore,

$$\varepsilon_a^{(2)} = \varepsilon_a^{(1)} = 1.26 + (0.1)b_{21} = 1.26 + (0.1)(1.04) \approx 1.4$$

The two-moment approximations for Leviathan Ltd. are summarized in Table 8.4, which is obtained from the one-moment results in Table 8.2 by multiplying the performance measures for the repair station by

$$\frac{\varepsilon_a^{(1)} + \varepsilon_s^{(1)}}{2} = \frac{1.4 + 2.0}{2} = 1.7$$

TABLE 8.4
Performance measures for the proposed repair facility using the second-moment approximation

performance measure	repair station	inspection station
ℓ_q (customers)	19.91	1.71
w_q (days)	3.59	0.30

and multiplying the performance measures for the inspection station by

$$\frac{\varepsilon_a^{(2)} + \varepsilon_s^{(2)}}{2} = \frac{1.4 + 0.75}{2} \approx 1.08$$

The expected delay and queue length at the repair station increase substantially, relative to the one-moment approximation, due to the variability in both the arrival and service processes surpassing that of an exponential distribution. However, the performance measures for the inspection station change little because the variability in the arrival process is offset by the low variability in the service process of the inspection station.

8.10.2.7   THE APPROXIMATION, REVISITED.   The derivation of $\varepsilon_a^{(j)}$ in the previous section illustrates how the approximations are used, but it is a cumbersome approach in practice. In Exercise 26 you are asked to show that the approximations (A1)–(A7) can be combined to give

$$b_{jk} = \sum_{i=1}^{m} r_{jk} c_{ij} b_{ij} + d_{jk} \tag{8.35}$$

where

$$c_{ij} = \frac{r_{ij} \lambda^{(i)}}{\lambda^{(j)}}$$

and

$$d_{jk} = r_{jk} \left(\frac{a_{0j}}{\lambda^{(j)}}\right) b_{0j} + (1 - r_{jk})$$

Taken over all $j, k = 1, 2, \ldots, m$, Equation (8.35) implies a system of m^2 equations in m^2 unknowns that can be solved to determine the b_{jk} and therefore the $\varepsilon_a^{(j)}$ via (A2).

8.11   EXERCISES

8.1. An industrial engineer (IE) observing the production of parts by a machining station recorded the following times (in minutes) to produce 10 parts: 2.3, 1.4, 8.0, 4.1, 2.2, 5.0, 1.1, 1.8, 2.8, 3.2. Estimate the production *rate* of the machining station from the available data. If a second, identical machining station is added to the factory, what is the production rate of the factory when both machines are busy? Suppose that the IE decides to model the time to produce a part as a random variable X having an exponential distribution. Write an expression for the cdf of this random variable, including a numerical value for its parameter.

8.2. Is it better to have one fast server or many slower servers? To answer the question make a plot of ℓ_q vs. ρ like Figure 8.3 for the M/M/s model with $s = 1, 2, 3, 4$.

8.3. Using the input distributions determined in Chapter 3, Exercise 23, and the queueing performance measures derived in Section 8.7.1, determine which proposed design for The Darker Image has the smaller long-run expected delay in queue, w_q. Carefully state your assumptions about the arrival and service processes in each case. Do you have any concerns about modeling the system in which both copiers are for general use as an M/M/2 queue?

8.4. A small ice-cream shop competes with several other ice-cream shops in a busy mall. If there are too many customers already in line at the shop, then potential customers will go elsewhere. Potential customers arrive at a rate of 20/hour. The probability that a customer will go elsewhere is $j/16$ when there are $j \leq 16$ customers already in the system, and 1, when there are $j > 16$ customers already in the system. The server at the shop can serve customers at a rate of 10 customers per hour.

(a) Over the long run, at what rate are customers lost? Carefully state your modeling assumptions.

(b) On average, how many customers are waiting to be served (not including the customer in service)?

(c) Suppose that the shop makes revenue of $0.50 per customer served and pays the server $4 per hour. What is the shop's long-run expected profit per hour (revenue − cost)?

8.5. At present the Industrial Engineering Department and the Management Science Department at Small State University each have one typist. Each typist can type 25 letters per day. IE requires an average of 20 letters per day to be typed, while MS requires an average of 15 letters per day to be typed.

(a) For IE and MS, determine the expected length of time from submitting a letter to be typed until it is completed. Carefully state your modeling assumptions.

(b) Suppose that the two typists are grouped into a typing pool; that is, each typist is available to type letters from both departments. For this arrangement, determine the expected length of time from submitting a letter to be typed until it is completed. Carefully state your modeling assumptions. Compare the results with part 5a.

8.6. The service counter at Southwest Montana Airlines has a single queue for waiting customers and two ticket agents. One of the agents is on duty at all times; the other agent goes on duty whenever the queue of customers becomes too long. Suppose that customer arrivals to the counter are well modeled as a Poisson process with rate 45/hour. The agents both work at rate 30 customers/hour, and the second agent goes on duty if there are 3 or more customers at the counter (including the ones being served). Service times are modeled as exponentially distributed.

(a) Develop a queueing model capable of answering the questions below. Carefully state all of your modeling approximations.

(b) Determine the fraction of time that the second agent is on duty.

(c) Determine the average length of the queue, *not* including those being served.

(d) Now treat the number of customers that trigger the second agent to go on duty as a decision variable. When should the second agent go on duty to keep the expected number in the system under 5?

8.7. In this exercise we introduce a simple relationship between the customer delays in a GI/G/1 queue.

Let D_i be the delay experienced by the ith customer arriving at the queue. Let X_i be the service time of the ith customer, and let G_i be the interarrival-time gap between customers $i - 1$ and i. Consider the recursive relationship

$$D_i = \max\{0, D_{i-1} + X_{i-1} - G_i\} \tag{8.36}$$

for $i = 1, 2, \ldots$. In words, the delay of the ith customer is the delay of the $(i-1)$st customer plus the service time of the $(i-1)$st customer minus the gap between the arrival of these two customers, provided this quantity is greater than 0.

Show that Equation (8.36) is in fact valid. Hint: Try an induction proof.

8.8. A restaurant is currently operating a single drive-through window. Orders are placed at an intercom at the back of the parking lot. After placing an order, the customer pulls up and waits in line at the drive-through window. Since filling the orders takes much longer than placing an order, queueing is caused by the process of filling orders only. The question faced by the management is whether to have one or two people working to fill the orders (orders will be filled one at a time, regardless). With one person filling orders, the average service time is 2 minutes/customer. With a second person it can be reduced to 1 minute and 15 seconds/customer. Cars arrive at a rate of 24 per hour.

(a) Determine the average delay and waiting time for the one- and two-person systems. Carefully state your modeling assumptions.

(b) Determine the percentage of time that the workers are completely idle for each system.

(c) Suppose that when there are more than three cars waiting for the drive-through window (including the one receiving their order), then any additional arrivals cannot reach the intercom. In other words, there is room for only three cars between the window and the intercom. Given this information, evaluate the one- and two-person systems in terms of their affect on the intercom.

(d) Suppose that when there are six cars waiting for the drive-through window (including the one receiving its order), then any additional arrivals cannot join the queue. In other words, the length of the drive-through lane is limited, so that when there are six cars there, then new arrivals must either park or are lost to the restaurant. Given this information, evaluate the one- and two-person systems in terms of their affect on the rate of cars lost to the drive-through window.

8.9. A bank is designing a new customer service that allows customers to access account information, transfer funds and pay bills by using their touch-tone telephone. The service will be completely automated and maintained by a computer. The bank needs to determine how many simultaneous connections ("ports") into the computer are needed so that only rarely will customers be unable to connect to the service when they try. However, it does not want to have excess capacity because of the cost of the ports.

The following data are available based on marketing projections and similar systems at other banks: The bank expects that approximately 10000 customers will use the service when it is fully available, and that a customer will connect about 0.7 times per day (in other words, about 4.9 times per seven-day week). Further, it is anticipated that customers will stay connected for about 2.5 minutes on average.

Provide the bank management with information, in the form of a table or graph, that it could use to make a decision about the number of ports vs. performance in terms of some measure of customers' ability to connect.

8.10. A software company named Bug-Me-Not offers a customer help line. Owners of Bug-Me-Not software can call and talk to a support person who answers questions. Presently there are two support people who handle calls individually, and the average call takes 3 minutes to complete. When both support people are busy, customers who call are placed in a "hold queue" in which they listen to music and information about other Bug-Me-Not products. To date the company has not observed any reneging, but they are worried that the introduction of a new product may significantly increase the arrival rate of calls, causing the delays to become excessive. Presently calls arrive at a rate of 20 per hour during peak times.

(a) Develop a model capable of answering the questions below; list your modeling approximations.

(b) How large can the arrival rate become before the company will be forced to add another support person just to keep up with calls?

(c) With just two support people, how large can the arrival rate become before the expected time a customer spends on hold exceeds 4 minutes?

(d) With just two support people, how large can the arrival rate become before the percentage of the time that there are more than five people in the hold queue exceeds 15% of the time?

(e) Modify your model from part 10a to include the fact that customers will only wait, on average, 5 minutes before reneging from the hold queue.

(f) What is the expected number in the hold queue for the refined model that includes reneging? (Hint: Approximate the result by truncating the infinite series expression.)

8.11. Consider a computer system with dial-up services for users with a modem and personal computer. There are a limited number of ports (connections to the computer) reserved for dial-up users; suppose that there are k ports. If a caller finds all k ports in use, then the system will allow the user to stay on the line and wait for one to open up. The waiting queue can handle m additional callers. If all ports are in use and the wait queue is full, then the caller gets a busy signal and must try again later. Callers in the wait queue may renege. The amount of time they are willing to wait is assumed to be exponentially distributed with mean $1/\gamma$ minutes.

The population of dial-up users is large enough to treat it as infinite, so that the arrival process for calls will be modeled as a Poisson process with rate

λ calls/minute. The amount of time a caller uses the computer is exponentially distributed with mean $1/\mu$ minutes.

(a) Develop a queueing model capable of answering the questions below by defining your state space and giving a transition-rate diagram.

(b) What would you compute to determine the expected number of callers who are waiting for a port (that is, in the wait queue)?

(c) What would you compute to determine the expected number of callers per hour that receive a busy signal?

(d) Suppose that the computer center has two options: Add one more port, or add five positions to the waiting queue. What quantities would you compute and compare for each option to determine which is better?

8.12. A computer center is equipped with three identical computers. The number of users at the center at any time is 20 (in other words there are 20 terminals). For each user the time for entering a program or preparing a job is exponentially distributed with mean 1 minute. Once a program is entered, it is submitted directly to the center for execution and is executed by the first available computer. The computer time per program is also exponentially distributed with rate 4 per minute. Assuming that the center is in operation on a full-time basis and neglecting the effect of computer down time, determine the following:

(a) The probability that a program is not executed immediately upon receipt at the center.

(b) The average time until a program is released from the center.

(c) The average number of programs awaiting execution.

(d) The expected number of idle computers.

(e) The percentage of the time all computers are simultaneously idle.

(f) The percentage of idle time per computer.

8.13. Use the recursive relationship in Exercise 7 to develop a spreadsheet simulation of the GI/G/1 queue. Let column A be a sequence of 100 pseudorandom numbers and column B the corresponding interarrival-time gaps (see Exercise 21, Chapter 3 for help on generating random variates in a spreadsheet). Let column C be another sequence of 100 pseudorandom numbers and column D the corresponding service times. Finally, let column E be the resulting customer delays; specifically E[i] = max(0, E[i-1] + D[i-1] - B[i]). For each exercise that follows, plot the sequence of delays in customer order.

(a) Let the interarrival-time gaps be exponentially distributed with parameter $\lambda = 0.5$. Let the service times be exponentially distributed with parameter $\mu = 1$. Recalculate the spreadsheet and the plot three times. Then try $\lambda = 0.8$. Describe how the characteristics of the plot change for the new value of λ.

(b) Repeat the previous exercise except that the interarrival-time gaps are now uniformly distributed on $[0, 4]$, then on $[0, 2.5]$. Do the characteristics of the plot change?

8.14. The purpose of this exercise is to develop intuition about formulas for the steady-state probabilities π_j of M/M-type queues.[4]

(a) Recall that for the M/M/1 queue $\pi_j = \pi_0 \rho^j$, for $j = 0, 1, 2, \ldots$, where $\rho = \lambda/\mu$ and $\pi_0 = 1 - \rho = 1/\sum_{i=0}^{\infty} \rho^i$. This is the *geometric distribution* with parameter ρ.

Show that for the M/M/1/c queue $\pi_j = \pi_0 \rho^j$, for $j = 0, 1, 2, \ldots, c$, where $\pi_0 = 1/\sum_{i=0}^{c} \rho^i$. This is the *truncated geometric distribution*, truncated at the queue capacity c. As c goes to infinity, this becomes the M/M/1 result.

(b) Recall that for the M/M/∞ queue $\pi_j = \pi_0 (\lambda/\mu)^j/j!$, for $j = 0, 1, 2, \ldots$, where

$$\pi_0 = e^{-\lambda/\mu} = \frac{1}{\sum_{i=0}^{\infty} (\lambda/\mu)^i / i!}$$

This is the *Poisson distribution* with parameter λ/μ.

Show that for the M/M/c/c queue $\pi_j = \pi_0 (\lambda/\mu)^j/j!$, for $j = 0, 1, 2, \ldots, c$, where

$$\pi_0 = \frac{1}{\sum_{i=0}^{c} (\lambda/\mu)^i / i!}$$

This is the *truncated Poisson distribution*, truncated at the capacity c. As c goes to infinity this becomes the M/M/∞ result.

(c) Now consider the M/M/s/c queue with $s \leq c$. Let $\rho = \lambda/(s\mu)$ and let L be a random variable representing the steady-state number of customers in the system (therefore $\ell = E[L]$). Show that

$$\Pr\{L = j | L \leq s\} = \frac{(\lambda/\mu)^j / j!}{\sum_{i=0}^{c} (\lambda/\mu)^i / i!}$$

for $j = 0, 1, \ldots, s$. In words, the steady-state distribution of number of customers in the system behaves like an M/M/s/s queue, conditional on there being s or fewer customers in the system. Then show that

$$\Pr\{L = s + j | L \geq s\} = \frac{\rho^j}{\sum_{i=0}^{c-s} \rho^i}$$

for $j = 1, 2, \ldots, c - s$. In words, the steady-state distribution of number of customers in the system behaves like an M/M/1/(c−s) queue, conditional on there being more than s customers in the system. Of course, as s approaches c, this becomes the M/M/c/c result, and as c goes to infinity, it becomes the M/M/s result.

8.15. In Section 8.7.1 the expected number of customers in an M/M/s queue was shown to be $\ell_q = \pi_s \rho/(1 - \rho)^2$, where $\rho = \lambda/(s\mu)$. Whitt proposed the

---

[4]Mike Taaffe of the University of Minnesota suggested this exercise.

following approximation to this formula:[5]

$$\ell_q \approx \left(1 + \sqrt{2\pi}\,\beta\,\Phi(\beta)\,e^{\beta^2/2}\right)^{-1} \left(\frac{\rho}{1-\rho}\right)$$

where $\beta = (1 - \rho)\sqrt{s}$ and $\Phi(\cdot)$ is the standard-normal cdf. This formula might be useful when s is very large so that computing π_s is tedious. Evaluate the accuracy of this approximation relative to the exact result for $0.5 \le \rho \le 0.95$ and $1 \le s \le 20$.

8.16. Suppose that the arrival process to some queueing system is best characterized as being time-dependent. Because it is more difficult to characterize and estimate a time-dependent arrival-rate function, as opposed to a constant arrival rate, the following strategy is sometimes employed: Model the arrival process as a time-stationary process with constant arrival rate equal to the *peak rate* that the system would experience, and examine how the system performs under peak load. This is a reasonable strategy in some situations, but there are also pitfalls as illustrated in this exercise.

Let the arrival process to the queueing system be modeled as a nonstationary Poisson arrival process with rate function

$$\lambda(t) = \begin{cases} t^2 & \text{if } 0 \le t < 10 \\ (20 - t)^2 & \text{if } 10 \le t < 20 \end{cases}$$

Suppose that this is an infinite-server system with service time *exactly* 1 time unit for all customers. Stated differently, each customer departs exactly 1 time unit after arriving. We could call this model an M(t)/D/∞ queue.

Our interest is in the expected number of customers in the system when the system is at peak load (time $t = 10$). Determine this performance measure for the nonstationary model and also for a stationary model with arrival rate $\lambda = \lambda(10) = 100$. Compare the results, explain why they are different and discuss how a steady-state analysis at peak arrival rate can be misleading. (Hint: For this M(t)/D/∞ queue the number in the system at any time t is just the number that arrived during the previous 1 time unit.)

8.17. Let W_q be a random variable representing the steady-state delay in queue of a customer arriving to an M/M/s queue (therefore $w_q = E[W_q]$). Show that

$$\Pr\{W_q > a | W_q > 0\} = e^{-(s\mu-\lambda)a}$$

for $a > 0$. (Hint: Try the M/M/1 case first. Let L be a random variable representing the number of customers in the system when a customer arrives. Therefore $\Pr\{L = j\} = \pi_j$. Derive an expression for $\Pr\{W_q > a | L = j\}$ for

[5]W. Whitt. 1992. Approximations for the GI/G/m queue. AT&T Bell Laboratories Technical Report, December 15, 1992.

$j = 0, 1, 2, \ldots$ by recalling that the sum of independent, exponentially distributed random variables is Erlang. Then apply the law of total probability to obtain $\Pr\{W_q > a\} = \sum_{n=0}^{\infty} \Pr\{W_q > a | L = j\} \pi_j$. Finally, use the definition of conditional probability directly.)

8.18. A bank wants to determine how many automated teller machines (ATMs) to install at a very busy site. It is willing to model the time between arrivals of customers as exponentially distributed with mean 3 minutes. The time a person spends at the ATM is uncertain, but can be modeled as an exponentially distributed random variable with mean 2 minutes. The bank's performance measure is the number of customers that must wait more than 5 minutes to begin using an ATM. It wants no more than 15% of all customers who must wait to have to wait more than 5 minutes. Of course, the bank does not want to install more ATMs than necessary. How many should it install? (Hint: You will need the result from Exercise 17.)

8.19. The GI/G/s approximation (8.32) is exact for the M/G/1 queue, which is a single-server queueing system with Poisson arrivals and a general service-time distribution.

(a) Show that for the case of the M/G/1 queue, (8.32) can be rewritten as

$$w_q(\lambda, 1, \mu, \varepsilon_s, 1) = \frac{\rho^2 + \lambda^2 \sigma^2}{2\lambda(1 - \rho)}$$

where σ^2 is the variance of the service-time distribution. This is the *Pollaczek-Khintchine formula*. Notice that the expected delay increases linearly in the variance of the service time.

(b) In an M/D/1 queue $\sigma^2 = 0$, while in an M/M/1 queue $\sigma^2 = 1/\mu^2$. Compute the expected delay for each of these queues and compare them. How much does the variability of service time contribute to the expected delay?

8.20. Perform a simulation study to evaluate the accuracy of the GI/G/s approximation (8.32). There are many possible distributions you could use to examine the effect of coefficients of variation different from 1. For example:

- To examine coefficients of variation less than 1, you might use the Erlang distribution with different numbers of phases n. To obtain an expected interarrival-time gap (or service time) of b, set $\lambda = n/b$ for the Erlang distribution with n phases. The coefficient of variation of the Erlang distribution with parameter $\lambda = n/b$ and n phases is $1/\sqrt{n}$, so the relative variability decreases as n increases. When $n = 1$ the Erlang distribution is the exponential distribution.
- The *lognormal distribution* has two parameters—the expected value and variance—that can be set to obtain any coefficient of variation you desire. See Law and Kelton (1991) for a description of, and methods for generating observations from, the lognormal distribution.
- The *hyper-exponential distribution* of order 2 describes the following random variable X: With probability $1 - p$, X is exponentially distributed with

parameter b_0 and with probability p, X is exponentially distributed with parameter b_1. If X has such a distribution, then

$$E[X] = \frac{b_0 p + b_1(1 - p)}{b_0 b_1}$$

and

$$Var[X] = \frac{b_0^2 p + b_1^2(1 - p) + (b_0 - b_1)^2 p(1 - p)}{(b_0 b_1)^2}$$

The hyper-exponential distribution has coefficient of variation greater than or equal to 1. A particularly easy case to work with is $p = 1/2$.

8.21. A central computer at Ohio State University receives electronic-mail messages from outside the university. This computer is called `osu.edu`, and it distributes the mail it receives to other computers on campus. Suppose that e-mail is distributed at a rate of 3 messages per second, with the actual time depending on the size of the message, which can vary widely. E-mail arrives to `osu.edu` at a rate of 2 messages per second during regular university hours, and is well modeled as a Poisson arrival process. Historically 20% of all incoming mail goes to the College of Engineering. The College of Engineering computer is called `eng.ohio-state.edu`. However, `eng.ohio-state.edu` also receives mail directly (without passing through `osu.edu`) at a rate of 1 message per second during regular business hours. The rate at which `eng.ohio-state.edu` can distribute messages is 2 per second. Help the College of Engineering do capacity planning by answering the following questions:

(a) What is the expected number of messages queued for processing at `osu.edu`? What is the expected delay for a message before it is processed?

(b) What is the expected number of messages queued for processing at `eng.ohio-state.edu`? What is the expected delay for a message before it is processed?

(c) If the average size of an e-mail message is 12K bytes, what is the expected number of bytes queued at `eng.ohio-state.edu`?

(d) E-mail users would like mail delivery to be nearly instantaneous. For practical purposes we could define this to mean 5 seconds from receipt by `osu.edu` until final delivery by `eng.ohio-state.edu`. In the next year the arrival rates to `osu.edu` and directly to `eng.ohio-state.edu` are expected to increase by the same percentage (for example, each by 12%). How much of a percentage increase can occur before the expected time from receipt by `osu.edu` until final delivery by `eng.ohio-state.edu` exceeds 5 seconds, given current computer processing speeds?

8.22. Show that Equation (8.28) satisfies the balance equations (8.24)–(8.26) and that $\sum_{i=0}^{\infty} \sum_{j=0}^{\infty} \pi_{(i,j)} = 1$.

8.23. Modify the simulation in Section 8.9.1 to model the proposed repair system.

8.24. Derive the balance equations for the proposed repair facility for Leviathan Ltd., and verify that the Jackson network solution does satisfy the balance equations.

8.25. For the proposed repair facility for Leviathan Ltd., how large can r be before the system becomes unstable?

8.26. Derive Equation (8.35). Hint: Start by substituting (A5) and then (A2) into (A7).

8.27. A production line consists of two single-machine stations in series. Jobs arrive to the first station according to a Poisson process with rate 20 jobs/hour. The service rate at each station is 30 jobs/hour. At each station, approximately 10% of the jobs are found to be defective and are scrapped.

(a) Approximate the buffer space needed for each station to ensure that all arriving jobs can be accommodated at least 95% of the time.

(b) Determine the expected number of jobs at station 2.

(c) Determine the utilization of the machine at station 2.

(d) Now suppose that there is a rework station associated with *each* machine station. Defective items from station 1 are reworked and then sent to station 2. Defective items from station 2 are reworked and then sent out. However, only 50% of the items sent to the rework stations can be salvaged; the others become scrap. The rework stations complete jobs at a rate of 4/hour. Develop a new model of this system, and answer the questions above.

(e) Suppose we pool the rework stations, meaning that there is one rework station with two servers, each working at rate 4/hour. Develop a new model, and again answer the first two questions.

8.28. Data Driven, Inc. offers another customer service, similar to the one described in Case 6.1, Chapter 6. There are still four transactions (log on, fetch, read and log off), and the estimated transaction matrix is

$$\mathbf{P} = \begin{pmatrix} 0 & 0.79 & 0.01 & 0.20 \\ 0 & 0.17 & 0.63 & 0.20 \\ 0 & 0.16 & 0.40 & 0.44 \\ 0 & 0 & 0 & 1 \end{pmatrix}$$

In addition, the expected durations for transactions have been estimated and are given in the following table:

transaction	expected duration (minutes)
log on	3
fetch	3
read	12
log off	0.6

Data Driven wants a model that translates log-on rate (in customers per hour) into an expected number of simultaneous users, because there is an upper limit of 500 simultaneous users on the entire system but no specific limit on number of users performing each transaction. Develop such a model, and use it to predict the expected number of simultaneous users when the log-on rate is 1000 customers per hour; then 1500 customers per hour. (Hint: Since computers are fast relative to customers, assume that the time for the computer to actually perform the transactions is negligible. Treat the system as a network of infinite server queues, one for each type of transaction.)

8.29. A job shop is comprised of five machine groups, each with a number of identical machines as summarized below:

Machine Group	Number of Machines in the Group
Casting Units	19
Lathes	3
Planers	4
Drill Presses	5
Shapers	16

Jobs are processed first-come-first-served, except at the planer where priority is given to jobs that have been on the shop floor the longest.

Jobs arrive at the shop somewhat irregularly. Industrial engineers working at the company collected data on interarrival-time gaps over 4 hours and obtained the following data: 10.2, 11.3, 22.2, 10.7, 8.7, 6.8, 9.8, 8.2, 19.5, 4.1, 13.2, 13.0, 3.8, 13.2, 18.0, 8.3, 10.0, 10.2, 13.1, 5.2 (times are in minutes). Looking back over company records for standard times and the types of jobs processed in the last two weeks, the following data were obtained:

job type	number of jobs last 2 weeks	sequence number	routing	standard operation time (minutes)
1	460	1	casting unit	125
		2	planer	35
		3	lathe	20
2	540	1	casting unit	235
		2	shaper	250
		3	drill press	50
		4	planer	30

Raw data for operation times are not available, but the actual operation times are considered to be "all over the map" by the machine operators.

The machine shop is typically in production 24 hours a day, and after a break of any kind (such as a holiday) production picks up where it left off. The company is interested in where it should add capacity (more machines to a group) if it was able to do so. The primary performance measures of interest are the flow time for each job type, the utilization of each machine group and the total work-in-process inventory. Your job is to make a recommendation and support it. (Hint: This system has features that we do not know how to model directly, so we must approximate them as closely as possible.)

8.30. Double Fault, Inc. produces tennis balls for sale to sporting goods stores. The engineers at Double Fault have proposed the following design for a new plant to package tennis balls: Cans of tennis balls travel down a conveyor, one at a time, spaced precisely 2 seconds apart. The cans travel on the conveyor for exactly 20 seconds until they reach the automatic packer. The packer produces 1 case out of 75 cans of balls and takes between 1 and 3 minutes to do it, with the actual time being well modeled as a uniformly distributed random variable. The packing process does not start until there are 75 cans available. There is adequate space for cans of balls waiting to be packed.

The completed cases wait to be moved by a forklift truck, one case per forklift. The time for the forklift to pick up a case and transport it to shipping is well modeled as a lognormally distributed random variable with a mean of 3 minutes and standard deviation of 1 minute. The time for the forklift to then return to pick up another case is exactly 1 minute.

Double Fault is interested in how many forklifts it will need to keep up with the production of cases of balls. The primary performance measures it is interested in are the number of cases waiting for forklifts, the utilization of the forklifts, and the time from when a case is produced until it has completed transport to shipping. Double Fault works one 12-hour shift each day. (Hint: This system has features that we do not know how to model directly, so we must approximate them as closely as possible.)

8.31. A small bank has three tellers, two tellers working inside the bank and one teller that services a single drive-up window. The bank is open from 9 a.m. until 4 p.m.

The time between arrivals of customers seeking to use the drive-up window is well modeled as an exponentially distributed random variable with mean 2 minutes. There is only enough space at the drive-up window for three cars to wait behind the one being served. Those drive-up customers who find no space available typically park their cars and attempt to receive service inside the bank; the time required to park and enter the bank is approximately 1 minute.

There are also customers who seek service inside the bank without ever attempting to use the drive-up window. The time between arrivals of such customers is modeled as exponentially distributed with mean 1 minute.

All customers who enter the bank join a single line to wait for service from the tellers.

The time to service a customer is a random variable having a normal distribution. Those customers who are served inside have a service time with mean 1.4 minutes and standard deviation 1.0 minute. Service at the drive-up window is typically a bit slower (mean 1.8 minutes with standard deviation 1.0 minute).

The bank is interested in the relative impact of adding an additional teller inside versus outside. Both options involve new construction, but adding another drive-up window is significantly more expensive. (Hint: This system has features that we do not know how to model directly, so we must approximate them as closely as possible.)

8.32. Read the article by Buchanan, J. and J. Scott. 1992. Vehicle utilization at Bay of Plenty Electricity. *Interfaces* **22**, 28–37. Suggest ways that their model could be modified to be more faithful to the situation they had.

8.33. Read the article by Agnihothri, S. R. and P. F. Taylor. 1991. Staffing a centralized appointment scheduling department in Lourdes Hospital. *Interfaces* **21**, 1–11. Describe the queueing approximation they present, and compare it to the GI/G/*s* approximation in Section 8.10.1.

8.34. Read the article by Srikar, B. N. and B. Vinod. 1989. Performance analysis and capacity planning of a landing gear shop. *Interfaces* **19**, 52–60. Discuss how their model differs from an open Jackson network (Section 8.9.3).

8.35. Read the article by Kolesar, P. 1984. Stalking the endangered CAT: A queueing analysis of congestion at automatic teller machines. *Interfaces* **14**, 16–26. Describe the difficulties Kolesar had in parameterizing the model and how Kolesar handled them.

CHAPTER
9

TOPICS IN
SIMULATION OF
STOCHASTIC
PROCESSES

This chapter completes our treatment of the simulation of stochastic processes by describing critical statistical issues that arise in the design and analysis of simulation experiments and discussing the use of rough-cut models prior to a simulation study. By doing this, the chapter provides a bridge from this book into a book or course covering the simulation of complex systems using a commercial-quality simulation language.

9.1 STATISTICAL ISSUES IN SIMULATION

Statistical analysis of sample paths plays a key role in simulation studies. We use statistics to make statements about performance measures associated with the model that generated the sample paths, performance measures that presumably cannot be derived via mathematical analysis.

This section describes certain critical statistical issues that arise in the design and analysis of simulation experiments. All of these issues center around error in simulation-based estimates. In Section 9.1.1 the error is due to bias, while in Sections 9.1.2 and 9.1.3 the error is due to the randomness represented by the pseudorandom-number generators. A more complete treatment of statistical issues in simulation at the same level as this chapter can be found in Nelson (1992). A comprehensive treatment at a more advanced level is in Law and Kelton (1991).

To highlight these critical issues, the examples we use are all stochastic processes that we *do not actually need to simulate* because we can derive the desired performance measures via mathematical analysis. By using such examples, we can clearly demonstrate the mistakes that occur when we do not account for design-and-analysis issues.

9.1.1 Initial-Condition Effects

One of the strengths of mathematical analysis is in deriving time-independent (long-run) performance measures of stochastic processes; see Chapters 5–8. Estimating such performance measures via simulation is difficult because simulated sample paths must be finite in length. In other words, we have to stop the simulation somewhere short of infinity. Here we describe the precise nature of the difficulty that this produces.

Recall that the steady-state probabilities π of an irreducible Markov chain with one-step transition matrix $\mathbf{P}$ are independent of the initial state of the process (Section 6.7.2). As time goes to infinity, the state in which the process started is irrelevant when making probability statements about a future state of the process. In contrast, a probability statement about a finite sample path of a Markov chain, such as the n-step transition probability $p_{ij}^{(n)}$, typically depends on the initial state. The following example illustrates the impact of this fact on simulation studies.

> **Case 9.1.** A food manufacturer plans to introduce a new potato chip, Box O' Spuds, into a local market that already has three strong competitors. Because the manufacturer makes a superior product, the marketing analysts believe that brand loyalty to Box O' Spuds will be high after people try it. However, it will be difficult to persuade people to switch from the three established brands. The marketing analysts would like to forecast the long-term market share for Box O' Spuds to determine whether it is worth entering the market.

Suppose the marketing analysts formulate a Markov chain model of customer brand switching in which the state space $\mathcal{M} = \{1, 2, 3, 4\}$ corresponds to which of the three established brands or Box O' Spuds, respectively, that a customer currently purchases. The time index is the number of bags of chips purchased. Based on market research and experience with other products, the one-step transition matrix the marketing analysts anticipate is

$$\mathbf{P} = \begin{pmatrix} 0.70 & 0.14 & 0.14 & 0.02 \\ 0.14 & 0.70 & 0.14 & 0.02 \\ 0.14 & 0.14 & 0.70 & 0.02 \\ 0.05 & 0.05 & 0.05 & 0.85 \end{pmatrix}$$

Given $\mathbf{P}$, the steady-state probabilities can be obtained by solving $\pi' = \pi'\mathbf{P}$ and $\pi'\mathbf{1} = 1$. To three decimal places they are

$$\pi \approx \begin{pmatrix} 0.294 \\ 0.294 \\ 0.294 \\ 0.118 \end{pmatrix}$$

Interpreting the steady-state probabilities as the long-run fraction of the time spent in each state, the long-run market share for Box O' Spuds is $\pi_4 \approx 0.118$, or roughly 12% of the market.

For the sake of illustration, suppose tha. the marketing analysts do not know how to compute π, but they do know how to conduct a simulation experiment. Then they could estimate π_4 by simulating customer brand switching and observing the long-run fraction of time that the process is in state 4, corresponding to Box O' Spuds. Since π_4 is independent of the initial state, they could take the initial state in the simulation to be any brand, for instance brand 1.

As in Chapter 6, let $F_{N|L=i}$ be the cdf of the next state of the Markov chain given that the last state was state i. Let the random variable X be the number of times the sample path enters state 4 in a simulation of m transitions (chip purchases). Then X/m is the fraction of time the process is in state 4, and it is an estimator of π_4. The simulation is as follows:

$e_0()$ **(first brand purchased)**

     $S_0 \leftarrow 1$ (initial brand is brand 1)

$e_1()$ **(next brand purchased)**

     $S_{n+1} \leftarrow F^{-1}_{N|L=S_n}(\texttt{random}())$ (next brand depends on current brand)

algorithm simulation
1. $n \leftarrow 0$ (initialize system-event counter)
     $e_0()$ (execute initial system event)
     $X \leftarrow 0$ (initialize counter for state 4)

2. $e_1()$ (update state of the system)
     $n \leftarrow n + 1$ (update event counter)

3. $X \leftarrow X + \mathcal{I}(S_n = 4)$ (add 1 to counter if brand 4)

4.   if $n < m$ then   (check length of sample path)
         repeat 2   (continue simulation)
     else
         $\widehat{\pi}_4 = X/m$ (estimate π_4)
         stop   (terminate simulation)
     endif

Is $\widehat{\pi}_4$ a reasonable estimator of π_4? One formal way to answer this question is to determine $E[\widehat{\pi}_4]$, the expected value of the estimator. Ideally, $E[\widehat{\pi}_4] = \pi_4$, in which case we say that the estimator is *unbiased* for π_4. Intuitively, an unbiased estimator is one whose distribution is centered on the quantity we want to estimate.

By definition,

$$E[\widehat{\pi}_4] = E\left[\frac{X}{m}\right] = E\left[\frac{1}{m}\sum_{n=1}^{m}\mathcal{I}(S_n = 4)\right] = \frac{1}{m}\sum_{n=1}^{m}E\left[\mathcal{I}(S_n = 4)\right]$$

But notice that

$$E[\mathcal{I}(S_n = 4)] = 0 \cdot \Pr\{S_n \neq 4\} + 1 \cdot \Pr\{S_n = 4\} = 1 \cdot p_4^{(n)} = p_4^{(n)}$$

where $p_i^{(n)} = \Pr\{S_n = i\}$ is the probability of the process being in state i at step n. Therefore,

$$E[\widehat{\pi}_4] = \frac{1}{m}\sum_{n=1}^{m}p_4^{(n)} \tag{9.1}$$

The model is a Markov chain, so we can easily evaluate the right-hand side of Equation (9.1).

Figure 9.1 shows a plot of $p_4^{(n)}$ for $n = 1, 2, \ldots, 20$, where the horizontal line is the true value of π_4. Notice that $p_4^{(n)}$ is converging to π_4 as n increases, as it should, but for small values of n it is far from the steady-state value. Substituting these values into Equation (9.1) gives $E[\widehat{\pi}_4] \approx 0.090 < 0.118$, when $m = 20$. Thus, $\widehat{\pi}_4$ is a *biased* estimator of π_4, which means that it estimates something other than π_4. Bias is one source of experiment error.

We have frequently employed replications to obtain better estimates, typically the more replications the better. Unfortunately, replications do not help the bias problem. If we make k independent replications and let $\widehat{\pi}_{4j}$ be the estimate

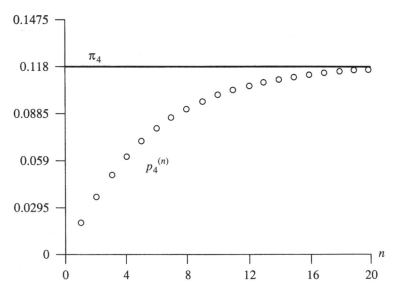

FIGURE 9.1
Plot of $p_4^{(n)}$ converging to π_4 for the Box O' Spuds example.

from replication j, then the sample average across k replications is

$$\bar{\pi}_4 = \frac{1}{k} \sum_{j=1}^{k} \widehat{\pi}_{4j}$$

which has expected value

$$E[\bar{\pi}_4] = E\left[\frac{1}{k} \sum_{j=1}^{k} \widehat{\pi}_{4j}\right] = \frac{1}{k} \sum_{j=1}^{k} E\left[\widehat{\pi}_{4j}\right] = \frac{1}{k}kE\left[\widehat{\pi}_4\right] = E\left[\widehat{\pi}_4\right] \approx 0.090$$

where $\widehat{\pi}_4$ is generic for any of the $\widehat{\pi}_{4j}$. The bias of the estimator is unchanged because the replications all have the same expected value (they are "identically distributed"). In fact, since replications reduce sampling error (see Section 9.1.2 below), the more replications we make, the better the estimate we have of the *wrong quantity*!

The bias of a sample average is unaffected by increasing the number of replications. If the number of replications is increased at the expense of reducing the length of each sample path, then increasing the number of replications may increase the bias.

The bias of an estimator like $\widehat{\pi}_4$ is due to the residual effects of the initial state of the stochastic process, because $p_4^{(n)}$ depends on how the initial state is chosen even though $\pi_4 = \lim_{n\to\infty} p_4^{(n)}$ does not. In this case $p_4^{(n)} = p_{14}^{(n)}$, because we started the process in state (brand) 1, so it is obvious that $p_4^{(n)}$ depends on the initial state. For that reason we refer to the problem as *initial-condition bias*. In theory the initial-condition bias can be eliminated by randomly sampling the initial state of the process from the steady-state distribution. For example, if we let the distribution of the initial state be $\mathbf{p} = \pi$ rather than forcing the initial state to be brand 1, then the n-step state probabilities are

$$\mathbf{p}^{(n)'} = \mathbf{p}'\mathbf{P}^n = \pi'\mathbf{P}^n = (\pi'\mathbf{P})\mathbf{P}^{n-1} = \pi'\mathbf{P}^{n-1} = \cdots = \pi' \qquad (9.2)$$

for all values of n. In other words, $\Pr\{S_n = j\} = \pi_j$ for *all* n and the bias disappears. Of course, to use this result we must already know the steady-state probabilities, which we would not know in a problem that really required simulation.

In realistic situations where we cannot generate the initial state from the steady-state distribution, there are two standard solutions to the problem of initial-condition bias: (a) Make the length of the sample path m as long as possible and (b) ignore (delete) some of the initial outputs $S_1, S_2, \ldots, S_d$, for $d < m$, to eliminate observations that are the most biased. For instance, if we ignore the first $d = 10$

transitions (chip purchases) in the brand-switching example and take

$$\widehat{\pi}_4 = \frac{1}{20 - 10} \sum_{n=11}^{20} \mathcal{I}(S_n = 4)$$

(the fraction of time the process is in state 4 for the transitions 11 through 20), then $\mathrm{E}[\widehat{\pi}_4] \approx 0.110$ which is closer to the true answer of 0.118.

When we have a plot like Figure 9.1, we can make a decision about how large m and d should be by looking at the plot. Such a plot is not available in practical applications, but we can *estimate* one via a simulation experiment and make our decisions based on the estimate. To illustrate the idea, suppose that we make k replications of length m of the brand-switching process, and let $S_{n,j}$ be the state of the system at chip purchase n on replication j. Then an estimator of $p_4^{(n)}$ is

$$\bar{p}_4^{(n)} = \frac{1}{k} \sum_{j=1}^{k} \mathcal{I}(S_{n,j} = 4)$$

which is just the fraction of the k replications for which the process was in state 4 at time n. A plot of $\bar{p}_4^{(n)}$ for $n = 1, 2, \ldots, m$ estimates the plot in Figure 9.1, with the estimate improving the larger k is. See Nelson (1992) and Law and Kelton (1991) for further refinements of this approach.

Table 9.1 contains the output from a simulation of $k = 20$ replications, each of $m = 20$ transitions. Every column is the sample path from one replication which represents a customer's initial brand (brand 1) and switching among the four brands. The estimator $\bar{p}_4^{(n)}$ is obtained by averaging across the $(n+1)$st row, after applying the indicator function $\mathcal{I}(S_{n,j} = 4)$ to map states $1, 2$ and 3 into a zero and state 4 into a one. For example, $\bar{p}_4^{(1)} = 1/20$, since there is only one "4" in the second row.

Figure 9.2 displays the true values of $p_4^{(n)}$ as ● (the same values as displayed in Figure 9.1), $\bar{p}_4^{(n)}$ using only the first $k = 10$ replications as ○, and $\bar{p}_4^{(n)}$ using all $k = 20$ replications as ∗. Notice that both sample plots overshoot the true values, but the $k = 20$ plot is closer. As k is increased further, the sample plot will converge to the true values.

Another approach that is sometimes recommended is to intelligently select the initial state of the process by using a mathematically tractable model as an approximation of the more general process. The most likely (largest probability) state of the approximating process can be used as the initial state for the simulation. While often helpful, notice that this approach is not a panacea: we initialized the Box O' Spuds simulation in state 1, which is (tied for) the most likely state of that process, and still encountered bias.

Initial-condition bias only arises when we attempt to estimate a long-run, time-independent performance measure via simulation. Such experiments are sometimes called *steady-state simulations*. There is no initial-condition bias when we are interested in time-dependent performance measures that are supposed to reflect the initial conditions. Simulation experiments performed to estimate

TABLE 9.1
Sample paths from a simulation of $k = 20$ replications of $m = 20$ chip purchases for the Box O' Spuds Markov chain

										replication									
1	2	3	4	5	6	7	8	9	10	11	12	13	14	15	16	17	18	19	20
1	1	1	1	1	1	1	1	1	1	1	1	1	1	1	1	1	1	1	1
1	1	2	3	2	1	3	1	1	1	1	1	1	1	1	1	3	1	1	4
1	1	1	1	1	2	1	3	1	3	1	1	2	1	1	1	1	1	1	4
1	1	1	3	1	1	1	1	1	1	1	3	2	1	1	1	1	2	1	4
1	3	1	4	1	1	1	1	2	1	3	1	3	1	3	1	1	1	1	3
1	1	2	4	1	3	1	1	1	1	1	3	3	1	2	1	1	1	1	1
1	1	1	4	1	1	1	1	1	1	1	2	1	1	1	1	1	2	2	2
1	1	2	4	1	1	3	1	1	1	1	3	1	1	1	1	1	1	1	3
3	1	2	4	1	1	1	1	1	1	1	3	1	1	2	4	1	1	3	1
1	1	2	4	2	4	1	1	1	1	1	1	1	1	1	4	1	2	1	1
1	1	2	4	2	4	1	2	1	1	1	1	1	2	1	4	1	1	3	3
2	1	1	4	4	4	1	3	1	3	1	4	1	1	1	3	1	1	1	3
3	1	1	4	4	4	3	3	2	1	2	4	1	1	1	2	1	1	1	1
1	1	1	4	4	4	2	1	1	1	1	4	1	1	3	1	3	1	1	1
2	2	2	4	4	4	3	3	1	1	2	4	2	1	2	1	1	1	3	2
1	1	4	4	4	4	2	3	1	3	1	4	1	2	1	1	3	1	1	1
1	1	4	4	4	4	1	3	1	1	1	4	1	1	1	2	2	1	1	1
1	1	4	4	4	4	2	3	1	1	1	4	3	3	2	1	3	1	1	1
1	1	4	4	4	4	2	2	2	3	3	4	1	1	3	1	1	2	2	1
1	4	4	4	4	4	1	1	2	1	1	2	1	1	1	1	1	1	1	1
3	4	2	4	4	4	3	1	1	2	1	3	1	1	2	3	1	1	1	2

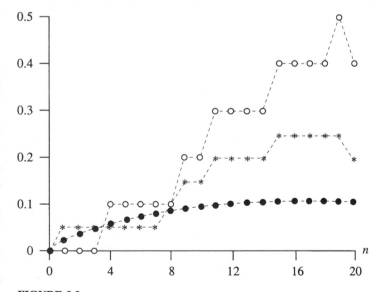

FIGURE 9.2
Plot of $p_4^{(n)}$ (represented by ●); of $\bar{p}_4^{(n)}$ using only the first $k = 10$ replications (represented by ○); and of $\bar{p}_4^{(n)}$ using all $k = 20$ replications (represented by *).

time-dependent performance measures are sometimes called *terminating simulations*, because each replication ends when the time horizon of interest has been simulated.

9.1.2   Measures of Error

An advantage of mathematical analysis over simulation is that no decision is required regarding the number of replications or the length of the sample path necessary to obtain a good estimate of system performance, because there is no error in a result achieved via mathematical analysis (except the numerical error when evaluating it on a calculator or computer). In a sense, mathematical analysis is equivalent to a simulation experiment that generates an infinite number of replications or an infinitely long sample path.

Simulation-based estimators, on the other hand, are subject to *sampling error* because they are functions of a finite number of sample paths, and the particular sample paths that are generated depend on the pseudorandom numbers. If the simulation experiment is appropriately conducted, then these estimators have good properties, such as being unbiased (see Section 9.1.1). Nevertheless, they are subject to sampling error, and it is important that the error is measured and controlled. The following key result from mathematical statistics is central to measuring and controlling sampling error:

Let $Z_1, Z_2, \ldots, Z_k$ be independent and time-stationary random variables with common expectation $\theta = \mathrm{E}[Z_i]$ and common standard deviation $\sigma = \sqrt{\mathrm{Var}[Z_i]}$. Let $\bar{Z} = \sum_{j=1}^{k} Z_j / k$. Then

$$\mathrm{E}[\bar{Z}] = \theta \tag{9.3}$$

and

$$\mathrm{se} \equiv \sqrt{\mathrm{Var}[\bar{Z}]} = \frac{\sigma}{\sqrt{k}} \tag{9.4}$$

In words, $\bar{Z}$ is an unbiased estimator of θ with standard error that decreases at rate $1/\sqrt{k}$.

This result suggests that we use the sample average, $\bar{Z}$, as an estimator of θ, and that we use the sample standard error, $\widehat{\mathrm{se}} = \widehat{\sigma}/\sqrt{k}$, as an estimator of the potential error of $\bar{Z}$, where $\widehat{\sigma}^2$ is the sample variance of the Z_i (see Equation (3.17)). In some contexts we can go further and construct a *confidence interval* for θ. One frequently used confidence interval is

$$\boxed{\bar{Z} \pm t_{1-\alpha/2, k-1}\, \widehat{\mathrm{se}}}$$

where $t_{1-\alpha/2,k-1}$ is the $1 - \alpha/2$ quantile of the t-distribution with $k - 1$ degrees of freedom. The "confidence" in a confidence interval is the probability that the procedure used to construct the interval will yield an interval that contains θ. The width of a confidence interval is often used as a measure of sampling error.

The following example illustrates in what situations (9.3)–(9.4) are applicable to simulation experiments.

Case 9.2. Code++, a small computer software company, has a 24-hour-a-day help line that is always staffed by a technician. Customers who call when the technician is busy with another customer are placed in a hold queue. Due to the pending introduction of a new software product, Code++ expects the rate of calls to increase by 25%. It wants to evaluate the effect that this increase will have on customer delay to determine if it needs to add technicians to staff the line. A management scientist at Code++ has modeled the help line as an M/M/1 queue (Chapter 8) with arrival rate $\lambda = 20$ calls per hour and service rate $\mu = 30$ calls per hour.

Since the management scientist has chosen an M/M/1 model, the steady-state expected customer delay when the traffic intensity is $\rho = (1.25\lambda)/\mu = 5/6$ can be calculated as

$$w_q = \frac{\rho^2}{\lambda(1 - \rho)} = \frac{1}{6} \text{ hours}$$
$$= 10 \text{ minutes}$$

Given this result, Code++ can decide whether an average 10-minute delay in the hold queue is an acceptable level of service.

For the sake of illustration, suppose that the management scientist is unaware of how to calculate the expected delay for an M/M/1 queue and therefore decides to estimate w_q via a simulation experiment. The experiment consists of k replications, each replication consisting of the delays of m customers.

Within replication j, let D_{ij} be the delay experienced by the ith simulated customer. Let X_{ij} be the service time of the ith customer on replication j, and let G_{ij} be the interarrival-time gap between customers $i - 1$ and i on replication j. In Chapter 7, Exercise 8, a simple recursive relationship among these random variables was presented, specifically

$$D_{ij} = \max\{0, D_{i-1,j} + X_{i-1,j} - G_{ij}\} \tag{9.5}$$

for $i = 1, 2, \ldots$. In words, the delay of the ith customer is the delay of the $(i - 1)$st customer plus the service time of the $(i - 1)$st customer minus the gap between the arrival of these two customers, provided this quantity is greater than 0. This recursion leads to a simple algorithm that does not require identifying system events:

algorithm code++

```
for j = 1 to k                                  (across-rep loop)
do
      D_{0j} ← 0                                 (initialize each rep)
      X_{0j} ← 0
      for i = 1 to m                             (within-rep loop)
      do
            G_{ij} ← F_G^{-1}(random())          (generate gap)
            D_{ij} ← max{0, D_{i-1,j} + X_{i-1,j} - G_{ij}}   (within-rep delay)
            X_{ij} ← F_X^{-1}(random())          (generate service time)
      enddo
      D̄_j ← (1/(m-d)) Σ_{i=d+1}^{m} D_{ij}       (within-rep average)
enddo
D̄ ← (1/k) Σ_{j=1}^{k} D̄_j                        (across-rep average)
```

We display the simulation in this compact form to emphasize the important distinction between *within-replication* statistics and *across-replication* statistics. The sample average $\bar{D}_j$ is a within-replication statistic, the average of the delays within replication j after deleting d delays to reduce initial-condition bias (see Section 9.1.1). Each $\bar{D}_j$ is an estimator of w_q. The sample average $\bar{D}$ is an across-replication statistic, since it is the average of the within-replication statistics $\{\bar{D}_1, \bar{D}_2, \ldots, \bar{D}_k\}$; it is also an estimator of w_q.

The following fundamental principle guides how we use $\bar{D}_j$ and $\bar{D}$ to estimate w_q:

Outputs across replications are typically independent and time stationary, while outputs within replications are typically neither independent nor time stationary.

The Code++ example illustrates this principle. Within a replication, $D_{1j}, D_{2j}, \ldots, D_{mj}$ are related by the recursive formula (9.5). Therefore, they are dependent, and the effect of the initial condition ($D_{0j} = 0$) makes them nonstationary. On the other hand, results across replications are based on different pseudorandom numbers; therefore they are independent. If the replications are executed identically except for the random numbers, then they are also time stationary ("identically distributed").

Consider Figure 9.3, which shows the delays from $k = 2$ replications of this simulation for $m = 100$ customers. Notice that within each replication large

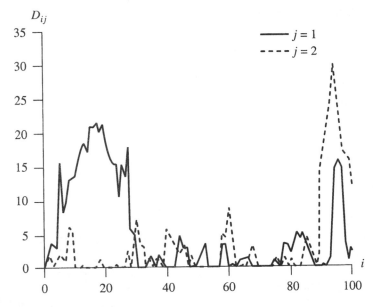

FIGURE 9.3
Sample paths of two replications of the Code++ simulation with plot points connected as a visual aid.

delays tend to be followed by large delays, while short delays tend to be followed by short delays. This is a manifestation of dependence within replications. And although the two replications are qualitatively similar, they are different in their detailed behavior. This is a manifestation of independence across replications.

A consequence of this principle is that the expectation and standard error results (9.3)–(9.4) typically apply only to across-replication sample averages, but not to within-replication sample averages. In fact, estimating the standard error of any one of the $\bar{D}_j$ is a very difficult problem, while estimating the standard error of $\bar{D}$ is straightforward: use $\hat{\sigma}/\sqrt{k}$ where $\bar{D}_j$ plays the role of Z_j, and

$$\hat{\sigma}^2 = \frac{1}{k-1} \sum_{i=1}^{k} (\bar{D}_j - \bar{D})^2 = \frac{1}{k-1} \left(\sum_{j=1}^{k} \bar{D}_j^2 - \frac{(\sum_{h=1}^{k} \bar{D}_h)^2}{k} \right)$$

The fact that the standard error decreases at rate $1/\sqrt{k}$ implies that sampling error decreases quite slowly as a function of the number of replications. For instance, it takes 4 times more replications to reduce the standard error by $1/2$, since $\sigma/\sqrt{4k} = (1/2)(\sigma/\sqrt{k})$.

Equation (9.4) can also be used to control sampling error: If an estimate $\hat{\sigma}$ is available based on experience with similar problems or some preliminary replications, then an estimate of the number of replications required to reduce the standard error below δ can be obtained by solving

$$\frac{\widehat{\sigma}}{\sqrt{k}} \leq \delta$$

for k. For example, suppose that $k = 10$ replications of the algorithm code++ yields a standard error for $\bar{D}$ of $\widehat{se} = 14.23/\sqrt{10} \approx 4.50$ minutes. If we want to estimate the expected delay to within a standard error of 2 minutes, we can solve

$$\frac{14.23}{\sqrt{k}} \leq 2$$

to find that we need $k = (14.23/2)^2 \approx 51$ replications in all.

9.1.3 Random Number Assignment

Many simulation studies are conducted to compare the performance of alternative system designs. In this section we argue that it is beneficial to assign the same random number seeds or streams to the simulation of each of the alternative systems. This approach is called using *common random numbers* (CRN).

The intuition behind CRN is that a fairer comparison among systems is achieved if they are subjected to the same experimental conditions, specifically the same source of randomness. CRN ensures this. The mathematical justification is given by the following key result from mathematical statistics:

Let $X_1, X_2, \ldots, X_k$ be independent and time-stationary random variables with common expectation $\theta_X = \mathrm{E}[X_i]$ and common standard deviation $\sigma_X = \sqrt{\mathrm{Var}[X_i]}$. Let $Z_1, Z_2, \ldots, Z_k$ be independent and time-stationary random variables with common expectation $\theta_Z = \mathrm{E}[Z_i]$ and common standard deviation $\sigma_Z = \sqrt{\mathrm{Var}[Z_i]}$. Let $\bar{X}$ and $\bar{Z}$ be the corresponding averages. Then

$$\mathrm{E}[\bar{X} - \bar{Z}] = \theta_X - \theta_Z \tag{9.6}$$

and

$$se \equiv \sqrt{\mathrm{Var}[\bar{X} - \bar{Z}]} = \frac{\sigma_{X-Z}}{\sqrt{k}} \tag{9.7}$$

where

$$\sigma_{X-Z} = \sqrt{\sigma_X^2 + \sigma_Z^2 - 2\sigma_{XZ}}$$

and

$$\sigma_{XZ} \equiv \mathrm{E}\left[(X - \theta_X)(Z - \theta_Z)\right]$$

*The quantity σ_{XZ} is called the **covariance** between X and Z and is also denoted $\mathrm{Cov}[X, Z]$. In words, $\bar{X} - \bar{Z}$ is an unbiased estimator of $\theta_X - \theta_Z$ with standard error that depends on the marginal variances of X and Z and their covariance.*

This result suggests that we use the difference of the averages, $\bar{X} - \bar{Z}$, as an estimator of the difference in expectations, $\theta_X - \theta_Z$, and that we use the sample standard error,

$$\widehat{se} \equiv \sqrt{\frac{\widehat{\sigma}_X^2 + \widehat{\sigma}_Z^2 - 2\widehat{\sigma}_{XZ}}{k}}$$

as an estimator of the potential error of $\bar{X} - \bar{Z}$, where $\widehat{\sigma}_X^2$ and $\widehat{\sigma}_Z^2$ are the sample variances of the X_i and the Z_i, respectively (Equation (3.17)), and

$$
\begin{aligned}
\widehat{\sigma}_{XZ} &\equiv \frac{1}{k-1} \sum_{j=1}^{k} (X_j - \bar{X})(Z_j - \bar{Z}) \\
&= \frac{1}{k-1} \left(\sum_{i=1}^{k} X_i Z_i - \frac{(\sum_{j=1}^{k} X_j)(\sum_{h=1}^{k} Z_h)}{k} \right)
\end{aligned}
\tag{9.8}
$$

is the sample covariance. In some contexts we can go further and construct a *confidence interval* for $\theta_X - \theta_Z$. One frequently used confidence interval is

$$\bar{X} - \bar{Z} \pm t_{1-\alpha/2, k-1}\, \widehat{se}$$

where $t_{1-\alpha/2, k-1}$ is the $1 - \alpha/2$ quantile of a t-distribution with $k - 1$ degrees of freedom.

The covariance is a measure of the dependence between two random variables. If X and Z are independent, then $\sigma_{XZ} = 0$. If X and Z tend to be large or small together, then $\sigma_{XZ} > 0$. The following example illustrates how CRN tends to make $\sigma_{XZ} > 0$, therefore reducing the standard error of $\bar{X} - \bar{Z}$.

Case 9.3. Todd and Diane plan to take out a mortgage from Humorous Money Company to purchase a home. The current fixed-interest rate on a 30-year loan is 8%. For no additional service charge Humorous Money will allow Todd and Diane to "lock in" the current rate of 8% (option 1). For a service charge of $50 they can take whatever the current rate happens to be in 4 weeks (option 2). For a service charge of $100 Humorous Money will let the rate drift for 4 weeks, but lock it in if it hits 8.2% or 7.8% (option 3). The sales agent says that option 3 protects Todd and Diane from getting stuck with a very high rate, while giving them a chance for a lower rate.

Interest rates can change weekly, and Diane believes that interest rates are more likely to go down than up. For the next 4 weeks she believes that the following model is a good representation of how the interest rate will change each week: The rate will go up 0.1% with probability 0.3, go down 0.1% with probability 0.4, and stay the same with probability 0.3. Under this model Todd and Diane want to decide which option is best for them.

To aid them in their decision, Todd and Diane could compute the expected interest rate under each option. If they lock in the current rate, then, of course, their expected rate is 8%. Let the random variable X represent their interest rate if they choose option 2, and let Z represent their interest rate if they choose option 3.

The possible values of X are the rates

$$\mathbf{r} = \begin{pmatrix} 8.4 \\ 8.3 \\ 8.2 \\ 8.1 \\ 8.0 \\ 7.9 \\ 7.8 \\ 7.7 \\ 7.6 \end{pmatrix}$$

since the rate changes at most 0.1% per week under Diane's model. Let r_i denote the ith element of $\mathbf{r}$, and let the state space $\mathcal{M} = \{1, 2, \ldots, 9\}$ correspond to the interest rates $\{8.4, 8.3, \ldots, 7.6\}$, respectively. If we let the time index be weeks, then the interest rate for the next 4 weeks can be modeled as a Markov chain with one-step transition matrix

$$\mathbf{P} = \begin{pmatrix} 1 & 0 & 0 & 0 & 0 & 0 & 0 & 0 & 0 \\ 0.3 & 0.3 & 0.4 & 0 & 0 & 0 & 0 & 0 & 0 \\ 0 & 0.3 & 0.3 & 0.4 & 0 & 0 & 0 & 0 & 0 \\ 0 & 0 & 0.3 & 0.3 & 0.4 & 0 & 0 & 0 & 0 \\ 0 & 0 & 0 & 0.3 & 0.3 & 0.4 & 0 & 0 & 0 \\ 0 & 0 & 0 & 0 & 0.3 & 0.3 & 0.4 & 0 & 0 \\ 0 & 0 & 0 & 0 & 0 & 0.3 & 0.3 & 0.4 & 0 \\ 0 & 0 & 0 & 0 & 0 & 0 & 0.3 & 0.3 & 0.4 \\ 0 & 0 & 0 & 0 & 0 & 0 & 0 & 0 & 1 \end{pmatrix} \qquad (9.9)$$

and initial-state distribution

$$\mathbf{p} = \begin{pmatrix} 0 \\ 0 \\ 0 \\ 0 \\ 1 \\ 0 \\ 0 \\ 0 \\ 0 \end{pmatrix}$$

The expected interest rate at the end of 4 weeks is

$$\theta_X = \mathrm{E}[X] = \sum_{i=1}^{9} r_i \Pr\{S_4 = i\} = \mathbf{r}'\mathbf{p}^{(4)} \approx 7.96$$

You should convince yourself that option 3 can also be modeled as a Markov chain with state space $\mathcal{M} = \{1, 2, 3, 4, 5\}$ corresponding to the interest rates

$$\mathbf{r} = \begin{pmatrix} 8.2 \\ 8.1 \\ 8 \\ 7.9 \\ 7.8 \end{pmatrix}$$

one-step transition matrix

$$\mathbf{P} = \begin{pmatrix} 1 & 0 & 0 & 0 & 0 \\ 0.3 & 0.3 & 0.4 & 0 & 0 \\ 0 & 0.3 & 0.3 & 0.4 & 0 \\ 0 & 0 & 0.3 & 0.3 & 0.4 \\ 0 & 0 & 0 & 0 & 1 \end{pmatrix} \qquad (9.10)$$

and initial-state distribution

$$\mathbf{p} = \begin{pmatrix} 0 \\ 0 \\ 1 \\ 0 \\ 0 \end{pmatrix}$$

For this model the expected interest rate is

$$\theta_Z = \mathrm{E}[Z] = \mathbf{r}'\mathbf{p}^{(4)} \approx 7.97$$

Therefore, in terms of expected interest rate, none of the options is appreciably different (perhaps Todd and Diane should next look at their risk of obtaining a higher rate under each option).

Suppose that instead of calculating θ_X and θ_Z, a simulation experiment was performed to estimate the difference $\theta_X - \theta_Z$. If this difference is negative then option 2 is better than option 3; if it is positive, then option 3 is better. The magnitude of $\theta_X - \theta_Z$ determines whether the difference is enough to be meaningful.

The simulation of these two Markov chains proceeds in the standard manner described in Chapter 6, with state transitions generated as

$$S_{n+1} \leftarrow F^{-1}_{N|L=S_n}(\texttt{random()})$$

where $F^{-1}_{N|L=S_n}$ corresponds to the cdf derived from the particular transition matrix, **P**. In the actual simulation a random number seed or stream must be assigned to $\texttt{random()}$. To simulate the two options independently, we assign different seeds or streams, say $\texttt{random(2)}$ for option 2 and $\texttt{random(3)}$ for option 3. By simulating them independently, we force $\sigma_{XY} = \mathrm{Cov}[X, Z] = 0$.

Suppose that we assign the same random number seed or stream to simulate each option, say $\texttt{random(2)}$. Then the same pseudorandom numbers are used to generate transitions for each option. If the sample path $8.0, 8.1, 8.0, 7.9, 7.9$ is

TABLE 9.2
Interest rates from ten replications, with and without CRN

replication	option 2 (stream 2)	option 3 (stream 3)	option 3 (stream 2)
1	8.0	7.8	8.0
2	8.2	7.9	8.2
3	8.0	7.8	8.0
4	8.2	7.9	8.2
5	8.1	7.8	8.1
6	7.9	8.0	7.8
7	8.1	7.8	8.1
8	8.0	8.2	8.0
9	8.1	8.0	8.1
10	7.7	7.8	7.8
sample average	8.03	7.90	8.03

generated for option 2, then the *same sample path* will be generated for option 3 because the transition probabilities among the states $\{8.1, 8.0, 7.9\}$ are the same for the two options. Only when the sample path for option 2 strays outside the range 7.9–8.1 will the simulated sample paths for the two options differ, but even then they will match until the time when option 2 moves outside 7.9–8.1. Thus, the week-4 interest rate for the two simulated options will tend to be large or small together, implying that $\sigma_{XY} > 0$. This is the goal CRN.

Table 9.2 shows the interest rates that Todd and Diane obtained on 10 replications of each option. Columns 2 and 3 resulted from simulating options 2 and 3 with different random number seeds or streams, while columns 2 and 4 resulted from simulating both options with the same seed or stream. The entries in columns 2 and 4 are mostly identical, but in the cases when they differ, they are both small together. This is a manifestation of positive covariance. Table 9.3 shows the sample means and standard errors computed from these data. Neither estimate matches the true difference of -0.01, but the estimate obtained using CRN is closer (0.00 compared to 0.13) with an 83% reduction in se. And in this problem the improvement is important, because an apparent interest rate reduction

TABLE 9.3
Sample-average differences and standard error

	$\bar{X} - \bar{Z}$	se
independent	0.13	0.06
CRN	0.00	0.01

of $0.13 \pm 2(0.06)$ could make a significant difference in payments and make option 3 appear worthwhile, but a reduction of $0.00 \pm 2(0.01)$ indicates (correctly) that option 3 is not an improvement over option 2.

The effect of CRN can be enhanced by *synchronizing* the random numbers, which means forcing the random numbers to be used for the same purpose in the simulation of each stochastic process. The primary technique for achieving synchronization is to assign a different random number seed or stream to each input process and then to use the same collection of seeds or streams across systems. For example, in the simulation of several alternative queueing systems, a separate stream can be assigned to generate interarrival-time gaps and service times. When common streams are used across systems, the same random numbers will generate interarrival-time gaps and service times for each alternative. This is the primary reason for having different seeds or streams available, and we have followed this convention in assigning seeds or streams throughout the book.

Positive covariance between the sample performance of two systems' simulations can often be induced by using CRN. The magnitude of the covariance can be increased by synchronizing the pseudorandom numbers.

9.2 ROUGH-CUT MODELING

We have presented a unified approach to stochastic modeling, analysis and simulation. The unifying principle is the discrete-event view of sample paths, which decomposes sample paths into system inputs and logic. Simulation is a general approach for analyzing the sample paths of such models, but we have emphasized recognizing those situations in which mathematical analysis is possible. Mathematical analysis, when feasible, is typically preferable to simulation analysis because of the lack of sampling error.

Textbook treatments of stochastic modeling—including this book—tend to give the impression that there is a unique best model of any real or conceptual system. This is not correct. For any modeling and analysis problem there are many reasonable models that differ in their scope and their level of detail. The salient test is whether a model is capable of answering the analysts' questions with enough precision to facilitate the decision at hand. In some cases a model that is amenable to mathematical analysis will suffice, but in other cases simulation analysis will be required. Even when simulation is required, there is a role for a mathematically solvable and typically less detailed model to provide a rough-cut analysis prior to simulating. Specifically:

- A rough-cut model can provide preliminary results quickly.
- A rough-cut model can provide an estimate of system performance to use as a gross check on the correctness of the simulation model.

- A rough-cut model's input parameters (for example, the number of servers or the arrival rate in a queueing model) can be varied to obtain a qualitative perception of system sensitivity to those parameters.
- A rough-cut model can provide bounds on system performance by parameterizing the model to represent extreme conditions (for example, overloaded and underloaded conditions).
- A rough-cut model can be used to determine initial conditions for a steady-state simulation.

For these reasons we recommend developing a rough-cut, and mathematically solvable, model before undertaking any simulation study. Sometimes the rough-cut model will reveal that a more sophisticated model is not necessary. More often it will provide some guidance regarding how to conduct, and what to expect from, the simulation study.

There are several ways in which a complex model can be simplified enough to make it mathematically solvable. For the models presented in this book, they include the following:

- Simplify the state space of the process by leaving out less critical details.
- Use steady-state (time-independent) performance measures to approximate time-dependent performance measures.
- Approximate general input distributions as exponential distributions with the same mean.
- Approximate the dependence in the process as satisfying the Markov property.
- Approximate a nonstationary process as a stationary process with the average rate, the maximum rate or the minimum rate.

The following cases illustrate these ideas:

> **Case 9.4.** There are a substantial number of processing steps required to resolve accident claims submitted to MEP Insurance Company's claims processing department. And the required steps vary depending on the claim. Sometimes an adjustor must be sent to examine the accident scene. Sometimes a police report must be obtained. Sometimes payments must be coordinated with another insurance company. And so on. Each step necessitates work by some group within the claims processing department. Any change in the law or policies that adds a new step or any change in organization that consolidates or reorganizes steps affects the entire department. MEP would like to be able to evaluate the impact of any changes prior to implementing them.

The processing steps that a claim requires depend almost entirely on the nature of the claim and the situation that demanded it. Nevertheless, a Markov chain model might be derived by treating each processing step as a state and forcing the probability distribution of the next processing step to depend only on the last processing step that a claim has completed. For example, there could be a fixed transition probability that a claim that required an adjustor's inspection will next require ordering a police report, independent of all previous processing. Changes in claim processing then correspond to adding or deleting states or altering the transition probabilities.

Case 9.5. Trucks arrive to the Krissek Paper Co. warehouse to pick up paper supplies. The warehouse has four loading bays and two forklift trucks to retrieve orders. Due to union rules, the truck drivers must load their own trucks; the forklifts only bring the items to the bay and then leave them for the drivers to load. When no bay is available, arriving trucks wait in the parking lot. Krissek Paper is interested in the relative merits of adding an additional forklift compared to adding an additional loading bay.

There is a complex interaction between the limited number of bays and forklifts. The forklifts can be idle because the trucks are all in the process of loading their supplies. On the other hand, some trucks can sit idle in the bays because all of the forklifts are busy. Capturing this interaction requires a detailed model.

A simple queueing analysis of each option can be obtained by simplifying the system. To examine the impact of adding a bay, treat the number of forklifts as unlimited (of course, the service time must include the time to retrieve and load the order). Similarly, to examine the impact of adding a forklift, treat the number of bays as unlimited. This is an optimistic analysis, since the interaction between bays and forklifts must lead to more delay, but it provides an idea of the best that can be expected from each option.

Case 9.6. Fraser Foods, a large grocery store chain, wants to determine an allocation of express checkout lanes and full-service checkout lanes that enhances customer satisfaction. At a selected store, data are available on the number of items purchased by customers and the number of customers that visit the store. Since the number of customers in the store and the number of items they purchase vary throughout the day, Fraser Foods is willing to consider strategies that change the number and type of checkout lanes during the day. Causal observation of the store reveals that customers sometime "jockey," meaning that they change checkout lanes if another lane is observed to be less crowded.

There are numerous complicated aspects of this system, including the time-varying load and the unpredictable customer behavior. All of these features can be incorporated into a simulation, but a rough-cut queueing analysis can be based on some reasonable simplifications.

We could treat the express checkout lanes and the full-service checkout lanes as essentially independent queueing systems, meaning that customers who are eligible for the express lanes (or some specified fraction of them) always choose the express lanes. This should be reasonable if there are enough express lanes to keep the queues from getting too long.

Each checkout lane is naturally thought of as a server with its own queue of customers. The effect of jockeying, however, is similar to what would occur if there were a single queue for all lanes, because jockeying keeps servers from being idle when there are customers waiting. Therefore, we could treat the store as (approximately) two independent queueing systems (one for express, one for full service), each system having a single queue but multiple servers.

The time-varying customer load can be handled by developing several models with constant loads, one model for each distinct load level experienced throughout the day. This approximation is reasonable if the load stays relatively constant for a substantial period of time so that a steady-state analysis is relevant. If load is constantly shifting, then perhaps the average load over a period might be used.[1]

> **Case 9.7.** A piece of military hardware is composed of a number of components that may fail over time and with use. Each of the component suppliers is required to provide data on the reliability of the component parts. Some suppliers provide extensive data, while others provide only a time-to-failure rating. When the hardware is assembled, the failure of some components can be tolerated, while the failure of others are fatal to the system. System failures can also be caused by certain combinations of component failures, and the failure of some components causes other components to fail more rapidly. Information about the reliability of the system as a whole is desired.

In this example a rough-cut model may be obtained by treating the time-to-failure of individual components as exponentially distributed random variables—even if we have data that indicate otherwise—by setting the parameter of the exponential distributions equal to the failure rate of the component (the reciprocal of the expected time to failure). We then approximate the system as a continuous-time Markov process in which the state of the system indicates which components are working and which have failed. The failure rates can be increased in certain states to model the fact that failure of some components hastens the failure of others.

9.3   EXERCISES

9.1. Ten replications of a simulation experiment were performed to estimate the expected time to produce a particular product. The sample mean time across the 10 replications was 137 minutes with standard deviation $\hat{\sigma} = 110$ minutes. How many more replications are needed to estimate the expected time to within ± 20 minutes?

9.2. Redo the simulation of Bit Bucket Computers, Case 4.1, Chapter 4, using the tools you learned in this chapter to conduct an appropriate simulation experiment.

9.3. Redo the simulation of Floppy Express, Case 7.1, Chapter 7, using the tools you learned in this chapter to conduct an appropriate simulation experiment.

---

[1]If the arrival rate is $\lambda(t)$ customers per time for $a \leq t \leq b$, then the average arrival rate is $\bar{\lambda} = \int_a^b \lambda(t)\, dt / (b - a)$.

9.4. Redo the simulation of Leviathan Ltd., Case 8.11, Chapter 8, using the tools you learned in this chapter to conduct an appropriate simulation experiment.

9.5. Rework Exercise 9, Chapter 4, using the tools you learned in this chapter to conduct an appropriate simulation experiment.

9.6. Rework Exercise 31, Chapter 8, using the tools you learned in this chapter to conduct an appropriate simulation experiment.

9.7. Rework Exercise 20, Chapter 8, using the tools you learned in this chapter to conduct an appropriate simulation experiment.

9.8. The repair and inspection stations in a small company work in the following way: Items arrive to be repaired and inspected. There are two repair technicians, and an incoming item is given to the technician with the smallest number of items waiting to be repaired. After repair, the item is inspected. There is one inspector. If the item passes inspection, it is released from the shop. If it fails inspection the first time, it is returned to the technician who has the smallest number of items waiting to be repaired and is repaired again. If an item fails inspection a second time, it is junked. The following models have been proposed (all times in minutes):

activity	distribution
time between arrivals	exponential mean 4
repair time	normal mean 6, standard deviation 2
inspection time	uniform minimum 2, maximum 4
fail inspection	probability 0.2

A consultant has proposed having the technicians all draw from a single queue of items to be repaired. The consultant claims that this will result in "a dramatic reduction in the flow time of items, and increase the utilization of the technicians." Your job is to conduct a study to check these claims.

9.9. Patients arrive at a hospital emergency room where they are treated and then depart (or are admitted to the hospital, which is effectively the same thing from the perspective of the emergency room). Upon entering the emergency room, patients are immediately evaluated based on the severity of their injury or illness and assigned to category A (most severe) and category B (less severe). Historically, 15% of arriving patients are category A. Category A patients are sent immediately to wait for an available bed, while category B patients are sent to a registration nurse to fill out forms. Once the category B patients have registered (filled out forms), they are also sent to wait for a bed. A doctor will treat each patient one at a time after they have been assigned a bed. The emergency room has three beds, two doctors and one registration nurse at all times.

Category B patients wait first-come-first-served to see the registration nurse. Once category B patients have received treatment, they leave the emergency room. However, when category A patients have received treatment, they must see the registration nurse to fill out forms before leaving.

When visiting the registration nurse, category B patients have priority over category A patients. Of those patients waiting for beds, category A patients have first priority. Of those patients in a bed waiting for a doctor, category A patients have first priority.

Management engineers have developed models for the various quantities. All times are in minutes.

quantity	distribution
time between arrivals	exponential, mean 20
time to register	uniform, minimum 5, maximum 9
time to treat category A	normal, mean 72, standard deviation 20
time to treat category B	normal, mean 25, standard deviation 6

The emergency room is open 24 hours a day, 7 days a week. The hospital administration is interested in the relative merits of adding additional doctors or additional beds. Evaluate the impact of each option.

9.10. A lawn and garden company sells bags of mulch (wood chips) to customers. A bag contains 3 cubic feet of mulch. The number of bags a customer wants differs from customer to customer, but can be modeled as a random variable having a Poisson distribution with mean 6 bags. If there is not enough mulch in stock to fill a customer's order, then the customer takes all of the available mulch. The time between arrivals of customers to the company can be modeled as an exponentially distributed random variable with mean 1 hour.

The company orders more mulch only when it runs out of mulch, since it takes just one day for the new mulch to arrive. But when it runs out, it does lose sales because potential customers can easily go elsewhere to buy mulch. The company is interested in determining a good value for s, the number of bags to order each time it orders. Although costs could perhaps be assigned, the company would first like an exploratory analysis of the trade offs between average stock level, number of lost customers, and number of orders placed as s changes. The time period of interest is one summer (90 days).

9.11. A small warehouse supplies local businesses with paper products. Trucks pull up to the loading bays; the driver places an order and then waits for the order to be gathered by the warehouse worker who drives a forklift truck. There are currently four loading bays, and when they are full, trucks wait in the parking lot next to the warehouse until a loading bay is available.

The warehouse is open from 7 a.m. to 4 p.m., Monday through Friday. Trucks arrive throughout the day. The warehouse has two forklift drivers who pick orders one at a time on a first-come-first-served basis. The truck drivers load their own orders and then leave the loading dock.

The warehouse company is interested in whether it would be more effective to add another forklift driver or another loading bay, or if either option is substantially better than the current service. The performance measures it is interested in include the average number of trucks waiting in the parking lot and the total time a truck spends at the warehouse. It also wants to know the utilization of the forklift drivers. Data collected on the warehouse operation show the following:

- The arrival rate of trucks is 6.5/hour, with the time between arrivals being exponentially distributed.
- The mean time for a forklift to fill an order is 15 minutes. The distribution is normal with standard deviation 1.5 minutes.
- The time required for a driver to load an order is well modeled as an Erlang random variable with mean 15 minutes and 3 phases.

9.12. Carefully prove Equation (9.2).

9.13. Perform a simulation experiment to construct the mean plot $\bar{p}_4^{(n)}$ for $n = 1, 2, \ldots, 20$ described in Section 9.1.1. Starting with $k = 10$ replications, gradually increase the number of replications until the empirical plot closely matches the true values $p_4^{(n)}$.

9.14. Consider the simulation experiment described in Section 9.1.2. First construct a mean plot to determine a deletion point d. Then estimate w_q using $\bar{D}$ and compute its standard error.

9.15. Estimate covariance between X and Z in the Humorous Money example.

9.16. Show that an equivalent method for calculating $\hat{\sigma}_X^2 + \hat{\sigma}_Z^2 - 2\hat{\sigma}_{XZ}$ is

$$\frac{1}{k-1} \sum_{i=1}^{k} (V_i - \bar{V})^2$$

where $V_i = X_i - Z_i$.

9.17. When we are interested in customer delay in steady-state queueing simulations, it is typically the case that setting the initial conditions underloaded (such as empty and idle) leads to a negative bias in the estimators, while setting the initial conditions overloaded (initially loading the queue with more customers than the expected number in the system) leads to a positive bias. This suggests that we could make multiple replications, half of them initialized from an underloaded state and half from an overloaded state. Discuss all advantages and disadvantages of this approach that you can see.

9.18. In Section 9.2 we showed that when we generate the initial state of a Markov chain from its steady-state distribution, π, then the process starts in steady state. In other words, there is no initial-condition bias. We also showed that when we deterministically select the initial state as $\{S_0 = 1\}$, there is bias. But notice that even if we generate the initial state from π, one possible outcome is $\{S_0 = 1\}$. This outcome has probability π_1. Explain why *generating* the initial state leads to an unbiased estimator (even though $\{S_0 = 1\}$ may occur) but *selecting* the initial state to be $\{S_0 = 1\}$ leads to a biased estimator.

9.19. The purpose of this exercise is to demonstrate that initial-condition bias may dissipate at different times for different performance measures.

Consider the Markov chain S_n with state space $\mathcal{M} = \{1, 2, 3\}$ and one-step transition matrix

$$\mathbf{P} = \begin{pmatrix} 0.7 & 0.1 & 0.2 \\ 0.1 & 0.8 & 0.1 \\ 0.4 & 0.1 & 0.5 \end{pmatrix}$$

Compute π, the steady-state distribution, and let S be a random variable having this distribution. In other words, $\Pr\{S = i\} = \pi_i$, $i = 1, 2, 3$. Compute $\Pr\{S = 3\}$, $E[S]$ and $\mathrm{Var}[S]$.

Now assume that $\{S_0 = 1\}$. Plot $p_{13}^{(n)} = \Pr\{S_n = 3 | S_0 = 1\}$, $E[S_n | S_0 = 1]$ and $\mathrm{Var}[S_n | S_0 = 1]$ as functions of n. How many time steps are required to be within 1% relative error of the limiting values for each quantity? Do the probability, expected value and variance converge to within 1% relative error of the limiting values at the same time step?

9.20. The purpose of this exercise is to demonstrate that initial-condition bias may dissipate at different times for systems that are logically similar.

An (r, s) inventory system involves the periodic review of the level of inventory of some discrete unit.[2] If the inventory position at a review is found to be below r units, then enough additional units are ordered to bring the inventory position up to s units. When the inventory position at a review is found to be above r units, no additional units are ordered. One possible goal is to select the values of r and s that minimize inventory cost.

Let $\{S_n; n = 0, 1, 2, \ldots\}$ represent the inventory position just after a review at period n, and let $\{X_n; n = 0, 1, 2, \ldots\}$ represent the demand for units of inventory in period t. Suppose the inventory position S_n changes in the following way:

$$S_{n+1} = \begin{cases} s, & S_n - X_n < r \\ S_n - X_n, & S_n - X_n \geq r \end{cases}$$

The initial inventory position is taken to be s, so that the inventory position is initially at its maximum. The demand $\{X_n; n = 0, 1, 2, \ldots\}$ is modeled as a sequence of independent, time-stationary Poisson random variables with mean 25 units.

In each period there are costs associated with the inventory position. If $S_n - X_n < r$, then in period $n + 1$ a cost of $32 + 3[s - (S_n - X_n)]$ is incurred, which is a fixed cost plus a per unit cost of bringing the inventory position up to s. In period $n + 1$, if $S_{n+1} \geq X_{n+1}$, then a holding cost of $S_{n+1} - X_{n+1}$ dollars is incurred; otherwise a shortage cost of $5(X_{n+1} - S_{n+1})$ dollars is incurred.

---

[2]This exercise is based on a model in L. W. Koenig and A.M. Law. 1985. A procedure for selecting a subset of size m containing the l best of k independent populations, with applications to simulation. *Communications in Statistics–Simulation and Computation* **14**, 719–734.

Let D_n be the cost incurred in period n. The four inventory policies under consideration are given in the following table. The performance measure of interest is the long-run expected cost per period of each inventory policy, with a smaller expected cost being preferred.

policy	r	s
1	20	40
2	20	80
3	40	60
4	40	100
5	60	100

(a) Simulate each inventory policy, and determine a deletion point using the mean-plot approach. Make sufficient replications so that you believe that the absolute bias is less than $20 at the deletion point.

(b) Using the deletion points determined above, perform a simulation experiment to find the inventory policy with minimum expected cost per period. Use common random numbers to sharpen your comparisons.

9.21. Read the article by Larson, R. C., M. F. Cahn and M. C. Shell. 1993. Improving the New York City arrest-to-arraignment system. *Interfaces* **23**, 76–96. Describe the combination of queueing and simulation analysis used.

APPENDIX

A

SIMULATION
PROGRAMMING
EXAMPLES

This appendix contains complete programs in Fortran, SLAM II (Pritsker Corporation), SIMAN IV (Systems Modeling Corporation) and GPSS/H (Wolverine Software Corporation) for the simulation of The Darker Image, as described in Section 4.6. In each case the program prints or plots the number of customers in the queue ($Y_{1,t}$) and the status of the server ($Y_{2,t}$) as a function of time, t, or gives time averages of these outputs.

A.1 FORTRAN

```
C**********************************************************
C     example Fortran simulation of The Darker Image
C
C     the simulation is written to mimic the algorithm
C     in chapter 4, rather than for efficiency
C
C     key variables:
C         queue = number of customers waiting in queue
C         avail = 0 if copier available, 1 otherwise
C         t = current event epoch (time)
C         clock(1) = clock for next customer arrival
C         clock(2) = clock for next customer finish
C         n = system event counter
C         i = system event indicator
C         seed(5) = array of random number seeds for
C                       function random
C         stream = random number stream 1,2,3,4,5
C
C     output:
C         simulated sample path of t, queue and avail
C         from time 0 to time 360 minutes
C
C**********************************************************
C     initialize five random number streams with seed
C     values 131,072 pseudorandom numbers apart
          integer*4 seed(5)
          data seed(1),seed(2),seed(3),seed(4),seed(5)/
      $ 748932582, 1985072130, 1631331038, 67377721,
      $ 366304404/
          common /seeds/seed
C
C     declare state variables, clocks and event epoch
          integer queue, avail, n, i
          real clock(2), t
          parameter(infty=1.e32)
          common /events/ clock(2), t
C
C STEP 1
          n = 0
          t = 0.0
          call start(queue, avail)
C
  2       if (t .lt. 360.) then
C
C STEP 2
              t = amin1(clock(1), clock(2))
              i = 1
```

```
              if (clock(2) .lt. clock(1)) then
                 i = 2
              endif
c STEP 3
              clock(i) = infty
c STEP 4
              if (i .eq. 1) then
                 call arrive(queue, avail)
              else
                 call finish(queue, avail)
              endif
              n = n + 1
c STEP 5
              write(6,*) n, t, queue, avail
              goto 2
           else
              stop
           endif
           end


           subroutine start(queue,avail)
c***************************************************
c   system event representing start of day
c***************************************************
           integer queue, avail
           real clock(2), t
           parameter(infty=1.e32)
           common /events/ clock(2), t
c
c   initialize state
           queue = 0
           avail = 0
c
c   set clock for first arrival, no finish
           clock(1) = expon(0.166667,1)
           clock(2) = infty
           return
           end
```

```
      subroutine arrive(queue,avail)
c***************************************************
c    system event representing customer arrives
c***************************************************
      integer queue, avail
      real clock(2), t
      common /events/ clock(2), t
c
c    if copier available put into use
c    otherwise one more waiting
      if (avail .eq. 0) then
         avail = 1
         clock(2) = t + expon(0.217391,2)
      else
         queue = queue + 1
      endif
c
c    set clock for next arrival
      clock(1) = t + expon(0.166667,1)
      return
      end


      subroutine finish(queue,avail)
c***************************************************
c    system event representing customer finishes
c***************************************************
      integer queue, avail
      real clock(2), t
      common /events/ clock(2), t
c
c    if no customers waiting make copier available
c    else one fewer customer waiting
      if (queue .eq. 0) then
         avail = 0
      else
         queue = queue - 1
         clock(2) = t + expon(0.217391,2)
      endif
      return
      end
```

```
      real function random(stream)
c*****************************************************
c    portable random number generator implementing the
c    recursion:  seed = 16807*seed  mod (2**(31) - 1)
c    using only 32 bits, including sign
c
c    code modified from UNIF in Bratley, Fox and
c    Schrage (1987), A GUIDE TO SIMULATION (2/e),
c    Springer-Verlag, New York, page 331
c
c    some compilers require the declaration:
c       integer*4 seed(), k1
c
c    input:
c       stream = an integer random number stream
c                that maps into an integer seed that is
c                greater than 0 and less than 2147483647
c
c    output:
c       seed(stream) = new seed value
c       random = a uniform pseudorandom number
c*****************************************************
c
      integer stream
      integer*4 seed(5), k1
      common /seeds/seed
c
      k1 = seed(stream)/127773
      seed(stream) = 16807*(seed(stream) - k1*127773)
     $               - k1*2836
      if (seed(stream) .lt. 0) then
         seed(stream) = seed(stream) + 2147483647
      endif
      random = seed(stream)*4.656612875e-10
      return
      end
```

```
      real function expon(lambda, stream)
c*******************************************************
c    function to generate a random variate from the
c    exponential distribution with mean =1/lambda
c    via inverse cdf method
c
c    input:
c       lambda = parameter of exponential distribution
c       stream = integer random number stream
c
c    output:
c       expon = random variate
c*******************************************************
c
      real lambda, u
      integer stream
c
      u = random(stream)
      expon = -alog(1.-u)/lambda
      return
      end
```

A.2 SLAM II

```
GEN,NELSON,DARKER IMAGE,4/13/94,1,NO,NO,YES,NO,YES,72;
LIMITS,1,1,100;
EQUIVALENCE,EXPON(6.0,3),GAP/EXPON(4.6,4),SERVICE;
INITIALIZE,0,360;                          1 day = 360 minutes
NETWORK;
      RESOURCE/1,COPIER(1),1;
      CREATE,GAP;                           generate customer
      AWAIT(1),COPIER;                      queue for copier
         ACTIVITY,SERVICE;                  service time
      FREE,COPIER;                          copier available
      TERMINATE;                            customer departs
      END;
RECORD,TNOW,T,,B,1.0;
VAR,NNQ(1),Q,NUMBER IN QUEUE;
VAR,NRUSE(1),A,SERVER UTILIZATION;
FIN;
```

A.3 SIMAN IV

```
BEGIN;
PROJECT,             DARKER IMAGE,NELSON,4/13/94;
QUEUES:              COPY QUEUE;
RESOURCES:           COPIER,1;
DSTATS:              NQ(COPY QUEUE),NUMBER IN QUEUE,
                     "QUEUE.DAT":
                     NR(COPIER),SERVER UTILIZATION,
                     "AVAIL.DAT";
DISTRIBUTIONS:       1,EXPO(6.0,3):      !interarrival gap
                     2,EXPO(4.6,4);      service time
REPLICATE,           ,,360.0;            1 day = 360 minutes
END;
```

```
BEGIN;
        CREATE,      ,ED(1):             !generate customer
                     ED(1);
        QUEUE,       COPY QUEUE;         queue for copier
        SEIZE:                           !use copier
                     COPIER;
        DELAY:                           !service time
                     ED(2);
        RELEASE:                         !copier available
                     COPIER:
                     DISPOSE;
END;
```

A.4 GPSS/H

```
        SIMULATE
        GENERATE     RVEXPO(3,6.0)       generate customer
        QUEUE        SERVICE             queue for copier
        SEIZE        COPIER              use copier
        DEPART       SERVICE
        ADVANCE      RVEXPO(4,4.6)       service time
        RELEASE      COPIER              copier available
        TERMINATE
        GENERATE     360                 1 day = 360 minutes
        TERMINATE    1
        START        1
        END
```

REFERENCES

Bhat, U. N. 1984. *Elements of applied stochastic processes*. Second edition. New York: John Wiley & Sons.

Bratley, P., B. L. Fox and L. E. Schrage. 1987. *A guide to simulation*. Second edition. New York: Springer-Verlag.

Çinlar, E. 1975. *Introduction to stochastic processes*. Englewood Cliffs, New Jersey: Prentice-Hall.

Clarke, A. B. and R. L. Disney (1985), *Probability and random processes: A first course with applications*, Second Edition, New York: John Wiley & Sons.

Glasserman, P. 1991. *Gradient estimation via perturbation analysis*. Boston: Kluwer Academic Publishers.

Law, A. M. and W. D. Kelton. 1991. *Simulation modeling & analysis*. New York: McGraw-Hill.

Nelson, B. L. 1992. Statistical analysis of simulation results. In *Handbook of industrial engineering* (G. Salvendy, ed.). New York: John Wiley.

Ravindran, A., D. T. Philips and J. J. Solberg. 1987. *Operations research: Principles and practice*. Second Edition. New York: John Wiley.

Schruben, L. 1992. *Event graph modeling using SIGMA*. Second Release. San Francisco: The Scientific Press.

Whitt, W. 1983. The queueing network analyzer. *The Bell System Technical Journal* **62**, 9, 2779–2815.

Winston, W. L. 1991. *Operations research: Applications and algorithms*. Boston: PWS-Kent.

INDEX

Statistics, 24, 37–43
within and across replications, 290
Steady-state probability:
of birth-death processes, 231–232
of Markov chains, 151, 153–154,
171–172
of Markov processes, 198, 200–201,
226–227
of semi-Markov processes, 211–212
Stock level, 175
Strong law of large numbers, 37, 116
System event, 12, 25, 70
System inputs, 10
System logic, 10
Stochastic process:
definition of, 4, 62–63, 73
generic model, 70, 72–73, 80

T

t distribution, 42
Time average, 68,
calculation of, 69, 72–73
definition of, 62
Time-stationary property, 38
of Markov chains, 134, 154–156
of Markov processes, 185, 208–210
Time study, 6, 7, 17
Traffic intensity, 232–233
Transaction (see Simulation)
Transient state, 149, 197–198
Transition probability:
diagram, 138–139
function, 189
initial state, 134, 135, 182–183

limiting, 147, 151, 197
matrix, 135, 170, 189
n step, 140–142, 145–147, 171
one step, 134, 170
parameterizing (see Markov chain)
steady state (see Steady-state probability)
Transition function (see Transition
probability: function)
Transition rate, 183, 186, 225
diagram, 186–187, 225
(See also Balance equations)

U

Unbiased estimator, 283, 288
Uniformization, 188, 191–192, 193–194
derivation of, 202–205
Utilization, 75, 233

V

Variance:
of Bernoulli, 31
of binomial, 85
definition of, 31
of Erlang, 86
of exponential, 31
of normal, 39
of Poisson, 118
sample, 39
of Weibull, 89

W

Waiting time, 233, 274–275
Whitt, W., 261, 265, 273–274, 315
Winston, W. L., 240, 315